Spices, Scents and Silk: Catalysts of World Trade

Dedication
To my Mom and Dad for giving me the keys to a happy, fulfilling life.

Spices, Scents and Silk: Catalysts of World Trade

James F. Hancock
Emeritus Professor Department of Horticulture
Michigan State University, USA

CABI is a trading name of CAB International

CABI
Nosworthy Way
Wallingford
Oxfordshire OX10 8DE
UK

Tel: +44 (0)1491 832111
Fax: +44 (0)1491 833508
E-mail: info@cabi.org
Website: www.cabi.org

CABI
WeWork
One Lincoln Street
24th Floor
Boston, MA 02111
USA

Tel: +1 (617)682-9015
E-mail: cabi-nao@cabi.org

© James F. Hancock 2021. All rights reserved. No part of this publication may be reproduced in any form or by any means, electronically, mechanically, by photocopying, recording or otherwise, without the prior permission of the copyright owners.

A catalogue record for this book is available from the British Library, London, UK.

Library of Congress Cataloging-in-Publication Data

Names: Hancock, James F., author.
Title: Spices, scents and silk : catalysts of world trade / James F. Hancock.
Description: Oxfordshire, UK ; Boston, MA : CAB International, [2021] | Includes bibliographical references and index. | Summary: "This book describes the central role that these exotic luxuries - spices, scents and silks - played in the lives of the ancients, traces the development of the great international trade networks that delivered them, and explores how the demand for such luxuries shaped the world"-- Provided by publisher.
Identifiers: LCCN 2021018831 (print) | LCCN 2021018832 (ebook) | ISBN 9781789249743 (hardback) | ISBN 9781789249750 (paperback) | ISBN 9781789249767 (ebook) | ISBN 9781789249774 (epub)
Subjects: LCSH: Commerce--History--To 500. | Commerce--History--Medieval, 500-1500. | Spice trade--History. | Incense industry--History. | Silk industry--History. | Trade routes--History.
Classification: LCC HF357 .H36 2021 (print) | LCC HF357 (ebook) | DDC 382.09--dc23
LC record available at https://lccn.loc.gov/2021018831
LC ebook record available at https://lccn.loc.gov/2021018832

References to Internet websites (URLs) were accurate at the time of writing.

ISBN-13: 9781789249743 (hardback)
9781789249750 (paperback)
9781789249767 (ePDF)
9781789249774 (ePub)

DOI: 10.1079/9781789249743.0000

Commissioning Editor: Rebecca Stubbs
Editorial Assistant: Emma McCann
Production Editor: Shankari Wilford

Typeset by SPi, Pondicherry, India

Contents

Preface		xiii
1	**Introduction**	1
2	**Early History of Scents, Spices and Silk**	5
	Setting the Stage – Exotic Luxuries of Antiquity	5
	Smoke of the Gods in Antiquity	5
	The Most Ancient of Spices: Cinnamon, Ginger and Pepper	8
	The European Spice of Antiquity: Saffron	12
	The Indonesian Spices: Clove, Nutmeg and Mace	14
	A Sightless Moth's Gift to the World: Silk	17
3	**Exotic Luxuries in Antiquity**	21
	Setting the Stage – Ancient Reports on Incense, Spice and Silk	21
	Use of Frankincense and Myrrh in Antiquity	21
	The Spices in Antiquity	24
	Silk in Antiquity	32
4	**Ancient Mediterranean Trade Links**	37
	Setting the Stage – Early Egyptian–Levantine Trade	37
	Minoans and Mycenaeans	38
	Invisible Commodities in Early Commerce between Egypt and the Levant	39
	Solomon and the Kingdom of Israel	39
	Phoenicians	40
	Emergence of the Greek City States	42
	Alexander and the City of Alexandria	43
	Egypt Under the Ptolemies	44
	Rome and Carthage Rise and Fight for Mediterranean Supremacy	45
	And Then the Romans Controlled Egypt	47
5	**Land of Punt and the Incense Routes**	50
	Setting the Stage – Land of Punt	50
	Red Sea Trade after Rameses III	52

	Canal of the Pharaohs	52
	The Rise of the Incense Kingdoms	53
	Domestication of the Camel	54
	Caravan Routes	55
	Ma'rib Dam	57
	The Sayhad Desert and Further Points North	57
	Profits Along the Way	59
	The Great Intermediaries: the Nabataeans	59
	Petra – Jewel of the Nabataeans	60
	Maritime Incense Trade	61
	Roman Invasion of the Incense Route	62
6	**Origins of the Spice Trade in the Indian Ocean**	**65**
	Setting the Stage – Central Role of Rivers	65
	Persian Gulf Routes	65
	To the Red Sea and Beyond	67
	Early Indonesian Seafarers	67
	Royal Road of the First Persian Empire	68
	Persian and Greek Explorations	70
	Arab Stranglehold on Egyptian Trade	71
	War Elephants and Red Sea Travel	72
	On to India	73
	The Roman Sea	73
	Periplus of the Erythraean Sea	75
	Rome's Breathtaking International Trade Network	77
7	**Silk Route Beginnings**	**80**
	Setting the Stage – Ancient Steppe Routes	80
	The Horse and the Balance of Power in Central Asia	82
	China Stretches Its Muscles	83
	Chinese Struggles with Mighty Xiongnu	84
	Adventures of Zhang Qian	85
	Han Chinese Take Control of Their Borderlands	86
	Silk Route Map	87
	The Engines of the Silk Routes	88
	The Merchants of the Silk Routes	89
	Cultural Diffusion along the Silk Routes	90
	Postscript – Discovery of the Buddhist Cave Complexes	91
8	**Silk Route Connections**	**94**
	Setting the Stage – Empires of the Middle East	94
	Roman Intrusions into the Middle East	95
	Palmyra Emerges as the Greatest of the Middle Eastern Trading Centres	95
	Zenobia Grabs Power	97
	Kushans Take the Centre of the Silk Routes	98

	Kushan Connections	99
	Parthia Controls the Terminus	101
	Sasanians Take Over	101
	Ebbs and Flows of the Silk Route	102
	A Plague Slows Trade	103
	Silk Trade after 400 CE	104
9	**Ancient South East Asian Maritime Trade**	**107**
	Setting the Stage – Indonesia Awakes	107
	Origin of Trade between India and South East Asia	107
	Maritime Trade of the Anuradhapura Kingdom	108
	Indianization of Indonesia	109
	China's Slow Entry into the South East Asia Trade Network	110
	Java Becomes the Nucleus of Indonesia	110
	The Chinese Pilgrims – Chroniclers of the Ancient Spice and Silk Routes	111
	Early Trade in the Outer Reaches of Indonesia	113
	The Golden Peninsula	114
	The First Great Trading Empire: Funan	115
	South East Asian Trading Spheres in the Early First Century CE	116
	European Connections	117
	The Two Ways to Rome	117
	The First Direct Contact between Rome and China	119
10	**Golden Age of Byzantium**	**122**
	Setting the Stage – Roman Power Shifts	122
	Constantine Builds His City	123
	The World Turns	124
	The Exotic Luxuries of Byzantium	124
	The Golden Age of the Eastern Roman Empire under Justinian	125
	Byzantine Attitudes about Trade	127
	The 'Dollar of the Middle Ages'	127
	Trading with the Enemy	128
	Aksum and Byzantium's Indian Ocean Connections	129
	Christians Surrounded by Muslims	130
	The Secret of Silk Escapes	131
	Justinian's Plague	131
	End of the Red Sea Portal	132
11	**Pan Islamica**	**135**
	Setting the Stage – Rapid Spread of Islam	135
	The Byzantines Redirect Their Trade	135
	A New Trading Empire Emerges in the Northern Steppes: the Khazar Khaganate	137
	Arab Agricultural Revolution	138
	The Centre of the Muslim World Shifts	138

	The Round City of Baghdad	139
	Islam and Medieval Medicine	140
	Spread of Islam across South East Asia	141
	Muslim Expansion Towards China	142
	Muslim Maritime Trade with South East Asia	142
	Muslim Sea Trade with China	143
12	**Spice Trade in the Dark Ages of Europe**	**146**
	Setting the Stage – Collapse of the Western Roman Empire	146
	Spice Use in Europe During the Dark Ages	146
	Level of Western Trade in the Early Medieval Age	147
	Mediterranean Trade in the Early Medieval Period	148
	Early Medieval Trade in Europe	149
	The Radhanites, Medieval Tycoons	150
	Rise of the Gotlanders	152
	Rus' Trade with the Muslims and Byzantines through Khazaria	153
	Rus' Attacks on the Islamic and Byzantine Worlds	154
13	**The Eastern Roman Empire and The Rise of Venice**	**157**
	Setting the Stage – Eastern Roman Empire Struggles to Control Italy	157
	Byzantine Empire after Justinian	157
	Islam in the Ninth Century	158
	Rule of Europe after the Fall of Rome	158
	Resurgence of Byzantine Empire in the Tenth and Eleventh Centuries	159
	Byzantine Trade in Silk and Spices with the Muslims	160
	The Rise of Venice	161
	Rise of Other Italian Mercantile States	163
	Western Maritime Trade in the High Middle Ages	164
	A Byzantine Cry for Help	165
	Influence of Crusader States on European Trade	165
	The Byzantine Empire Struggles with the Rising Power of Venice	167
	The Fourth Crusade and the Sacking of Constantinople	168
14	**Medieval Shifts in the Balance of Power**	**172**
	Setting the Stage – The Late Medieval European Economy	172
	Spices in Medieval Cuisine	173
	Spices in Medieval Medicine	174
	Silk in Medieval Europe	175
	The World System in the Thirteenth Century	176
	The Venetian Trading Empire	177
	The Catalonian Trade Networks	178
	The Hanseatic League	178
	Internal European Trade and the Champagne Fairs	179
	Genghis Khan Reopens the Silk Route	180
	End of the Crusader States and Muslim Trade	183

	The Alternative Black Sea Route	184
	Venice Moves into the Black Sea	185
	Periodic Confrontations with the Mongols	186
	Black Death Ravages the World	186
15	**Monsoon Islam**	**189**
	Setting the Stage – Ottoman Takeover of the Middle East	189
	The Sea that Fed the Ottoman Spice Routes	190
	Ibn Majid and Indian Ocean Navigation	191
	Tomé Pires	191
	India at the Centre	192
	Aden: Ottoman Portal to Indian Ocean Trade	194
	Ormuz: Safavid Portal to Indian Ocean Trade	195
	Swahili Coast of Africa	196
	Strait of Malacca and the City of Malacca	196
	Sumatra and Java	198
	Ceylon	198
	Moluccas	199
	Far East Asian Connections	201
	Early Chinese Ming and the Indian Ocean	201
	Voyages of Zheng He	202
	Chinese Return to Isolationism	203
16	**Portuguese Discovery and Conquest**	**206**
	Setting the Stage – Medieval European Knowledge of the Spice Trade	206
	Emergence of Portugal	207
	Treaty of Tordesillas	208
	Vasco da Gama Discovers an Unknown World	209
	The Portuguese Conquest of India Begins	212
	The Pace Quickens	213
	The First Viceroy of Portuguese India	214
	Socotra and Ormuz	215
	The Portuguese–Mamluk Naval War	215
	Albuquerque Takes Over	217
	The Portuguese in the Bandas and Moluccas	218
	Portuguese Move into Ceylon	219
17	**The Portuguese Build an Empire**	**222**
	Setting the Stage – Long-Term Effects of Albuquerque's Battles	222
	The Cartaz System	223
	The India Run	224
	Importance of Indian Cotton in the Spice Trade	225
	Sixteenth-Century Shifts in European Spice Trade	226
	The Portuguese Distribution Network in Europe	226
	Portuguese Spice Profits	227

	The Mughals and Portuguese in India	228
	The Last Frontier: Portugal and China	229
	Dutch and English Look for Alternative Routes to the Spices	230
	Henry Hudson Searches Both Ways	233
18	**The Spanish Build Their Empire**	**235**
	Setting the Stage – Spain Stretches Its Muscles Across the Atlantic	235
	Magellan Finds Another Route to the Pacific	236
	The Fate of Magellan's Ships and Crew	237
	Spanish Struggle to Keep a Foothold in the Spice Islands	238
	The Spanish Explore the Pacific	239
	Andrés de Urdaneta Finds the Way to New Spain	240
	Manila Galleons	241
	Spanish Establish a Silk Industry in New Spain	242
	English Sea Dogs Growl at Spanish	243
	English Sea Dog Activity Against Portuguese	244
19	**The Dutch and English Conquest of South East Asia**	**247**
	Setting the Stage – The Rise of the Dutch and English Empires	247
	Ralph Fitch	248
	Jan Huygen van Linschoten	249
	The First Dutch Expeditions to the East Indies	250
	Jacob van Heemskerck and the Law of Prize and Booty	251
	English East India Company	252
	Dutch East India Company (Verenigde Oost-Indische Compagnie)	253
	The Dutch Establish an East Asian Trading Network	254
	First English–Dutch Confrontations	255
	The Ill-Fated Missions of Admiral Pieter Willemszoon Verhoeff	256
	Brouwer Route	257
	Jan Pieterszoon Coen	258
	Dutch Finally Take Portuguese Malacca	259
	The VOC in the North Moluccas	260
	East India Companies' Effect on European Trade	261
	Atlantic Trade of Europeans	261
20	**Age of Expansion**	**264**
	Setting the Stage – the EIC and VOC Move into India	264
	Mughals of India in the Seventeenth Century	265
	EIC Entry into India	265
	Early Dutch Spread into India	266
	East India Companies Spread Out from Surat	267
	Country Trade	270
	The Port of al-Makha (Mocha)	271
	Affairs at Home and the East India Trade	272
	VOC Gets into Trade with Japan	273
	The Dutch and the Pirates	274
	Dutch Take Over Ceylon	275

21	**The Ottoman and Safavid Silk Trade**	**278**
	Setting the Stage – Ottomans in the Centre	278
	Silk Production and Movement in Safavid Persia	279
	France and the Ottomans	280
	Enter the Levant Company	281
	Dutch in the Levant	282
	French Impact Grows in the Levant	283
	Raw Silk Around the Horn	284
	The VOC Gets into the Silk Market	285
	Bengali Silk Trade	285
	Luxury Silks in Europe	286
	Death of the Worms	287
22	**End of the Spice Era**	**289**
	Setting the Stage – European Tastes Change	289
	The Shift Away from Spices	290
	The Shift to Cotton	290
	Cotton Politics	291
	British Weave Their Own	292
	The Shift to Tea	292
	Opium Wars	293
	The EIC Moves Heavily into India	294
	VOC Loses Steam	294
	Last Century of the EIC – From Traders to Rulers	295
	European Competitors in India	296
	End of the EIC	297
	Epilogue – Spice Trade after the East India Companies	297

Index   **301**

Preface

This book is about how scents, spices and silk catalysed world trade. It has essentially three parts. The first three chapters introduce the exotic luxuries that came to have the greatest impact on human societies, including their origins, culture and uses. The next 12 chapters describe how trade routes evolved in antiquity to deliver scents, spices and silks to the Western world. The last seven chapters discuss the Renaissance period after the Portuguese discovered the route around the Cape and the Europeans began going after their own spices and silks. In my account, I use a blend of primary literature and historical fiction to produce what I believe is an expansive, sometimes amusing narrative that is rooted in scientific fact. I quote others freely in this book, as I don't believe in paraphrasing what others have already said well.

This book is meant to concentrate on the trade routes and not on the rise and fall of nations, although I do periodically review world events to maintain a historical timeline. I also sometimes go off on tangential topics such as camels, elephants, city structure and ship construction where they seem relevant to the story of trade. I have tried hard to give equal weight to both the Western and Eastern worlds, but fear the book does have a Eurocentric outlook, as Europe was often the end point of the trade networks.

Three books have influenced me greatly: William Bernstein's *A Splendid Exchange: How Trade Shaped the World* (Grove Press, New York, 2008); L. Paine's *The Sea & Civilization: A Maritime History of the World* (Vintage Books, New York, 2013); and Peter Frankopan's *The Silk Roads: A New History of the World* (Bloomsbury, New York, 2015). Bernstein's book served as a model for mine with its sweeping discussion of the centrality of trade in world affairs. Paine's book provides an incredibly detailed, seamless history of how sea travel evolved. Frankopan's treatise impressed me with its unabashed declaration that world history does not revolve around the West. I have also greatly benefited from the historical essays found in the *Ancient History Encyclopedia* (now *World History Encyclopedia*) and *ThoughtCo*, in particular those descriptions of Mark Cartwright, who helped fill in many holes in my knowledge of world history.

I want to thank my wife Ann once again for her enthusiastic support of all I do and the rich environment that surrounds us. My friend Jim Varmecky offered many insights during our weekly discussions of 'all things important'. The CABI editorial staff was phenomenal. Rachael Russell for her early enthusiasm in this project and Rebecca Stubbs for her leadership during the editorial and production process. Emma McCann and Shankari Wilford played key roles in making the book beautiful. Gill Watling gave the manuscript an uncommonly thoughtful read and suggested a number of important edits. I greatly appreciated her general 'scrutinize-everything-that-moves' tendency.

Introduction

In our grocery stores and shops you can purchase an almost boundless array of products that came from all over the globe – blueberries from South America, electronics and appliances from China, tomatoes and strawberries from Mexico, wines from France, rugs from Persia, cooking oil from Malaysia, coffee and bananas from Central America and T-shirts from Bangladesh – the list is seemingly endless.

For millennia in the ancient world only a very select handful of goods were transported great distances for sale – the most widespread became silk, the spices and incense. Produced in remote corners of South East Asia and China, these commodities were carried for thousands of miles across land and sea to deliver them to their markets. Their origin from distant, mystical locations added to their status and allure. From the dawn of civilization to at least until the modern era, one of the exotic luxuries from the Far East was always at the centre of the world stage.

The first three great civilizations emerged about 5000 years ago: Egypt along the Nile River in northern Africa; Sumer between the Tigris and Euphrates Rivers in today's Iraq; and Harappa in the Indus Valley in what is now Pakistan. From almost their inception, these early societies became addicted to the luxury products of far-off lands and established long-reaching trade networks. As the massive Chinese, Persian, Hellenistic and Roman empires subsequently emerged, the acquisition and trade of exotic goods continued to be a focal point. India and Indonesia became hotbeds of commerce.

International trade in spices, scents and silk pushed humans to explore and then travel to the far corners of the earth. The great powers of the world fought mightily for the kingdoms where silk, spices and scents were produced. Ultimately, European lust for the spices led to the discovery of the sea route from Europe around Africa to South East Asia. The New World was accidentally discovered by Columbus in his quest for spices.

Among the most sought-after luxury products in the ancient world were silk cloth, the incenses frankincense and myrrh, and the spices cinnamon, pepper, ginger, clove, saffron, nutmeg and mace. What made trade in these luxury products so remarkable was that the plants producing them grew in very restricted areas of the world, distant from the wealthy civilizations of northern Africa, Greece and Europe (Fig. 1.1). Silk was spun only in China by domesticated moths fed leaves of mulberry trees. Frankincense and myrrh could only be obtained in a narrow band of south-west Arabia and coastal Africa. Pepper was limited to north-east and western India, cassia to southern Vietnam and the eastern Himalayas, ginger to south-eastern China, cinnamon to tiny Ceylon (today's Sri Lanka), and cloves, nutmeg and mace to the even tinier Molucca (or Spice) Islands in Indonesia. Only saffron was native to Europe.

Once the Western world discovered the intoxicating properties of these products, their procurement became a dominant force in the world economy. Traders carried silk, spices and the scent resins for thousands of miles by land and sea to satisfy the desires of the West. There were many other important trade commodities such as grain, slaves, porcelain, precious metals, pearls and gemstones, but the low bulk and high profits linked to the scents, silks and spices were particularly enticing. These luxuries could be carried on the backs of camels or in the holds of ships for months on end, arriving at their final destination in nearly perfect condition. The fact that they came from some distant, mysterious location made them all the more desirable. Nothing else compared with their possible profit returns.

This book tells the story of how the scents, spices and silks catalysed world trade. It describes the central role that these exotic luxuries came to play in the lives of the ancients and how those products got into their hands. It traces the

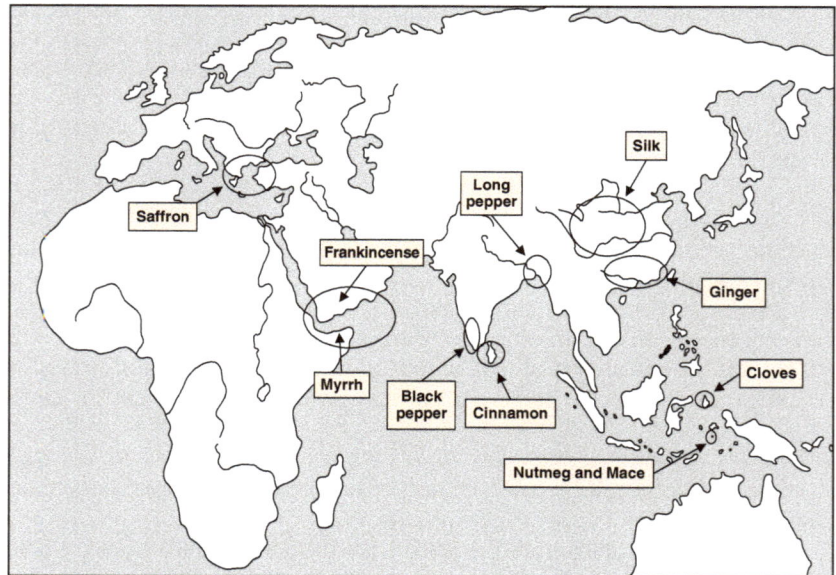

Fig. 1.1. Geographical origins of the silk, spices and scents that shaped the world.

development of the great international trade networks that delivered spices, scents and silk and how the demand for such luxuries shaped the world.

International long-distance trade began when the ancient Egyptians' lust for frankincense and myrrh led their pharaohs to send massive trade expeditions south-east to a place they called the 'Land of Punt'. Huge ships were built and launched down the Nile, disassembled and dragged across the desert to the Red Sea and then sailed to southern Arabia and Somalia in search of frankincense, myrrh and other exotic riches.

Eventually, a great trade route evolved across Arabia to serve the insatiable desire for incense in Egypt and the other ancient societies of Mesopotamia. What came to be called the 'Incense Route' wove 1000 miles from southern Arabia along the Red Sea coast and west through the Sinai to trade frankincense and myrrh. Cinnamon came to be included in these cargoes, after being carried another 2000 miles across the Indian Ocean and Arabian Sea by mariners from South East Asia. This Incense Route flourished most mightily between the seventh century BCE and the second century CE but had roots that went back as far as 1200 BCE.

For the first 600 years of the Incense Route, the primary consumers were the Dynasties of Egypt and the Kingdom of Assyria. As the Greek civilization grew in power across the Mediterranean it became the focus of the incense trade, first passing through Gaza and then Alexandria after Egypt's conquest by Alexander. When Rome achieved its world dominance and subjugated Egypt, it became the epicentre of the incense trade, also through Alexandria.

The original movement of cinnamon, from India and East Asia to Arabia, was the beginning of a global network of trade that later came to be called the 'Spice Routes'. For the first 2000 years of this trading sphere, the Arabs served as the sole middlemen in the spice trade to Egypt and kept them completely in the dark as to the origin of the spices. It wasn't until about 120 BCE that this stranglehold on the spice trade came crashing down, when a shipwrecked Indian sailor washed up on the Red Sea coast of Egypt and taught the Egyptian Greeks how to ride the monsoon winds across the Arabian Sea to India. Few Greek ships actually made this trip, but when the Romans took over Egypt in 30 BCE it was not long before about 120 ships were leaving Red Sea ports annually to load their holds with spices in India.

When the Romans began sailing to Indian ports, a dazzling array of spices came to their attention. By 1 CE, a full-blown trading network was operating across the seas of the Far East and India stood at the centre of this world. Indian dhows were sailing south through the Indian Ocean to Indonesia where they traded pepper for cloves and nutmeg. Chinese junks plied the South China Sea and reached as far as the Spice Islands, Indonesia and Ceylon. They dropped off their ginger, cassia, silk and porcelain at these stops in return for cloves, nutmeg and pepper. With the advent of the Roman trade, a maritime trading route of 9000 miles had evolved that stretched all the way from Rome, across the Mediterranean to northern Africa, through the Indian Ocean to Indonesia and on to China, with India at its centre.

In the Roman period, spices from South East Asia were also travelling through the Persian Gulf to North Africa and the Mediterranean. Exotic goods from India and Indonesia were carried first by Indian and then by Persian sailors to the central Persian Gulf ports where they were moved by land-based caravans north towards the Mediterranean. Magnificent trading centres like Petra and Palmyra evolved in Syria to facilitate this trade. Rome and Persia fought long and hard for centuries to control these trade routes.

In the first century CE, another great commodity had entered the world market: silk. The Romans had fallen head over heels for this precious commodity. Tracing its discovery to antiquity in north-eastern China, a vast trading network evolved across the mountains and deserts of China through Central Asia to the Persian Parthian Empire. This 'Silk Route' travelled over 4000 miles across the wide geography covering China, Central Asia and the Middle East.

After Rome's collapse in 250 CE, the epicentre of world trade shifted first to the Byzantine Empire and then to Europe as it emerged from the Dark Ages. Venice became a worldwide trading powerhouse. The Muslims gained control of the Spice Routes as they took over the Middle East and a large part of South East Asia. Europeans fought the Crusades in large part to maintain a portal to the spice trade. Ultimately the Portuguese found the route to the spices of South East Asia, and they entered a period of conquest and subjugation in India and Indonesia to control that trade. The other half of the world came under the purveys of the Spanish, who traded silver from Bolivia for Chinese silk and then shipped it half-way around the world across the Isthmus of Panama on to Europe. The Dutch followed by the English burst on to the scene next, took the lucrative spice routes from the Portuguese and then battled each other mightily for dominance. They held on to their winnings dearly until European passions shifted from highly spiced foods to sugar and tea.

Human desire for the exotic drove the extensive trade routes that carried the scent resins, spices and silk. Someone in a far-off place had something that someone wanted, and opportunists were willing to deliver it for a profit. People wanted fragrances to surround them, and the best scents came from far away. They wanted powerful aromas to adorn their foods and heal their bodies, and the most desirable were produced in distant lands. They wanted to wrap their bodies in fine cloth, and the finest textiles came from places thousands of miles away. Once people discovered the intoxicating properties of these products, their procurement became a dominant force in the world economy. Herein lies the story of how scents, spices and silk shaped our world.

Early History of Scents, Spices and Silk

2

Setting the Stage – Exotic Luxuries of Antiquity

The insatiable desire of humans for the exotic catalysed the formation of extensive trade networks that moved incense, spices and silk in antiquity. The great early civilizations of Mesopotamia, Egypt, the Indus Valley and China, from their beginnings, were addicted to the luxury products of far-off lands. Merchants carried these products for thousands of miles across land and sea to reach their ultimate destinations.

Three great streams of travel evolved over time across the world: the Incense, Silk and Spice Routes. The Incense Route was the first, linking the aromatics' growing regions of Arabia with Egypt, Babylon and Rome. The Silk Route followed to become the longest overland trade route across the world – travelling some 7,000 kilometres (4,350 miles) across the mountains, deserts and steppes of Central Asia, joining western China with Persia, the Levant and Europe. The Spice Route developed as a series of sea lanes across the China and Indian seas that came to fully link the East with the West through the Arabian and Red Seas.

This story is about the major commodities moved across these routes. Among the most sought-after luxury products in the ancient world were silk cloth, the aromatics frankincense and myrrh, and the spices cinnamon, pepper, ginger, clove, saffron, nutmeg and mace. Numerous other luxury commodities, such as cumin, cardamom and sandalwood, were also widely traded in antiquity, but the above 'top ten' are the ones that played a central role in the economies of both Eastern and Western civilizations.

Smoke of the Gods in Antiquity

No one really knows why people began to use the resins of frankincense and myrrh. Watt and Seller speculate:

> Was it ancient man, who, back in the distant past, first discovered the haunting fragrance of frankincense and the smoky notes of myrrh? When he threw the

twigs on the camp fires did their aromas lull him to sleep? Is it possible too that their preservative properties may first have been observed when insects were found in the resin masses perfectly preserved. Perhaps when hands were grazed and cut gathering wood, their wound healing properties were first noticed.

(Watt and Seller, 1996, p. 1)

In the tiny slice of the earth now represented by Yemen, many thousands of years ago, a few small kingdoms emerged that became fabulously wealthy by trading frankincense and myrrh to Mesopotamia, Egypt, the Levant and the Mediterranean. When frankincense was at its highest demand in the second century CE, Greece and Rome were importing more than 3000 tons annually. The scents wafting from these resins became the hallmark of civilization, dominating the world economy for 4000 years as 'the premier luxury products of antiquity' (Bernstein, 2008, p. 59). Their enormous popularity came from their symbolism in religious ceremony and their practicality in masking the odours of the ancient world.

Natural history of frankincense and myrrh

The trees producing frankincense and myrrh are quite restricted in geography (Watt and Sellar, 1996; Singer, 2006). Frankincense comes from *Boswellia sacra* trees that grow in the hills, plains and valleys of Dhofar in southern Oman and the Hadramawt region in eastern Yemen. The most favoured myrrh comes from trees of *Commiphora myrrha* found in southern Oman and Yemen, as well as Ethiopia and Somalia.

The incense trees stand like wizened sentinels in the desert. Frankincense trees are short and squat, reaching no more than about 4.5 metres (15 feet) high (Fig. 2.1). They have a papery peeling bark which varies in colour from white to reddish. Multiple trunks emerge out of a disc-like base. Trees of *C. myrrha* are shorter than *B. sacra*, reaching a height of about 2.7 metres (9 feet). They have a single truck, with greyish white bark and their branches are knotted and covered with nasty thorns. The resin, exuded by slashing the bark, hardens into what are called tears (Fig. 2.2).

Both species have very exacting requirements and demand a limestone soil that is never soggy with moisture. The climate of southern Arabia has the perfect moisture pattern for them – from the end of May until September the region is subject to the Khareef, a monsoon wind blowing in from East Africa that delivers a fine mist rather than rain.

The region where the scent resins grow was first settled between 6000 and 4000 BCE by Neolithic farmers originating from the Levant (Crassard and Drechsler, 2013). They brought with them sheep, goats and cattle, as well as the whole crop assemblage domesticated in the Near East. When the agriculture of these ancient settlers was producing dependable harvests, a merchant class emerged that traded frankincense and myrrh to enrich themselves. As will be described later, they established broad trading networks with Egypt, Mesopotamia, the Levant and eventually Greece and Rome (Lawler, 2010).

Fig. 2.1. Frankincense tree standing like a sentinel in arid Oman. (Source: Land of Frankincense (Oman). From UNESCO as part of a GLAM-Wiki partnership. Author: Giovanni Boccardi. WikiMedia Commons.)

Fig. 2.2. Dried resin of the frankincense tree. (Source: snotch. Licensed under the Creative Commons Attribution-Share Alike 3.0 Unported.)

Ancient culture of frankincense and myrrh

Small family units have collected resins from clumps of frankincense and myrrh trees growing wild in wadis since antiquity. The Roman philosopher and chronicler, Pliny the Elder (23–79 CE), provided details on their harvest methods, which are essentially unchanged today. The collection of resins took place:

> about the rising of the Dog-star [during the hottest days of summer] ... They made an incision where the bark appears to be fullest of juice and distended to its thinnest; and the bark is loosened with a blow but not removed. From the incision, a greasy foam spurts out, which gradually coagulates and thickens, being received on a mat of palm leaves ... the residue clinging to the tree is scraped off with an iron tool, and consequently contains fragments of bark.
>
> (Bostock and Riley, 1855–57, 12.32)

The families harvesting the incense resins maintained almost sacred, hereditary control of these trees. Pliny reported that 'there are not more than 3,000 families who retain the right of trading frankincense as a hereditary property, and consequently the members of these families are called sacred ... no one is known to plunder his neighbor' (Bostock and Riley, 1855–57, 12.32).

The Most Ancient of Spices: Cinnamon, Ginger and Pepper

As with frankincense and myrrh, the use of spices runs deep into the antiquity of human society. They were used in medicines and food long before there were written records. Plainly defined, the word *spice* refers to an aromatic, dried part of a plant used for the seasoning and flavouring of food. However, spices represent so much more. As Andrew Dalby describes in his book, *Dangerous Tastes: The Story of Spices*:

> They are natural products from a single limited region that are in demand and fetch a high price, far beyond their place of origin, for their flavor and odor. These powerful, pleasurable, sensual aromatics have been used in foods, drinks, scented oils and waxes, perfumes and cosmetics, drugs; in these various forms, they have served human beings as appetizers, digestives, antiseptics, therapeutics, tonics and aphrodisiacs.
>
> (Dalby, 2000, p. 16)

The first spices to find their way to the West via international trade were cinnamon, ginger and pepper. All are oft mentioned in the early natural history writings of the Greeks, Persians and Romans. Cinnamon was native to Ceylon and China and was first brought to southern Arabia 3000 years ago before being carried by camel caravan with incense to Egypt – a remarkable journey of over 4000 miles. Ginger from China and pepper from India were also being transported across the Indian Ocean and Arabian Sea by the last century BCE and were widely used in Roman cuisine by 50 BCE.

Natural history and ancient culture of cinnamon

What we call cinnamon today comes from two species in the laurel family: *Cinnamomum verum* or *Cinnamomum zealanicum* (true or Ceylon cinnamon); and *Cinnamomum cassia* (cassia, bastard cinnamon or Chinese cinnamon). True cinnamon was restricted to Ceylon, while cassia had a broader native range across southern China and Indochina.

The spice cinnamon comes from the dried bark of both trees (Fig. 2.3). They produce a similar spicy, aromatic condiment but their flavour intensity varies greatly. As far back as the second century CE, the Greek physician/author Galen observed that best cassia differed little from the lowest-quality cinnamon and that cassia could be substituted for the second, if a double weight of it was used. Today, almost all of what is called cinnamon in the USA is really cassia.

There are no ancient records of how cinnamon was grown and harvested, but an excellent account of traditional culture is found in Garcia de Orta's remarkable *Colloquies on the Simples & Drugs of India*, published in 1563. de Orta was a Portuguese, Sephardic Jewish physician, who worked for decades for the elite in the Portuguese colony of Goa, India. In what must be an accurate description of age-old practices, he tells us that:

> The trees are about the size of olives or rather smaller, the branches are numerous and not crooked, but somewhat straight. The flowers are white, the fruit black and round, larger than a myrtle, or between that and a nut. The canela [cinnamon in Portuguese] is the second bark of the tree: for it has two barks like the cork tree, which has bark and shell ... First, they take off the outer bark and clean the other. The outer bark, cut in squares, is then thrown on the ground. When on the ground it rolls itself up in a round form, so as to look like the bark of a stick which it is not. For the poles or sticks are the size of a man's thigh. The thickest of the bark is the thickness of a finger. It takes a vermilion colour, or that which is given when burnt by the sun; or more like ashes mixed with red wine, very little of the cinder and a great deal of the wine ... This year's bark is taken and leaving the tree for three years it renews its bark.
>
> (de Orta, 1563, pp. 129–130)

Natural history and ancient culture of ginger

Ginger, *Zingiber officinale*, is thought to have been domesticated in India and southern China, but surprisingly is no longer found in the wild (Nayar and

Fig. 2.3. Dried bark strips, bark powder and flowers of the cinnamon tree. (Source: Simon A. Eugster. Licensed under the Creative Commons Attribution-Share Alike 3.0 Unported.)

Ravindran, 1995). The spice comes from its pungent, knobby underground stem or rhizome (called 'hands') (Fig. 2.4).

Ginger was one of the crops carried by the Austronesians as they spread from the coast of south-eastern China and Taiwan across the Malay Archipelago to as far as Madagascar and to Easter Island (Dalby, 2000; Greenhill and Gray, 2005). What is called the 'Lapita cultural complex' began its island diaspora about 5500 to 6000 years ago and ended it about 950 years ago. The people carried with them on outrigger canoes pigs, dogs, chickens, breadfruit, banana, yam, taro, sweet potato and ginger. Ginger probably added zest to their prepared foods.

There are also no really ancient accounts of ginger culture. Like cinnamon, the origin and culture of ginger remained a mystery in the Western world for over a thousand years. Arab traders were more than content to keep that mystery alive to keep profits high. The first Westerner to pinpoint where ginger was grown was Rabbi Benjamin Tudella, who visited the west coast of India between 1159 and 1173 CE. He referred to its presence, along with cinnamon and 'many other spices', but left us no mention of how it was grown.

In a limited eyewitness report on the ginger harvest in the mid-1500s, Garcia de Orta tells us that in India:

> It is gathered in December and January, dried and covered with clay in holes to prevent it from decaying. It is also enclosed in clay to make it weigh more and to keep it fresh, preserving its natural humidity. Besides, if it is not well covered with clay the worms eat it. It is also more humid and has a better taste.
>
> (de Orta, 1563, p. 399)

Marco Polo was astonished to find ginger cultivation in China during his travels in the late thirteenth century. In his famous travelogue, he describes how after 20 days on mountainous roads he emerged into an immense valley on the north edge of Sichuan that 'produces such a vast quantity of ginger that it is exported throughout the whole of Cathay, bringing in great profit' (Dalby, 2000, p. 24). Sadly, he also made no mention of how it was grown.

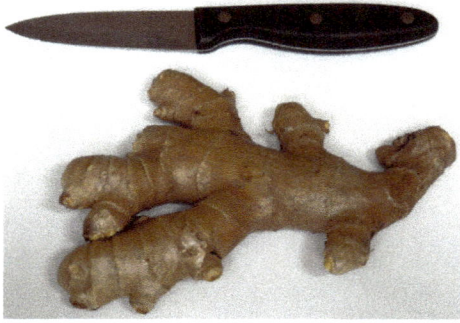

Fig. 2.4. Fresh ginger rhizome. (Source: Frank C. Müller, Baden-Baden. Licensed under the Creative Commons Attribution-Share Alike 3.0 Unported.)

Natural history and ancient culture of pepper

Pepper grows native in the tropical forests of India. There are two domesticated species of pepper: long pepper (*Piper longum*) found in the north-east (Fig. 2.5) and black pepper (*Piper nigrum*) located along the Malabar Coast in the south-west. Of the two, long pepper was the most popular in the ancient Western world. Its popularity dropped dramatically during the Middle Ages in Europe, and is now largely forgotten (Dalby, 2000; Shaffer, 2013). For close to 2000 years, however, long pepper was one of India's most important exports.

The earliest record of black pepper cultivation was also made by Garcia de Orta, who wrote:

> The tree of the pepper is planted at the foot of another tree, generally at the foot of a palm or cachou tree. It has a small root, and grows as its supporting tree grows, climbing round and embracing it. The leaves are not numerous, nor large, smaller than an orange leaf, green, and sharp pointed, burning a little almost like betel. It grows in bunches like grapes, and only differs in the pepper being smaller in the grains, and the bunches being smaller, and always green at the time that the pepper dries. The crop is in its perfection in the middle of January.
>
> In Malabar, the plant is of two kinds, one being the black pepper and the other white; and besides these there is another in Bengal called the long pepper ... You will now see that these three trees are different, namely those of long, black, and white pepper ... The tree of the long is no more like that of the black pepper than a bean is like an egg. If you do not want to believe me, believe in these three seeds, that one is of long, the other of black, the other of white pepper. The black and white pepper trees are very-like each other, and only the people of the country can tell them apart, just as we cannot tell the black from the white vines unless they are bearing grapes. If you do not want to believe me, believe in these three seeds, that one is of long, the other of black, the other of white pepper.
>
> (de Orta, 1563, p. 399)

Fig. 2.5. Dried long pepper catkins. (Source: Lemmikkipuu. Licensed under the Creative Commons Attribution-Share Alike 3.0 Unported.)

Unfortunately, de Orta was wrong about black and white pepper being from two separate species. They are in fact from the same species, *P. nigrum*. Black pepper is produced by picking the berries green before they mature and drying them in the sun until they become black and wrinkly. White pepper is produced by allowing the berries to ripen, soaking them in water and then rubbing off their outer skins before drying the remaining seed.

There are not any early published records of the cultivation of long pepper, but it is likely that it was grown at the base of trees, like black pepper. The whole catkins were harvested whole when they were blackish-green at maturity and then dried in the sun.

The European Spice of Antiquity: Saffron

Saffron holds a unique place in the history of world trade in that it was the only spice that achieved international stature that was not from Indochina or South East Asia. It was probably first gathered by the Minoans of Greece (Caiola, 2004; Caiola and Canini, 2010). The earliest known image of saffron dates to the second millennium BCE and is found on a Minoan palace fresco on the Greek island of Santorini. In radiant colours it depicts the flowers being picked by young girls and monkeys. There were two saffron-growing Minoan settlements on Santorini, Thera and Acrotiri, which were ultimately destroyed by a massive earthquake and volcanic eruption between 1645 and 1500 BCE. Saffron harvests were subsequently curtailed when much of the island sank below the sea, but the ash produced by the volcano entombed the saffron frescoes and helped preserve them.

Later Greek records tell of voyages to Cilicia, on the south-eastern coast of Asia Minor, where merchants sought what was considered the finest saffron threads. There is also strong evidence that the ancient Persians were cultivating saffron by the tenth century BCE.

The saffron species now cultivated across the world, *Crocus sativus*, has dramatically longer styles (called 'threads') than the wild forms (Fig. 2.6) (Caiola, 2004; Nemati *et al.*, 2018). Modern DNA analysis has shown that *C. sativus* was derived from a wild species, *Crocus cartwrightianus*, which is found only in Greece. While the DNA content of these two species is similar, they vary dramatically from each other in one important genetic characteristic: *C. sativus* is triploid, meaning it carries one more set of chromosomes than *C. cartwrightianus*. The first individual of saffron must have originated through a genetic mutation that altered the chromosome number of a gamete that subsequently fertilized a normal one to produce a triploid individual that survived. How often this type of event occurred is, of course, not known but it must have been exceedingly rare.

Perhaps an even more amazing event is that this exceptional individual was noted by some ancient saffron collector. Some incredibly lucky and observant ancient must have recognized a single individual with dramatically longer

Fig. 2.6. Brilliantly coloured saffron flower with its exceptionally long styles. (Source: Serpico. Licensed under the Creative Commons Attribution-Share Alike 3.0 Unported.)

threads in a field of hundreds or thousands of others and had the good sense to propagate it. This happy accident was then passed on through the ages.

Any chronicle of this event is long lost in the mist of time, but it is very similar to the happy accident that produced bread wheat about 8000 years ago in the Fertile Crescent of Mesopotamia (Hancock, 2012). The bread wheat (*Triticum aestivum*) has six sets of chromosomes and emerged when emmer wheat (*Triticum dicoccum*) with four sets of chromosomes accidently crossed with a wild goat grass (*Aegilops tauschii*) with two sets and somebody recognized it.

There is also considerable mystery as to what saffron the ancient Cilicians in Asia Minor and the Persians were growing, as referenced by Herodotus and Pliny (Caiola, 2004; Caiola and Canini, 2010). Since *C. sativus* is native only to Greece, the early production of saffron in Asia Minor and Mesopotamia must have come from either different wild species of *Crocus* in those regions or cultivated fields of the Greek cultigen introduced by Aegean traders. It seems unlikely that the ancient Greek merchants would have freely given away their precious source of wealth, so the people of Asia Minor and Mesopotamia probably first domesticated their own local species and then later abandoned them when they finally got their hands on the superior Greek source.

On the wall of the Minos Palace at Knossos (Crete) is an extraordinary fresco depicting the harvest of crocus dating back to 1700–1600 BCE (Sarpaki, 2000). The painting covers two adjacent walls and features a central, beautifully dressed female who is almost life-sized and surrounded by animals and a landscape dominated by crocus. Scholars have interpreted the activities depicted in these frescoes 'as representing various activities including fertility rituals, initiation and/or marriage ceremonies, autumnal rebirth, and local industry' (Ferrence and Bendersky, 2004, p. 204). The fresco clearly signifies the central importance of the crocus and saffron to the daily lives of the early Greeks and its mystical qualities. We learn that the crocus was harvested in the wild from the rocky and mountainous habitat of Crete. The collection of styles into baskets was coordinated by high-ranking central authorities and likely had important healing properties.

As Ferrence and Bendersky describe the painting:

> it depicts an elevated central female figure, almost life-size, surrounded by animals and young girls in a crocus-filled landscape. This beautifully dressed and coifed woman is seated on a stack of cushions on a multicolored elevated tripartite platform. She wears many pieces of jewelry, including necklaces of beads in the shape of waterfowl and dragonflies, and a bodice decorated with crocus flowers. To the right of the woman, a leashed griffin stands with its forepaws on the upper level of her platform ... On her left, a blue monkey steps up to the lower level of her platform while it extends a mass of crocus stigmas (taken from the basket in front of it) toward the woman, who reaches her right hand in the direction of the cluster. A young girl looks directly at the elevated woman as she empties crocus blooms from her small hand basket into a much larger basket on the ground to the left of the monkey. Another young girl to the right of the griffin strides toward the woman while she carries a hand basket on her shoulder. On an adjacent wall, two more young females converse with each other as they delicately pluck blooms from crocus clumps in an uneven rocky landscape.
>
> (Ferrence and Bendersky, 2004, p. 202)

In another fresco they describe:

> three female figures in a rocky landscape with crocus clumps face toward the adjoining wall, which depicts an architectural structure crowned by a set of Horns of Consecration streaked with blood or decorated with crocus stigmas. The figure on the left wears a diaphanous blouse ornamented with crocus flowers. She extends a necklace in her left hand as she walks in the direction of the architectural structure. The central figure is seated on a rocky outcrop with her left hand holding her head or adjusting the leafy spray in her hair as she reaches with her right hand down toward her bleeding left foot. Two crocus stigmas are falling next to her foot. The younger female figure on the right is wrapping herself in a full-length veil as she steps away from the architectural structure, and turns her head backward to face it.
>
> (Ferrence and Bendersky, 2004, pp. 203–204)

These frescos leave us with no details on how the crocuses were processed after collection and utilized in daily lives, but they do leave us with a strong sense of the central cultural and economic importance of saffron to the Minoans.

The Indonesian Spices: Clove, Nutmeg and Mace

Clove, nutmeg and mace were last of the spices to achieve worldwide fame, being touted widely across the world by the first century CE for their medical properties and as healthy, tasty additions to foods. The roots of their trade, however, reach back much more deeply. Found exclusively on the tiny Maluku (Molucca or Spice) Islands of Indonesia (Fig. 2.7), they became important medicines in ancient China and India, being carried in outrigger canoes to

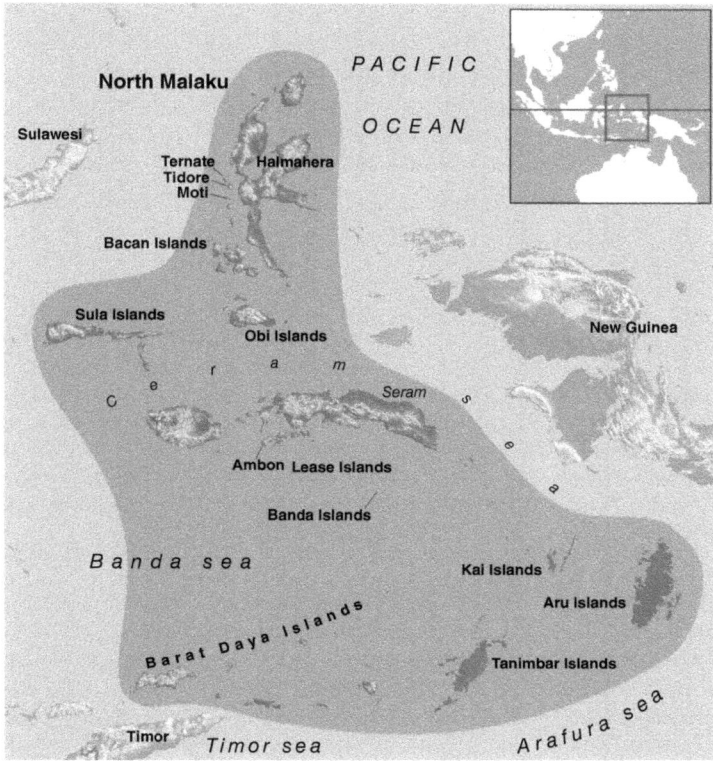

Fig. 2.7. Location of the Maluku (Molucca) Islands (shaded area) in South East Asia. (Redrawn from Lencer, CC BY-SA 3.0 <https://creativecommons.org/licenses/by-sa/3.0>, via Wikimedia Commons. File: Maluku Locator Topography.png)

Malaysia and the larger islands of Indonesia for trade with Chinese and Indian mariners. From there they ultimately found their way to Western Europe.

Cloves originated on only a few tiny, volcanic Maluku Islands located in the East Indian Archipelago (Ambon, Ternate, Tidore and the Bacan Islands). The English expression 'cloves' refers to the dried, unopened buds of the evergreen tree, *Syzygium caryophyllata* (also known as *Eugenia aromatica*) in the myrtle family. The clove tree is a small, straight-trunked tree that grows to about 9–12 metres (30–40 feet) tall (Rosengarten, 1969).

The earliest eyewitness accounts of clove cultivation did not appear until the 1500s (Donkin, 2003). Antonio Pigafetta (1480–1534), chronicler of the first circumnavigation of the world by Magellan, wrote (Donkin, 2003, p. 6): '[cloves] grow only in the mountains [up to about 600 metres] ... and if one of these trees is planted in the low ground ... it dies; we saw almost every day a cloud descends and encircle first one of these mountains and then the other, whereby the cloves become more perfect.' The first chronicler of tropical medicine and ethnobotany, Garcia de Orta, in his *Colloquies on the Simples & Drugs of India* (1563, p. 248), relates that 'they do not grow very near the sea, but a cannon shot distant from it, though on islands surrounded by the sea.'

The most complete, early description of the clove harvest is found in Antonio Galvao's *Historia das Moluccas* (1544):

> [The islanders] eat before they harvest [the clove] because it causes a strong nausea. As soon as it begins to ripen, they gather it; for if they let it attain ripeness, it becomes woody and falls without being of any use. The harvesters climb up the trees and take with them a rope and a pole. They throw the rope down, and those who are standing there tie a basket to it, and it is hoisted up. And they fasten it with some cord around their shoulders, and thus it stays on their back. They pick the clove with their hands, breaking the ends of the boughs bearing it, and throw it into the saloi. Where their hand cannot reach, they substitute the pole for it; and when the basket is filled they send it back down the rope. They bring it to their houses, and they put it to dry on mats in the sun or on reeds in the smoke as [one does with] chestnuts. Cloves that have not been properly dried lose weight by dehydration in transit to the disadvantage of the merchant.
>
> (Donkin, 2003, pp. 6–7)

The nutmeg tree is native to sheltered valleys on the hot tropical Banda Islands in the Maluku Islands of Indonesia (Rosengarten, 1969). Nutmeg and mace come from a large tropical tree (12–18 metres or 40–60 feet tall), *Myristica fragrans*, of the *Myristicaceae* family. The nutmegs are the dark reddish-brown seeds within the fruits. These seeds are surrounded by a deep red, fleshy net-like membrane or aril, which is mace (Fig. 2.8). The nutmeg tree also has an extremely limited native range like clove and is found only on the Banda Islands.

Fig. 2.8. Reddish-brown nutmeg seed surrounded by a deep red, net-like membrane of mace. (Source: Anton. Licensed under the Creative Commons Attribution-Share Alike 3.0 Unported.)

Garcia de Orta described nutmeg and mace as follows:

> It is like a pear tree, or, to be more exact, like a small peach tree. The rind is hard, the outer skin being harder than green pears. Removing the thick rind, there is a very fine rind like that which encircles our chesnuts. This goes round the nut. The nut is like a small gall nut. The delicate skin which encircles it is the mace ... You must know that when the nutmeg begins to swell, it breaks the first rind, as our chesnuts burst their prickly covering, and the mace becomes very red, appearing like fine gram. It is the most beautiful sight in the world when the trees are loaded ... When the nutmeg is dried it does without the mace, which changes in colour from red to a pale orange.
>
> (de Orta, 1563, pp. 272–273)

Today, as was also likely done in the past, the harvest of nutmeg begins about six months after flowering and continues for the rest of the year. Fruits are collected from the ground after the fruits burst open, or are collected with long, pronged poles into baskets. The nuts are removed from the husks, and then the mace is peeled from the nutmegs. The two are dried separately for several weeks. To prevent insect damage, the nutmegs were traditionally treated with lime.

Asian spice popularity in both cuisine and medicine reached its historical peak during the Middle Ages in Europe. Spices were not only thought to be healthy, but also were widely used to enrich the natural qualities of food (Freedman, 2008).

A Sightless Moth's Gift to the World: Silk

Probably 5000 years ago in China, someone figured out how to collect and weave the silk threads of the sightless moth *Bombyx mori*, a native of China. The ancient Chinese fed their larvae on a diet of the leaves of mulberry trees and learned to plunge their cocoons (Fig. 2.9) into boiling water at the appropriate time to ensure the release of the full length of silk strands. They

Fig. 2.9. A cut silkworm cocoon. (Made available by Pixabay under the Creative Commons CC0 1.0 Universal Public Domain Dedication.)

also developed the process called 'reeling', where long single strands of silk are collected on bobbins.

The importance of silk as a worldwide object of trade cannot be underestimated. Its worldwide demand spawned the great Silk Routes that wove all across China and Central Asia through treacherous mountains and deserts to Mesopotamia and then Mediterranean ports. Ancient Chinese dynasties kept warring barbarians at bay by paying huge ransoms of silk cloth. The Romans became greatly enamoured with silk clothes, with aristocratic men and women wearing the sheerest of fabrics. Silk came to represent the magnificent power of the Christian God as the flowing purple robes worn by the orthodox clergy of Byzantium. Silk remained one of the most sought-after luxuries of antiquity until the first silkworms were spirited out of China and great Persian and Byzantine industries arose in the fourth century.

Archaeological evidence reveals that silk was being produced in China as far back as 2750 BCE, during the realm of Huangdi, the 'Yellow Emperor' (Kuhn, 1984). Half a silkworm cocoon was unearthed along the Yellow River in Shanxi Province in northern China that was dated to between 2600 and 2300 BCE, and a group of silk ribbons, threads and fabric fragments of about the same age was found at Qianshanyang in Zhejiang Province. Silk strands have also been found on bronze artefacts from the Indus Valley dating to 2450 BCE, suggesting the Harrapans may have independently learned to produce silk (Good *et al.*, 2009).

There are many Chinese, Japanese and Western books that repeat a long-standing tradition that the Yellow Emperor's wife, Lady Leizu (Hsi Ling Shih), introduced silkworm rearing and invented the loom. As Kuhn relates:

> The Shih-chi (Record of the Historian) informs us that 'Huang-ti [Huangdi] dwelt on the hillock of Hsien-yüan and married a woman from Hsi-ling whom he made his principal wife.' Regrettably, the Shih-chi does not supply its source of reference. The story of lady His-ling [Leizu] is usually told as follows: 'The empress, known as the lady of His-ling, wife of the famous emperor, Huang-ti, encouraged the cultivation of the mulberry tree, the rearing of the worms and the reeling of the silk. This empress is said to have devoted herself personally to the care of the silkworms, and she is by the Chinese credited with the invention of the loom. From ancient antiquity she was (regarded as) the First Sericulturalist.'
>
> (Kuhn, 1984, p. 216)

Perhaps the most charming legend concerning the Lady Leizu's discovery of silk is that she was sitting under the shade of a mulberry tree, enjoying a cup of herbal tea, when a silkworm cocoon accidentally dropped into her steaming cup. The angry empress fished the cocoon out of her cup and found that it had started to unwind a thread that appeared almost endless. In an 'aha moment', Lady Leizu deduced that the thread could be used to make a delicate, soft yarn and crafted the whole silk-making strategy.

The discovery of silk had a dramatic effect on the clothing worn in China. K'ung Yin-ta (524–648 CE) relates in the oldest Chinese classic, *I Ching* (Yi Jing) (*Book of Change*), that:

> Before Huang-ti's [Huangdi] time clothing was made from the skins of birds and animals, but as time went by people increased and animals were few, causing great hardship. So Huang-ti ordained that clothing should be made from silk and hemp cloth. This is how the spiritual ruler changed matters for the peoples benefit.
>
> (Kuhn, 1984, p. 217)

Ancient Chinese sources such as *The Classic of History*, and *The Book of Rites* by Confucius give many details about ancient Chinese sericulture. Silk reeling and spinning were a cottage industry of women, while weaving and embroidery were done in targeted workshops as well as in the home. All the women in the household of silk-producing provinces spent most of their day, for six months each year, in the feeding and care of silkworms and to the spinning, weaving, dyeing and embroidering of silk. By the fifth century BCE, at least six Chinese provinces were producing silk.

Cartwright tells us that:

> weavers were usually women, and it was also their responsibility to make sure the silk worms were well fed on their favorite diet of chopped mulberry leaves and that they were sufficiently warm enough to spin thread for their cocoons. The industry became such a vital source of income for families that land dedicated to the cultivation of mulberry bushes was even made exempt from reforms which otherwise took away agricultural land from peasant ownership and mulberry plots became the only land that it was possible for farmers to claim hereditary ownership of. Mencius, the Confucian philosopher, advocated the smallest of land holdings always set aside a plot to plant mulberry. As demand grew, then the state and those with enough capital to do so set up large workshops where both men and women worked. Great aristocratic houses had their own private silk production team with several hundred workers employed in producing silk for the estate's needs and for resale.
>
> (Cartwright, 2017)

References

Bernstein, W.J. (2008) *A Splendid Exchange: How Trade Shaped the World*. Grove Press, New York.

Bostock, J. and Riley, H.T. (1855–57) *The Natural History of Pliny. Translated with Copious Notes and Illustrations*, Vol. 3. Harry G. Bohn, London.

Caiola, M.G. (2004) Saffron reproductive biology. *Acta Horticulturae* 650, 25–37.

Caiola, M.G. and Canini, A. (2010) Looking for saffron's (*Crocus sativus*) parents. *Functional Plant Science and Biotechnology* 4(Sp. Iss. 2), 1–14.

Cartwright, M. (2017) Silk in antiquity. *Ancient History Encyclopedia*. Available at: https://www.ancient.eu/Silk/ (accessed 27 January 2020).

Crassard, R. and Drechsler, P. (2013) Towards new paradigms: multiple pathways for the Arabian Neolithic. *Arabian Archeology and Epigraphy* 24, 3–8.

Dalby, A. (2000) *Dangerous Tastes: The Story of Spices*. University of California Press, Berkeley, California.

de Orta, G. (1563) *Colloquies on the Simples & Drugs of India*. New edition (Lisbon, 1895) edited and annotated by the Conde de Ficalho, translated by Sir Clements Markham. Henry Sotheran and Co., London.

Donkin, R.A. (2003) *Between East and West: The Moluccas and the Traffic in Spices up to the Arrival of Europeans*. American Philosophical Society, Philadelphia, Pennsylvania.

Ferrence, S.C. and Bendersky, G. (2004) Therapy with saffron and the goddess at Thera. *Perspectives in Biology and Medicine* 47(2), 199–226.

Freeman, P. (2008) *Out of the East: Spices and the Medieval Imagination*. Yale University Press, New Haven, Connecticut and London.

Good, I.L., Kenoyer, J.M. and Meadow, R.H. (2009) New evidence for early silk in the Indus civilization. *Archaeometry* 51(3), 457–466.

Greenhill, S.J. and Gray, R.D. (2005) Testing population dispersal hypotheses: Pacific settlement, phylogenetic trees, and Austronesian languages. In: Mace, R., Holden, C. and Shennan, S. (eds) *The Evolution of Cultural Diversity: Phylogenetic Approaches*. UCL Press, London, pp. 31–52.

Hancock, J.F. (2012) *Plant Evolution and The Origin of Crop Species*. CAB International, Wallingford, UK.

Kuhn, D. (1984) Tracing a Chinese legend: in search of the identity of the 'first sericulturalist'. *T'oung Pao* 70(4/5), second series, 213–245.

Lawler, A. (2010) A forgotten corridor rediscovered. *Science* 328, 1092–1097.

Nayar, N.M. and Ravindran, P.N. (1995) Herb spices. In: Smartt, J. and Simmonds, N.W. (eds) *Evolution of Crop Plants*. Longman Scientific and Technical, Harlow, UK, pp. 491–495.

Nemati, N., Blattner, F.R., Kerndorff, H., Erol, O. and Harpke, D. (2018) Phylogeny of the saffron-crocus species group, *Crocus* series *Crocus* (Iridaceae). *Molecular Phylogenetics and Evolution* 127, 891–897.

Rosengarten, F. (1969) *The Book of Spices*. Jove Publishing, Inc., New York.

Sarpaki, A. (2000) Plants chosen to be depicted on Theran wall paintings: tentative interpretations. In: Sherratt, S. (ed.) *The Wall Paintings of Thera: Proceedings of the First International Symposium, Petros M. Nomikos Conference Centre, Thera, Hellas, 30 August–4 September 1997*, Vol. 2. Thera Foundation, Akrotiri, Greece, pp. 657–680.

Shaffer, M. (2013) *Pepper: A History of the World's Most Influential Spice*. Thomas Dunne Books, St Martin's Press, New York.

Singer, C. (2006) The incense kingdoms of Yemen: an outline history of the South Arabian incense trade. In: Peacock, D.P.S. and Williams, D.F. (eds) *Food for the Gods: New Light on the Ancient Incense Trade*. EBSCO Publishing, Ipswich, Massachusetts, pp. 4–27.

Watt, M. and Sellar, W. (1996) *Frankincense and Myrrh: Through the Ages, and A Complete Guide to Their Use in Herbalism and Aromatherapy Today*. C.W. Daniel Company Limited, Saffron Walden, UK.

3

Exotic Luxuries in Antiquity

Setting the Stage – Ancient Reports on Incense, Spice and Silk

The antediluvian literature is filled with reports of the central role of spices and scents in the ancient world. Frankincense, myrrh, cinnamon, ginger and pepper are described by all the great natural historians including Herodotus (482–425 BCE), Theophrastus (371–287 BCE), Dioscorides (40–90 CE) and Pliny (23–79 CE). In the records of all the early civilizations of Assyria, Babylon, Egypt, Persia, Greece and Rome the widespread use of frankincense and myrrh is well documented. These fragrances played a central role in daily life, from consecrating temples and masking funeral odours to personal perfumery and treating illness. Cinnamon was also burned in massive quantities in divine worship. Pepper, cloves, nutmeg and ginger were very important in the ancient Indian Ayurvedic and Chinese herbals from 300 BCE. The cookbook attributed to the famous Roman gourmet Apicius (37–14 BCE) describes the liberal use of pepper on almost everything eaten. Clove, nutmeg and mace play a prominent role in the medical texts of India and China by the eighth century CE.

Once the Silk Route was established, this luxurious textile also took the world by storm. The first Romans to see silk may have been the legions of Marcus Licinius Crassus, Governor of Syria, at the Battle of Carrhae near the Euphrates River in 53 BCE. It is said that the soldiers were so startled by the bright silken banners of the Parthian troops that they fled in panic. Within decades, however, Chinese silks (supplied by the Parthians) were being widely worn by the rich and noble families of Rome and had become what Bernstein (2008) calls the 'luxury product par excellence'.

Use of Frankincense and Myrrh in Antiquity

Frankincense was widely used primarily as incense, burned to produce a sweet fragrance that filled the air. Its aroma flowed from shrines devoted to great deities and it was freely burned at banquets, funerals, sporting games and weddings. It was also commonly burned in public places to disguise the common reeks emitted in the crowded cities from sweat, dung, sewage and piles of rubbish. The sweet scent of incense was everywhere. Herodotus wrote in 450 BCE of Athens, Greece, 'the whole country is scented with them and exhales an odor marvelously sweet' (Nabhan, 2014, p. 18). The streets of Greece and Rome were said to be filled with fragrant smoke wafting from burners at the base of the emperor statuary that adorned the streets. Great braziers of incense were burned at Roman amphitheatres to cover the stench of the blood and gore left sizzling in the hot sun.

Scent resins were liberally burned during worship and at funerals, and when the powerful died, huge quantities were combusted in their honour. It was considered an odour of spiritual fragrance that glorified God and ensured that evil spirits would not rain down upon mankind pestilence, famine and disease. During the New Kingdom in Egypt (sixteenth to eleventh century BCE), religious services were preceded by large processions of priests stepping backward carrying long-handled, falcon-topped censors of burning incense. Incense was burned on temple altars at night to make sure that the sun-god Re travelled again the next morning across the Egyptian sky. Balls of incense were found in the tomb of King Tut (Tutankhamun) (1347–1339 BCE), presumably to be burned in his afterlife.

For thousands of years the Egyptians used frankincense and myrrh in embalming their dead. The ancient process began by removing the brain and viscera, then washing the skull and packing the body cavity with myrrh. The head was then annointed with frankincense and the body rubbed with a fragrant solution of myrrh, juniper and cinnamon. The mortuary priests would solemnly chant in the background as they plied their trade, their volume being modulated by the importance of the deceased. Burning frankincense filled the air surrounding them.

The Greeks and Romans, who cremated their dead, honoured their deceased and signified the transition between the states of life and death by burning prodigious quantities of incense at their funerals. After a loud processional with mimes, musicians and even professional mourners, the bodies were incinerated in an immense funeral pyre of wood scented with spices (primarily frankincense and cinnamon). The wealthier the dead, the more elaborate was the ceremony. Pliny reported that at the funeral of Nero's consort Poppaea, a full year's supply of Rome's incense was consumed; so much so that the economy of the empire was endangered. Frankincense and myrrh were widely used in the ancient world as perfumes and in cosmetics. The royalty in Greece and Rome would immerse themselves in aromatic smoke before lovemaking. In the Bible, there are references to the use of incense resins to perfume royal garments, the human body, as well as couches and beds. Egyptian temple and tomb walls show depictions of 'unguent cones' made of fat scented with myrrh and other aromatics. These were placed on the heads of guests at great banquets to slowly melt and drip on to their shoulders and upper body. The Deir el-Bahri complex of mortuary temples and tombs on the Nile has inscriptions describing how the female pharaoh Hatshepsut perfumed herself by rubbing myrrh on her legs. The ancient Greek philosopher Theophrastus describes in his treatise *On Odors* how perfumes with myrrh-oil were best for women because 'their strength and substantial character do not easily evaporate and are not easily made to disperse, and a lasting perfume is what women require' (*Loeb Classical Library* 79, 365).

Frankincense and myrrh were used in medicines to cure everything from ulcers to depression to infertility. Assyrian cuneiform texts describe how myrrh was effective in treating ailments of the eyes, ears, mouth, nose and anus. It was used in mouthwashes, enemas and poultices, and was burned in a censor placed at the head of a patient's bed. A mixture of wine and myrrh was given to Jesus just before his crucifixion (Mark 15:23), likely as an analgesic or painkiller. Greco-Roman *materia medica* are filled with descriptions of how frankincense could be used to stop bleeding, cleanse and heal wounds, serve as an antidote for hemlock poisoning and treat a number of ailments including chest pains, haemorrhoids, bladder stones, paralysed limbs and haemorrhages.

Tributes of scent

The ancient literature is replete with stories about tributes being paid in frankincense and myrrh. The first great historian, Herodotus (fifth century BCE), described an annual tribute of 1000 talents of frankincense, a weight equivalent to almost 45,400 kilograms (100,000 pounds), being provided to Babylonian King Nebuchadnezzar (634–562 BCE) to burn at the altar in the great Temple of Bel (Temple of Baal). Herodotus also reported that the Arabs of Gaza paid the Persian king Darius an annual tribute of almost the same amount of frankincense to honour his conquests and keep their autonomy.

The Greek biographer Plutarch (75 CE) records that when Gaza fell to Alexander the Great in 315 BCE, he sent a whole shipload of plundered frankincense and myrrh to his old tutor Leonidas 'in remembrance of the hope which that teacher had inspired his boyhood'. It seems that:

> as Alexander was one day sacrificing and taking incense with both hands to throw upon the altar-fire, Leonides said to him: 'Alexander, when thou hast conquered the spice-bearing regions thou canst be thus lavish with thine incense; now, however, use sparingly what thou hast.' Accordingly, Alexander now wrote him: 'I have sent thee myrrh and frankincense in abundance, that thou mayest stop dealing parsimoniously with the gods.'
> (*Delphi Complete Works of Plutarch*, 2013, Alexander 6, 7 and 8)

Probably the most famous tribute of frankincense and myrrh is recounted in the Bible's Gospel of Matthew (2:1–12). It tells how 'wise men' had seen a 'star in the east' and Herod the King of Judea was told by his priests that the star marked the birthplace of a 'ruler who will Shepard my people of Israel'. Herod sends the wise men in the direction of Bethlehem and tells them:

Go and search carefully for the young Child, and when you have found Him, bring back word to me, that I may come and worship Him also. When they heard the king, they departed; and behold, the star which they had seen in the East went before them, till it came and stood over where the young Child was. When they saw the star, they rejoiced with exceedingly great joy. And when they had come into the house, they saw the young Child with Mary His mother, and fell down and worshipped Him. And when they had opened their treasures, they presented gifts to him: gold, frankincense, and myrrh.

(Matt. 2:8–11)

The Spices in Antiquity

Cinnamon

Cassia and cinnamon were used as medicine in China and India dating back thousands of years. Cinnamon has played an important part in the Ayurvedic medicine of India for at least 3000 years. The first authentic record of cassia use in China is in the *Ch'u Ssu* (*Elegies of Ch'u*) written in the fourth century BCE (Rosengarten, 1969). An earlier mention of cinnamon dates to 2700 BCE in the alleged herbal of the Chinese emperor Shennong (Shen-nung); however, most scholars feel an unknown author wrote this herbal in the first century CE. Shennong could not have authored it, as there was no written language in China until the twelfth century BCE.

Cassia and cinnamon were utilized in all the ancient civilizations of Assyria, Babylon, Egypt, Persia, Greece and Rome. It has already been described how during the New Kingdom in Egypt, religious services were preceded by large processions of priests burning incense and cinnamon. How incense and spice were burned in temple altars at night to keep the gods happy. How for thousands of years the Egyptians, Greeks and Romans rubbed an aromatic solution of myrrh, juniper and cinnamon on the bodies of their dead and then incinerated them in immense funeral pyres of wood scented with frankincense and cinnamon.

In the Book of Exodus, the story is told of the Lord telling Moses:

Take the finest spices: of liquid myrrh 500 shekels [11 grams], and of sweet-smelling cinnamon half as much, that is, 250, and 250 of aromatic cane, and 500 of cassia, according to the shekel of the sanctuary, and a hin [3.5 litres] of olive oil. And you shall make of these a sacred anointing oil blended as by the perfumer; it shall be a holy anointing oil.

(Exod. 30:23–25)

In ancient Rome, cinnamon was so expensive that it was used primarily in medicine and perfumes and was little used as a spice. It is not mentioned at all in the cookbook of *Apicius*. Using figures provided by Pliny, Dalby (2001, p. 42) calculated that 'an ounce of cinnamon (the quantity we might buy for kitchen use) would have cost between sixty and a hundred denarii – several days' pay for an army centurion, at least a month's pay for a laborer.'

Cinnamon's most famous use in Roman medicine was in *theriac*, a supposed universal antidote to everything from snakebite to epilepsy (Tibi, 2006). Its composition varied widely depending on what drugs and aromatic substances were available, but at its base was opium. The great Greek physician Galen prepared a daily dose for Marcus Aurelius, adjusting the dose of opium depending on whether the emperor wanted to feel good or just get a good night's sleep.

There are numerous references in the ancient literature of cinnamon being used in intoxicating perfumes by the wealthy (Dalby, 2000). In the Book of Proverbs (7:17–18), the story is told about an adulteress who uses scents to entice her lovers: 'I have sprinkled my bed with myrrh, with aloeswood and with cinnamon, come let us drink deep of love until morning, and abandon ourselves to delight.' In the second-century CE Roman novel, *The Metamorphoses of Apuleius*, a rapturous lover relates: 'And now, with responsive desire waxing with mine into an equality of love, exhaling from her open mouth the odour of cinnamon, she ravished me with the nectareous touch of her tongue.' During another seduction, an alluring temptress admits: 'The lovely face of your brother, my husband, still lingers in my eyes, the cinnamon odour of his ambrosial body still haunts my nostrils' (Apuleius, *Metamorphoses*, 2.10).

Ginger

In ancient India, ginger was highly valued for its medicinal properties. It played an important role in the ancient Ayurvedic texts *Charka Samhita* written in the third century BCE, *Sushruta Samhita* (third century CE) and *Ashtanga Hridayam* (seventh century CE). Ginger was considered *mahabbeshaj*, which means literally 'the great cure'; it was considered panacea for a number of ailments (Ravindran and Babu, 2016). Ginger was thought to enhance digestion by improving assimilation and movement of nutrients to body tissues. In one Ayurvedic verse, everyone is recommended to eat fresh ginger before meals. It was considered a universal remedy for a wide range of ailments including elephantiasis, gout, haemorrhage, vertigo, skin conditions, arthritic joint pain and hyperactivity.

Ginger was just as important in Chinese medicine as in India, being used to treat a wide range of ailments from cholera to sexual impotency. The first written Chinese reference to dried ginger is found in the alleged herbal of the Chinese emperor Shennong. The use of ginger in China as medicine is also recorded by Tao Hongjing in 500 CE in his work *Mingyi bielu* (*Miscellaneous Records of Famous Physicians*) and *Bencao jizhu* (*Collection of Commentaries on the Classics of Materia Medica*). The Chinese monk Faxian (Sen, 2006), who went on an incredible journey through Central Asia to India, wrote in 406 CE that sea traders from South East Asia carried potted plants of ginger on their voyages to ward off scurvy.

Confucius is said to have eaten ginger at every meal (Slingerland, 2003). He approved of its being eaten during periods of fasting or sacrificial worship when other pungent foods were prohibited. He did, however, recommend that it be consumed in moderation to not increase the internal heat of the body.

Ginger also became an important component of ancient Greek and Roman medicine. Dioscorides (40–90 CE) describes it in his *De Materia Medica* as an exotic medicine from distant lands whose effect 'is warming, digestive, gently laxative, appetizing: it helps in cases of cataract, and is an ingredient in antidotes against poison' (Dalby, 2000, p. 22). *De Materia Medica* was the leading pharmacological text for 1500 years.

In the Roman recipe book of *Apicius*, ginger is used in many dishes, although not as much as pepper. In 'another style for roasts', *Apicius* instructs:

> Take 6 scruples [a scruple is a historical unit of weight of about 20 grains, used by apothecaries] of parsley, of laser just as many, 6 of ginger, 5 laurel berries, 6 scruples of preserved laser root, Cyprian rush 6, 6 of origany, a little costmary, 3 scruples of chamomile or pellitory, 6 scruples of celery seed, 12 scruples of pepper, and broth and oil as much as it will take up.
>
> (Vehling, 1936, p. 164)

Apicius describes a 'salt for many ills' that 'are used against indigestion, to move the bowels, against all illness, against pestilence as well as for the prevention of colds' and suggests that 'They're very gentle indeed and healthier than you would expect. Make them in this manner: 1 lb. of common salt ground, 2 lbs. of ammoniac salt, ground, 3 ounces white pepper, 2 ounces ginger, 1 ounce of Aminean bryony, 1 of thyme seed and 1 of celery seed' (Vehling, 1936, p. 67).

Pliny gives a rather detailed description of how ginger differs from pepper in his first-century *Natural History*:

> The root of the pepper-tree is not, as some people have thought, the same as the substance called ginger, or by others zinpiberi, although it has a similar flavour. Ginger is grown on farms in Arabia and Cave-dwellers-Country; it is a small plant with a white root. The plant is liable to decay very quickly, in spite of its extreme pungency. Its price is six denarii a pound. It is easy to adulterate long pepper with Alexandrian mustard. Long pepper is sold at 15 denarii a pound, white pepper at 7, and black at 4. Both pepper and ginger grow wild in their own countries, and nevertheless they are bought by weight like gold or silver. Italy also now possesses a pepper-tree that grows larger than a myrtle, which it somewhat resembles. Its grains have the same pungency as that believed to belong to myrtle-pepper, but when dried it lacks the ripeness that the other has, and consequently has not the same.
>
> (Dalby, 2000, p. 89)

Pepper

Since antiquity, pepper has always been the most important of the spices (Toussaint-Samat, 1992; Dalby, 2000; Shaffer, 2013). It is not known when it was first consumed by humans in food, but there are numerous records of its medical use in India that date back at least 3000 years. Pepper was a key component in the ancient Ayurvedic system of medicine, to stimulate appetite, aid digestion, reduce pain and cure coughs, dysentery and fevers (Vijayan and Thampuran, 2000). It was described as *katu* (pungent), *tikta* (bitter), *usbnaveerya* (energy restoring) and was thought to subdue *vatta* (biological functions controlled by the central and autonomous nervous systems) and *kapha* (heat regulation and formation of human fluids like mucus). In Sanskrit, black pepper is called *maricha* or *marica*, which refers to its facility in dispersing poisons.

The Romans were the first to use copious quantities of pepper in food. The wealthy used it liberally in almost everything eaten. In the cookbook attributed to the famous Roman gourmet Apicius, known for his opulent living and commitment to his belly, pepper is included in 349 of 469 recipes (Turner, 2005). In the first century CE, the Roman encyclopaedist Pliny described the allure and value of pepper as:

> Why do we like it so much? Some foods attract by sweetness, some by their appearance, but neither the pod nor the berry of pepper has anything to be said for it. We only want it for its bite – and we will go to India to get it. Who was the first person who was willing to try it on his viands, or in his greed for an appetite was not content merely to be hungry?
>
> (Dalby, 2000, p. 89)

The Byzantines, rich and poor, also required their food to be richly seasoned with pepper as was the Roman habit for centuries. In spite of the repeated interference of the Sasanians to limit their South East Asian trade, the Byzantines clung mightily to their Red Sea links, to a great extent to maintain their source of pepper. Chinese ginger and Indonesian nutmeg also became important in their diets, moving from the ancient Romans' medicine cabinets to the Byzantine kitchen.

As pepper's popularity as a spice surged, its medical use remained strong. This is documented clearly in *The Syriac Book of Medicines* written in the fifth century. Jack Turner reviews this work in his book *Spice* and marvels that:

> pepper alone is prescribed for a bewildering array of diseases; to be poured into the ears for earache and paralysis; for sore joints and excretory organs; for abscesses of the mouth and pustules in the throat; for general debility, blackening, and numbing of the teeth; for cancer of the mouth, toothache, gangrene, and stinking secretions; for a lost voice or a frog in the throat; for coughing up pus, for lung diseases; mixed with jackal fat for chest and internal pains; as a soporific; for heart disease and a weak stomach; for constipation and sunburn; for sleeplessness, insect bites, bad burps, and poor digestion; for a cold stomach, shivers, and worms; for a hard liver, a sore liver, wind and dysentery; for jaundice, hard spleen, loose bowels, dropsy, hernias and a general evil condition.
>
> (Turner, 2005, p. 160)

Black and long peppers also found their way to China in antiquity, although black pepper was far more important (Toussaint-Samat, 1992; Shaffer, 2013). There is written evidence that pepper was being traded overland from India to the Szechwan Province by the second century BCE (Simoons, 1990). Pepper was also mentioned in histories of the Han Dynasty (25–220 CE) published in the fifth century and a Tang Dynasty account four centuries later.

Pepper probably was first brought from India for medicinal purposes, but it did not take long for it to become important in the spicing of food. When Marco Polo travelled to China in 1271, he found pepper to be a major component of Chinese cooking and trade in pepper had become a major economic force (Shaffer, 2013). He was told by a customs official that just the city of Hang-zhou consumed 43 cartloads a day, each weighing 223 pounds (101 kilograms). Staggering quantities of pepper were being transported by Chinese junks from Java and Sumatra, as many as 5000 to 6000 baskets in each. By the Sung Dynasty (1271–1367 CE), it was standard for South Asian diplomats to bring tributes of pepper to the Chinese rulers.

Pepper was one of the driving forces for the historic sea voyages of Zheng He's 'Treasure Fleet of the Dragon Throne' in the early 1400s (Shaffer, 2013), which are described more fully in Chapter 15 (this book). In seven expeditions, Zheng He roamed far and wide from China to Korea and Japan, and through the Indian Ocean to India, the Persian Gulf and the east coast of Africa. The voyages were on a massive scale, composed of hundreds of ships and many thousand sailors, medical officers and soldiers.

Saffron

Many ancient myths and legends centre on the crocus plant that produces saffron (Hoffman, 2004; Caiola and Canini, 2010). The Greek god Zeus supposedly slept in a bed of saffron to enhance his amorous behaviour. Ovid tells the story of a young shepherd named Crocus, who falls deeply in love with a beautiful nymph named Smilax. The Greek gods are so impressed with Crocus' deep devotion to her that they grant him immortality by turning him into a saffron flower; its radiant orange styles representing his undying passion for her. In another version of this story, the goddess Hera is jealous of their passion and to keep them apart turns Crocus into a flower and Smilax into a vine. In still another story, Hermes loves but accidently kills Crocus and where the youth's blood is spilled to the ground, the first crocus flowers spring up.

Saffron came to play a central role in the lives of the wealthy in all the classical civilizations. The ancestral Greeks sprinkled and spiced many dishes with saffron. 'They served brilliant saffron-colored breads with the wine course after the main meal and used the golden spice in fish dishes' (Hoffman, 2004, p. 240). Alexander the Great is said to have soaked in saffron baths to sooth his battered body, drank saffron tea to refresh his soul and ate saffron rice to fuel his body. Willard tells us that:

> In a hot, sweaty country, what was most attractive about saffron after its property as a dye, was its scent. Its clean sharpness, clinging lightly to everything it touches, as alluring and elegant as a thin veil, would have been hard to resist. Soon saffron was being spread across the bed at night, freshening sheets and pillows, inducing a tranquil sleep. Persians swore that a cup of saffron tea removed their melancholy; a pouch of it worn on a string around the neck and dangling above the heart would rekindle love. During the burning months, when hot breezes brushed across the shade less plateau, saffron and sandalwood were stirred into water that was left in a bowl beside the front stoop to wash the dust and heat from parched bodies.
>
> (Willard, 2002, p. 18)

Many early Greek merchants made a fortune trading in saffron-pigmented textiles. As Dewan relates:

> Saffron-coloured clothing is well-attested in Classical Greece where the plant was the primary dye used to produce yellow cloth and pigments. The powerful yellow pigment of the styles is both water-soluble and resilient to light, colouring up to 100,000 times its volume when diluted ... In Egypt, Theban tomb reliefs depict Aegeans in procession carrying textiles to be presented as tribute while Aegean-style wall-paintings reflect artistic motifs derived from Minoan textiles. In similar fashion, Assyrian kings had Phoenician traders supply their courts with saffron-dyed materials, presumably taking pride in both the quality of the fabric and the implications of its valuable colour.
>
> (Dewan, 2015, p. 45)

The Persians also came to make great use of saffron in textiles – the brilliant yellows of their early carpets and funeral shrouds came from the tiny threads gathered from saffron.

The ancient Greeks and Romans particularly prized saffron in perfumes (Willard, 2002). The dried spice emits a pleasing, sensuous smell and its powerful pigment served as a natural colouring for the perfumes. Cleopatra was said to have added a quarter-cup of saffron to her baths to enhance her lovemaking. The wealthiest Romans took daily saffron baths, used it as mascara, blended it into their wines, hung it as potpourri in their halls and along streets and offered it to their gods in worship. The professional courtesans that the Greeks called *hetaerae* revered saffron as a perfume.

The classical Greek physicians Hippocrates (fifth to fourth century BCE), Erasistratus (fourth to third century BCE), Diocles (third century BCE) and Dioscorides (first century CE) all recommended saffron for a wide range of ailments, including eye diseases (painful eye, corneal disease and cataract, purulent eye infection), earache, toothache and ulcers (skin, mouth, genitalia) (Mousavi and Bathaie, 2011). Herodotus and Pliny the Elder rated Assyrian and Babylonian saffron as best to treat gastrointestinal and renal ailments.

Sumerian, Assyrian and Egyptian healers also used saffron in many medicinal ways, including pastes that were spread over ailing organs and mixtures dissolved in beer that were drunk. It was used to treat a wide range of complaints from stomach aches to urinary disorders, sexual dysfunction, infections, inflammation and diarrhoea. Dewan (2015, p. 48) suggests that 'amongst medicinal plants known in the Near East and Mediterranean, saffron can claim the largest number of applications, with 90 ethno medical parallels.'

Saffron was not native to Indochina, but likely arrived there in antiquity, perhaps carried by Persian traders on the Silk Routes. In India, saffron became an almost indispensable ingredient in many recipes of rice and sweets; it came to be used in Ayurvedic medicine and in religious rituals. Its golden-yellow properties as a fabric dye came to be revered by Buddhists. After the historical Buddha's death, his attendant monks decreed saffron as the official colour for Buddhist robes and mantles. Saffron is mentioned repeatedly in ancient Chinese medical texts, including the *Bencao gangmu* (*Compendium of Materia Medica*), which documents thousands of phytochemical-based medical treatments for various disorders.

Cloves

The first mention of cloves in the Chinese literature is made during the Han period, about the third century BCE (Rosengarten, 1969; Ptak, 1993). The spice called *hi-sho-hiang* ('bird's tongue') was first used as a breath freshener – officers of the court were required to place cloves in their mouth before discussions with their sovereign. Cloves were much more widely used in medicines than food preparation. They were considered an internal warming herb, which helped dispel cold and warm the body. They were used as tonics and stimulants and were prescribed as a digestive aid and antiseptic. Cloves were used to treat a wide range of ailments including intestine distress, impotence, diarrhoea, vomiting and cholera. They were made into a poultice to treat cracked nipples, scorpion stings, toothaches and any abscess that caused pain.

Cloves also played an important role in early Indian society, although they arrived several centuries later in India than China. Cloves became just as popular in Ayurvedic medicine as traditional Chinese medicine, being used to treat a wide range of problems including colds, asthma, indigestion, vomiting, toothache, laryngitis, low blood pressure and impotence. In the ancient Sanskrit text, *Charaka Saṃhita* (first century CE). It is stated that 'one who wants clear, fresh, fragrant breath must keep nutmegs and cloves in the mouth' (Dalby, 2000, p. 50).

The Roman Pliny was the first to describe cloves in the West in his *Natural History* (c. 70 CE) when he records that 'there is also in India a grain resembling that of pepper but larger and more fragile, called caryophyllom, which is reported to grow on the Indian lotus tree; it is imported here for the sake of its scent' (Dalby, 2000, p. 50). Cloves are not mentioned by Theophrastis (c.287 BCE) or Apicius (c. 30 CE). Emperor Constantine the Great is said to have presented St Silvester, the bishop of Rome (314–335 CE), with gold and silver vessels filled with incense and spices, including 150 pounds (68 kilograms) of cloves. The Greek physician, Paul of Aegina, wrote in the fifth century (Dalby, 2000, p. 50): 'It is the nature of some flower of some tree, woody, black, almost thick as a finger; reputed aromatic, sour, bitterish, hot and dry in the third degree; excellent in relishes and other prescriptions.' The eminent Byzantine physician, Alexander of Tralles, recommended cloves for seasickness, gout and appetite stimulation in his sixth-century *Twelve Books on Medicine*.

Nutmeg and mace

The use of nutmeg followed the pattern of the other spices, being used first in medicines and then as food. Nutmeg is frequently in the oldest scriptures of Hinduism in India, the *Vedas* texts, composed between 1500 and 1000 BCE (Weil, 1965). It was recommended for improved digestion and was prescribed for headache, neural problems, fevers from colds, bad breath and digestive problems. Later Indian texts described nutmeg as an important medicine for cardiac complaints, consumption, asthma, toothaches, dysentery, flatulence and rheumatism.

Nutmeg's arrival in China was much later than in India, as the first reference of what could have been nutmeg doesn't appear until the third century CE in Ji Han's *Nanfang caomu zhuang* (*Record of Southern Plants and Trees*). In it, he mentions a fragrant spice that comes from a tree whose flowers are coloured like a lotus (Dalby, 2000). Nutmeg isn't commonly mentioned in the Chinese literature until the eighth century. In traditional Chinese medicine, nutmeg has long been used to treat diarrhoea, dysentery, abdominal pain and bloating, reduced appetite and indigestion.

Nutmeg was largely unknown to the West until the fifth or sixth century CE (Dalby, 2000). Pliny wrote about a tree he called *comacum*, which had a fragrant nut, but it is not certain if he was really referring to nutmeg. Dioscorides also vaguely refers to a red bark of unknown origin called *macir*. The first clear references to nutmeg and mace aren't found until the Byzantine medical texts of the sixth century, which refer to a red bark, *macis* (mace), and a musky nut, *nux muscata* (nutmeg).

Nutmeg played a prominent role in the ninth-century medical texts written by the Arab physician, Isaac ibn Amran. These works, written in Arabic and translated into Hebrew, Latin and Spanish, were the foundation of the medical curriculum of medieval Europe. He 'recommended nutmeg for a variety of disorders ... mainly restricted to those of the digestive organs, from the mouth to the stomach to the intestines, to the liver and spleen, as well as for freckles and skin blotches' (Weil, 1965 quoting Warburg, 1897). He prescribed nutmeg to control pain, vomiting, blood disorders and as an aphrodisiac.

Spices and the humoral theory of Galen

Spices came to play an important role in medicine rooted in the 1000-year-old humoral theory of the Greek physicians Hippocrates (460–377 BCE) and Galen (129–210 CE) (Nam, 2014). They had proposed that our health and personality are dependent on the harmonious balance between four humours: *black bile*, *blood*, *phlegm* and *yellow bile*, and that foods and spices could be used to affect that balance.

Galen assigned the humours different levels of warmth and humidity, in sync with what the ancient philosopher Empedocles (493–433 BCE) considered the four fundamental elements of the universe: earth, water, fire and air (Stelmack and Stalikas, 1991; Freedman, 2008). The dominant humour *blood* was thought to be made in the liver and was attributed the qualities of air – warm and moist. The humour *black bile* was thought to be made in the spleen and was attributed the qualities of earth – cold and dry. The humour *yellow bile* was thought to be made in the gall bladder and was attributed the qualities of fire – warm and dry. The humour *phlegm* was thought to be made in the brain and lungs and was attributed the qualities of water – cold and moist.

In the humoral theory, foods and herbs affected the balance of humours in four different degrees:

> First degree – slightly affects metabolism but produces no overt sensation.
> Second degree – strongly affects metabolism but temporarily.
> Third degree – strongly effects metabolism permanently.
> Fourth degree – stops metabolic function (poisons).

The humoral theory was accepted almost without question by Greek and Byzantine doctors for a thousand years. It became influential in the Islamic world in the ninth and tenth centuries when major works of Galen were translated into Arabic. Muslim physicians advanced Galen's techniques by developing syrup-based medicines to treat illness and restore proper balance between the humours. An important component of these medicines were the spices – pepper, cinnamon, ginger, cloves, nutmeg, saffron and mace.

Silk in Antiquity

When silk was first discovered in the third millennium BCE, its use was reserved for the emperor and his inner circle. The emperor wore a robe of white silk and his principal wife and the heir to the throne wore yellow. As time went on, fancy-patterned silks became the major status symbol of the royalty. Officials, courtiers and priests were distinguished from the cotton- or plain-silk-clothed commoners by their brightly coloured and exquisitely embroidered silk robes.

By the second century BCE, silk had become the centrepiece of all Chinese society. Silk was used to make clothing of long robes, gowns and jackets for anyone who could afford them (Fig. 3.1). Silk was used to make Buddhist mandala offerings, hand fans, screens, military and funeral banners, wall hangings and writing media. During the Han Dynasty, silk even became a form of currency: taxes were paid by farmers in grain and silk, civil servants were paid in silk cloth and subjects were rewarded with bolts of silk.

Silk production became the subject of numerous Chinese poems and songs. The philosopher Master Xun wrote in the third century BCE (Cartwright, 2017):

> How naked its external form,
> Yet it continually transforms like a spirit.
> Its achievement covers the world,
> For it has created ornament for a myriad generation.
> Ritual ceremonies and musical performances are completed through it;
> Noble and humble are distinguished with it;
> Young and old rely on it;
> For with it alone can one survive.

Fig. 3.1. A painted silk scroll depicting Taizong, second emperor during the Tang Dynasty of China, r. 626–649 CE. (National Palace Museum, Taipei. Public domain. Uploaded by Mark Cartwright for *Ancient History Encyclopedia*.)

Silk was used internally to pay taxes and became the primary commodity traded with the rest of the world. Great amounts of silk were given to tributary states in appreciation for their loyalty to demonstrate the Chinese emperor's great wealth and largesse. In 25 BCE alone, the Han gave 20,000 rolls of silk cloth as gifts (Cartwright, 2017). Massive amounts of silk were traded with the nomadic people of the steppes in return for tens of thousands of horses.

Life-sized statues were built to honour patron saints of sericulture and the process of silk production became an object of worship. During the Han and Sung periods, the empress performed an annual silk ceremony. Pigs and sheep would be sacrificed in memory of the 'First Sericulturalist' (Lady Leizu) and the empress would participate in the gathering of mulberry leaves to feed the larvae. A silkworm palace was built in the third century CE that was supervised by the empress.

Chinese silk became a greatly desired luxury product in trade with the Persians, Greeks and Romans. As Richter relates:

> [The ancient Greek historians] Herodotos and Xenophon repeatedly tell us that the Persians wore the 'Medic dress' and that this was considered precious, luxurious, and beautiful. This Medic dress is described as of silk by the [Byzantine historian] Procopius: 'This is the silk of which they are accustomed to make the garments which of old the Greeks called Medic but which at the present time they call Seric,' as well as by [the early Christian theologist] Tertullian: 'Alexander conquered the Medic people and was conquered by the Medic dress. He stripped his breast, adorned with the emblems of armor, and covered it with transparent tissue; and, as it were, softening, quenched it, all panting with the labor of warfare, in floating silk.'
>
> (Richter, 1929, pp. 32–33)

The evidence of silk clothing in Greece is sketchy, but Richter continues:

> [the] second half of the fifth century was a time of unprecedented wealth for Greece and especially Athens. What more natural than that some of the luxurious habits of Persia should be adopted and that silk should be imported from the East to delight the women by its soft texture and the sculptors by its lovely folds and clinging quality?
>
> (Richter, 1929, p. 33)

Romans fell head over heels for what Bernstein (2008, p. 2) describes as 'the gentle, almost weightless caress of silk on bare skin'. It became so popular that the Senate periodically issued proclamations (mostly in vain) to prohibit the wearing of silk on both economic and moral grounds. The importation of silk was costing the Roman treasury concerning quantities of gold and the thin, translucent silk clothes were thought by many to be decadent and immoral. The poet Juvenal, writing around 110 CE, was appalled 'by luxury loving woman who find the thinnest of thin robes too hot for them; whose delicate flesh is chafed by the finest silk tissue' (Bernstein, 2008, p. 2). The Roman philosopher Seneca the Younger wrote in his *Da Beneficiis*: 'I see there raiments [clothing] of silk – if that can be called raiment, which provides nothing that could possibly afford protection for the body, or indeed modesty, so that, when a woman wears it, she can scarcely, with a clear conscience, swear that she is not naked' (*Loeb Classical Library* 310, 479).

From emperor to emperor, attitudes about wearing of silk swung widely, from the more severe moralists who restricted its use to those like Caligula (37–47 CE) who not only encouraged females to wear it, but also enjoyed displaying it on their own persons. Elagabalus, who ruled Rome from AD 218 to 222 and shocked the nation with his scandalous behaviour on many fronts, was the first Western leader to wear clothes made entirely of silk. Over time, the pendulum swung against the moralists and by AD 380 the use of silk was generally accepted by the elite, its wear having spread to anyone who could afford it, with no class distinctions.

References

Bernstein, W.J. (2008) *A Splendid Exchange: How Trade Shaped the World*. Grove Press, New York.
Caiola, M.G. and Canini, A. (2010) Looking for saffron's (*Crocus sativus*) parents. *Functional Plant Science and Biotechnology* 4(Sp. Iss. 2), 1–14.
Cartwright, M. (2017) Silk in antiquity. *Ancient History Encyclopedia*. Available at: https://www.ancient.eu/Silk/ (accessed 27 January 2021).
Dalby, A. (2000) *Dangerous Tastes: The Story of Spices*. University of California Press, Berkeley, California.
Dalby, A. (2001) Christopher Columbus, Gonzalo Pizarro, and the search for cinnamon. *Gastronomica* 1, 40–49.
Dewan, R. (2015) Bronze Age flower power: the Minoan use and social significance of saffron and crocus flowers. *Chronika V*, 42–55.
Freedman, P. (2008) *Out of the East: Spices and the Medieval Imagination*. Yale University Press, New Haven, Connecticut.
Hoffman, S. (2004) *The Olive and Caper: Adventures in Greek Cooking*. Workman Publishing Company, New York.
Mousavi, S.Z. and Bathaie, S.Z. (2011) Historical uses of saffron: identifying potential new avenues for modern research. *Avicenna Journal of Phytomedicine* 1(2), 57–66.
Nabhan, G.P. (2014) *Cumin, Camels and Caravans: A Spice Odyssey*. University of California Press, Berkeley, California.
Nam, J.K. (2014) Medieval European medicine and Asian spices. *Ui Sahak* 23(2), 319–342.
Ptak, R. (1993) China and the trade in cloves, circa 960–1435. *Journal of the American Oriental Society* 113, 1–13.
Ravindran, P.N. and Babu, K.N. (2016) *Ginger: The Genus Zingiber*. CRC Press, Boca Raton, Florida.
Richter, G.M.A. (1929) Silk in Greece. *American Journal of Archaeology* 33(1), 27–33.
Rosengarten, F. (1969) *The Book of Spices*. Jove Publishing, Inc., New York.
Sen, T. (2006) The travel records of Chinese pilgrims Faxian, Xuanzang, and Yijing. *Education about Asia* 11, 24–33.

Shaffer, M. (2013) *Pepper: A History of the World's Most Influential Spice*. Thomas Dunne Books, St Martin's Press, New York.

Simoons, F.J. (1990) *Food in China: A Cultural and Historical Inquiry*. CRC Press, Boca Raton, Florida.

Slingerland, E. (2003) *Analects: With Selections from Traditional Commentaries*. Hackett Publishing, Cambridge.

Stelmack, R.M. and Stalikas, A. (1991) Galen and the humour theory of temperament. *Personality and Individual Differences* 12, 255–263.

Tibi, S. (2006) *The Medicinal Use of Opium in Ninth-century Baghdad*, Vol. 5. EJ Brill Publishing Company, Leiden, the Netherlands.

Toussaint-Samat, M. (1992) *History of Food*, tr. A. Bell. Blackwell Publishing, Malden, Massachusetts.

Turner, J. (2005) *The History of a Temptation; Spice*. Vintage Books, New York.

Vehling, J.D. (1936) *Apicius: De Re Coquinaria*. Walter Hill, Chicago, Illinois.

Vijayan, K.K. and Thampuran, R.V.A. (2000) Pharmacology, toxicology and clinical applications of black pepper. In: Ravindran, P.N. (ed.) *Black Pepper, Piper nigram*. Harwood Academic Publishers, Amsterdam, pp. 455–466.

Warburg, O. (1897) *The Nutmeg; Their History, Botany, Culture, Trade and Exploitation, as well as Their Falsifications and Surrogates. It is also a contribution to the cultural history of the Banda Islands*. W. Engelmann, Leipzig, Germany (in German).

Weil, A.T. (1965) Nutmeg as a narcotic. *Economic Botany* 19(3), 194–217.

Willard, P. (2002) *Secrets of Saffron: The Vagabond Life of the World's Most Seductive Spice*. Beacon Press, Boston, Massachusetts.

4 Ancient Mediterranean Trade Links

Setting the Stage – Early Egyptian–Levantine Trade

The dynastic kingdoms of ancient Egypt were in early contact with the eastern Mediterranean coast, as evidenced by jars, flasks and pitchers from Syria and Palestine being found in the tombs of Egyptian pharaohs as far back as 3500 BCE (Casson, 1991; Paine, 2013). By the third millennium BCE, sea trade was brisk between Egypt and the Levantine littoral, the eastern shore of the Mediterranean from Palestine to the edge of Anatolia. One of the signature accomplishments of Pharaoh Snefru, ruler of Egypt in 2600 BCE, was the import of cedar logs from Gebal (later Greek Byblos) in today's Lebanon. A scribe of his left the first written record of Mediterranean Sea trade in hieroglyphics on the Palermo Stone, in what Casson (1991, p. 6) calls 'the world's first articulate record of large scale overseas commerce'. The scribe relates how 'a fleet of 40 ships slipped their moorings, sailed out of a Phoenician harbor, and shaped a course for Egypt to bring there a shipment of Lebanese cedar'.

Byblos may be the oldest city in the world, having been inhabited for over 7000 years (Paine, 2013). By 3000 BCE this city had 'become the single most important shipping port for timber to Egypt and elsewhere' (Mark, 2009). Byblos grew vastly wealthy from trade with the Egyptians. It was here that shipbuilding was first perfected, and the Phoenicians became the 'princes of the sea', as referred to in Ezekiel. Through Byblos and other Levantine trading centres, Egypt became well connected to the civilized world beyond, including Mesopotamia and the Aegean littoral.

The peak of Egyptian–Syrian trade occurred during the reign of Thutmose III (1481–1425 BCE). Casson (1991, p. 15) lists among the trade items moving to Egypt: 'ponderous timber' from Lebanon, Syrian-built chariots, cattle and silver from Asia Minor, wine and olive oil from Cyprus, 'swarms of Asiatic and Semitic slaves' and perfumes of Arabia and copper from Cyprus. In return, Egypt sent 'gold that was mined in Nubia', papyrus writing paper, linen textiles and 'the fine products of its workshops – beads, faience, scarabs, figurines'.

In Egyptian Memphis, a foreign merchants' quarter sprang up with temples to their own gods.

Minoans and Mycenaeans

The Minoans from the island of Crete were the most active Mediterranean traders during this period (Kelder *et al.*, 2018). The lightly built Egyptian ships probably ranged through the Red Sea to Punt and along the Mediterranean coast to the Levant but went no further. The sturdier ships of the Minoan traders sailed far and wide across the western Mediterranean, actively trading with Greece, Cyprus, Asia Minor, Egypt and the Levant. Casson (1991, p. 20) describes these ships from an ancient painted frieze: 'The galleys are graceful, slender vehicles, with handsomely decorated hulls and prows that jut forward in gentle, elegant curve. The rig consists of a broad squaresail with yard along the head and boom along the foot … the ships are driven by manpower, a line of 20 odd paddlers.'

The Minoan civilization had its glory years from about 1800 to 1500 BCE, until the Mycenaean Greeks colonized the various islands of the Aegean archipelago and took over the Mediterranean trading empire (Paine, 2013; Cartwright, 2019). Prominent city centres spread across Greece and the Aegean Sea including Mycenae, Tiryns (the oldest centre), Pylos, Thebes, Midea, Gla, Orchomenos, Argos, Sparta, Nichoria and Athens. These centres were likely politically independent but shared the Greek language and writing, and cultural aspects such as architecture, frescoes, pottery, jewellery and weapons.

The Mycenaeans also traded in their own 'home grown' spice: saffron (Dewan, 2015). The ancient records of Egypt are replete with references to saffron in perfumes and ointments, which they most likely obtained from Crete. Saffron was also widely traded across the Mediterranean by the Phoenicians, whose customers ranged from perfumers to physicians and cooks scattered all across the civilized world. Pat Willard in her book *Secrets of Saffron: The Vagabond Life of the World's Most Seductive Spice* tells us that the Phoenician 'ships were packed to the oars with saffron, a particularly prized commodity for them because there were so many uses for it. In almost every port where the Phoenicians dropped anchor, someone would be willing to pay for a bag of saffron' (Willard, 2002, p. 58). The Phoenicians also had large dye works operating in Sidon and Tyre, where robes were finished by dipping them in a saffron wash.

The Mycenaeans produced a wide array of additional goods for trade such as figs, grapes, wine, raisins, honey, assorted vegetables and herbs. Evidence of the broad scope of their trade is found in their pottery which has been found as far afield as Egypt, Mesopotamia, the Levant, Anatolia, Sicily and Cyprus. The Mycenaean civilization flourished until about 1200 BCE when it dissolved after invasions from 'The Sea Peoples', a mysterious group who also harassed the Egyptians, Hittites and Phoenicians.

Invisible Commodities in Early Commerce between Egypt and the Levant

Egyptian desire for large quantities of timber may have been the primary driver of their Mediterranean trade but that connection with Syrian merchants also:

> plugged Egypt more directly into the main trunk line of the long-distance commercial network that ran from Anatolia to Afghanistan. The cities of the Levantine coast became emporia for products bound for Egypt, ranging from silver to precious stones to perfumes and oils ... Palestine became a commercial cul-de-sac.
>
> (Smith, 2009, p. 41)

Through its Levantine intermediaries, Egypt was fully connected in trade with the Assyrians and Babylonians of Mesopotamia. This connection provided an additional link to the spices of South East Asia through their Persian Gulf trading networks.

There is little direct evidence of the Assyrians and Babylonians carrying on spice trade with the Indian subcontinent; however, there is strong evidence that spices from South Asia must have found their way to Egypt (Gilboa and Namdar, 2015). Grains of *Piper nigrum* (black pepper) were found in the abdomen and nasal cavities of the mummy of Rameses II (1279–1213 BCE) which was probably used in the mummification process. Black pepper in this period could only have come from southern India. Another piece of botanical evidence of ancient spice trade comes from the seventh century BCE in the form of a flower of *Cassia cassia* found in the sanctuary of Hera at Samos in ancient Greece.

The strongest evidence that there was early spice trade between East Asia and the West comes from the traces of cinnamaldehyde found in Iron Age Phoenician clay flasks dating to the eleventh to tenth century BCE (Namdar *et al.*, 2013). These flasks were the containers used commonly by the Phoenicians to ship liquids. Since cinnamaldehyde is only produced in large quantities by the plant genus *Cinnamomum*, it had to have come to the Levant via trade through the Persian Gulf with the Indian subcontinent or Ceylon. Once in Syria, it is highly probable that these spices were shipped further west to Cyprus and other Mediterranean destinations, comprising what Gilboa and Namdar (2015, p. 279) called 'one of the invisible commodities in Mediterranean commerce'.

Solomon and the Kingdom of Israel

Born around 1010 BCE, Solomon reigned for 40 years in one of the most prosperous periods in Israel's history (Mark, 2018a). His kingdom was located in the Levant between the great Mediterranean shipping ports of Gaza, Tyre and Sidon. It stood at the centre of the Incense and Spice Routes coming out of Arabia, India and Africa and as a result prospered mightily. According to the scriptures, 'The king made silver and gold as common in Jerusalem as stones, and he made cedars as abundant as the sycamores, which are in the lowland' (2 Chr. 1). King Solomon was the third and last king in the ancient, united, Kingdom of Israel and was renowned as a prophet, writer and builder in both the Hebrew Scriptures and the Qur'an.

The success of Solomon:

> was achieved through ingenious reforms and innovations such as the improvement of defense measures; the expansion of the royal court; the financial windfall from more sophisticated taxation, labor conscriptions of Canaanites and Israelites, tributes and gifts from foreign countries under the influence of Solomon; and a land and sea trading system that utilized a powerful navy and army to protect assets and trade routes.
>
> (Knox, 2017)

Solomon was a brilliant diplomat, forging strong alliances with the powerful nations surrounding him. Often related is the story of the Queen of Sheba who visited the Kingdom of Israel, fell under his aura of power and may have given him a child. He was famous for cementing partnerships with royal marriages and readily accepted gifts of concubines. His royal household may have consisted of 700 wives and 300 concubines at its peak.

Phoenicians

The Phoenician city states of the Levant took over control of Mediterranean trade when the Mycenaean civilization collapsed. By the ninth century BCE, the Phoenicians had established an extensive maritime trading network spanning the length and breadth of the Mediterranean (Fig. 4.1) (Paine, 2013; Cartwright, 2016). They traded with the Greek islands, across all southern Europe, up to

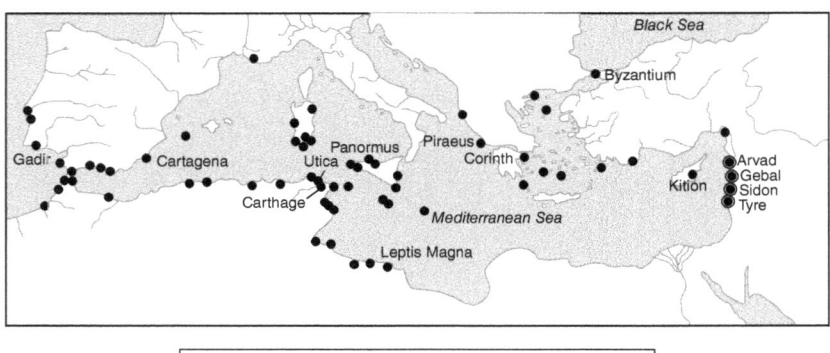

Fig. 4.1. Phoenician trade centres in the Mediterranean. (Redrawn from O. Mustafin, translated by Frantishak, CC BY-SA 3.0 <http://creativecommons.org/licenses/by-sa/3.0/>, via Wikimedia Commons. Wikimedia Commons. File: Phoenician colonisation be.PNG)

the British Isles and down the Atlantic coast of Africa. The Phoenicians established bustling ports in Tyre and Sidon (now Lebanon), Carthage (Tunisia), Gadir (Cádiz) and Cartagena (Spain), Piraeus and Corinth (Greece), Byzantium, and Marseille (France). Carthage became the greatest Phoenician centre in the western Mediterranean (Casson, 1991). They reached Arabia and India by the Red Sea and were tapped into Mesopotamia by overland caravan.

The Phoenicians dealt not only with what they themselves needed and produced, but also acted as middlemen for everything to everyone. They made enormous profits selling commodities bought at cheap prices at their source but fetching high prices elsewhere. Phoenician trade is widely referenced in the ancient literature 'from Mesopotamian reliefs to the works of Homer and Herodotus, from Egyptian tomb art to the Book of Ezekiel in the Bible' (Cartwright, 2016).

The Phoenicians built two kinds of ships, merchant ships and war vessels. For trading, they used ships with rounded hulls, curved sterns and a huge rectangular sail in the centre. These had an oar-like blade attached to the port side of the ship that was used to steer. Warships were longer and narrower than cargo ships and carried large numbers of soldiers. They had two sails, coverings on the deck to hide officers and at their bow was a forecastle, from which soldiers fired arrows or hurled catapults during battle. They were propelled by dozens of oarsmen, making possible rapid bursts of speed. The bow was fitted with a 'rostrum', a bronze tip used to ram other ships. The Phoenician ships were decorated extensively with detailed carvings and paintings. They had painted eyes to help the ship 'see' and frighten their enemies, and horses' heads that honoured Yamm, their god of the sea.

The Phoenicians were active exporters of wood, textiles and glass. They harvested aromatic cedar and fir from their own forests and sold it to Mesopotamia and Egypt for a handsome profit. The timber was carried directly to Egypt by ship and to Mesopotamia by overland caravan before being floated down the Euphrates River. They purchased wool from Damascus and Arabia, linen yarn from Egypt and silk from Persia, and transformed them into richly coloured clothes and carpets that were the rage throughout the ancient world. In particular, they were known for their cloth dyed purple using fluid extracted from the murex shellfish caught in baited traps off the coast. Huge piles of the shells have been uncovered near Sidon and Tyre and the shellfish almost became extinct on the coasts of Phoenicia. Their textiles were also acclaimed for their fine, intricate embroidery. The Phoenicians produced and traded glassware, specializing in transparent glass that was used to manufacture mirrors and plates. They also manufactured colourful semi-transparent coloured glass (blue, yellow, green and brown) which was used to adorn clothing.

The Phoenicians also were active in the trade of metals, ivory, precious stones, animals and spice. As Cartwright describes:

> The Phoenicians imported metals, especially copper from Cyprus, silver and iron from Spain, and gold from Ethiopia (and possibly Anatolia). This raw material was

transformed into ornate vessels and art objects in Phoenician workshops and then exported. Tin (from Britain), lead (Scilly Isles and Spain), and brassware were also traded, the latter principally coming from Spain. Ivory was imported from either Punt or India, as was ebony, both coming to Phoenicia via Arabia. Amber came either from the Baltic or Adriatic coast and was used in Phoenician jewellery. Embroidered linen and grain were imported from Egypt and fine, worked cloth from Mesopotamia. Grain, barley, honey, and oak timber used for oars on Phoenician ships, came from Palestine.

Phoenician markets also traded in slaves (from Cilicia and Phrygia but also captured by the Phoenicians themselves), sheep (Arabia), horses and mules (Armenia), goats, wool (Damascus and Arabia), coral, perfumes (Judah and Israel), agate, and precious stones such as emeralds (from Syria and Sheba). Spices came from the Arabian Peninsula (some coming from distant India) and included cinnamon, calamus, cassia, ladanum, frankincense, and myrrh.

(Cartwright, 2016)

In a nutshell, the Phoenicians were the world's first trading superpower. They moved all kinds of goods across all distant horizons of the Mediterranean to wherever the best price could be fetched. Their trade network would eventually be eclipsed by the Greeks, but for 500 years they dominated world trade. Phoenician-style ships and navigation tools would remain in use long after their empire had faded.

Emergence of the Greek City States

In the seventh and eighth centuries BCE, the Greek world slowly rose from the Dark Age that descended on it after the collapse of the Mycenaean civilization. A conglomeration of strong, independent cities emerged across the peninsula of Greece and on the islands of the Aegean Sea, Crete, Cyprus and Rhodes, as well as the western coast of Asia Minor. Greece's tapestry of islands and mountainous terrain made it very difficult for people to travel very far by land and each city (*polis*) developed independently from one another. Over 1000 *poleis* arose in the Greek world, with the most important being Athens, Sparta, Corinth, Thebes, Syracuse, Aegina, Rhodes, Argos, Eretria and Elis (Cartwright, 2013). The term 'Greece' represents a culture and language, not a nation. The *poleis*, in an ever-changing array of alliances, were in an almost continual state of warfare; only great outside threats brought them together in a unified whole.

Like the Phoenicians before them, these states began establishing trading colonies all across the Mediterranean and Black Seas until they had essentially wrestled control of Western trade from the Phoenicians. As Casson (1991, p. 67) describes it: 'The Greeks, in a series of bursts of activity, settled themselves the length and breadth of the Mediterranean and Black Sea, like frogs in a pond as Plato put it'. Between 750 and 550 BCE, they established some 250 colonies that still exist today. They injected Western civilization at multiple points in the sea of barbarism that surrounded the Mediterranean and Black Seas.

The Greek merchants and sailors picked these sites for their commercial potential, but they also served as a lifeline for the hard scrabble peasant farmers who had been barely eking a living from the poor soils of Greece (Bernstein, 2008). They emigrated to build farms on what must have seemed a paradise of rich soils. The rocky weak soils of Greece were sufficient to produce olive oil and wine but could not produce enough grain to feed the burgeoning populations.

Before the sixth century BCE, the Greeks relied on Egypt for their staple grain but as their populations continued to grow, they came to rely more on their own colonies. The city states of Sparta, Corinth and Megara located in the southern Peloponnese peninsula obtained the bulk of their grain from Sicily, while Athens and the other Aegean states depended on the Bug and Dnieper river valleys on the Black Sea's northern coast for their grain. Millions of bushels of grain passed annually through these trade routes.

The sea routes to these breadbaskets were extremely vulnerable to rival city states and pirates. They threaded through restricted channels and straits filled with islands. The route between the Aegean and Black Seas was particularly vulnerable, threading its way through the narrow straits of Hellespont (today the Dardanelles) on one side of the Sea of Marmara and the Bosporus on the other. For centuries, the great Athenian Empire and its allies battled mightily to keep control of the northern grain routes, fighting colonization and invasions of other Greek states like Sparta and the powerful Persian Empire.

Alexander and the City of Alexandria

Alexander turned the balance of power in the world upside down when he took control of most of western Asia and north-east Africa in the third century BCE. He founded the city of Alexandria in 332–331 BCE to serve as his link between Egypt and the rest of civilization surrounding the Mediterranean Sea. There are two charming legends associated with Alexander's decision to found the great city. In one, Alexander gets advice on where the city should be located in a dream from the great Greek author Homer. In the other, birds steal the first plans for the city that had been outlined in grain by Homer's architect. It is not too surprising that Homer played a central role in Alexander's dreams – as a child he had been captivated by Homer's *Iliad* and the character of Achilles, the exemplar of manly virtues.

The Greek author Plutarch of Chaeronea describes the stories in sections 26.4–10 of his *Life of Alexander*:

> They say that after his conquest of Egypt, Alexander resolved to found and leave behind him a large and populous Greek city which would bear his name. On the advice of his architects he was about to measure out and enclose a certain site, when during the night, as he was sleeping, he saw a remarkable vision. He thought he could see a man with very white hair and of venerable appearance standing beside him and speaking these lines: Then there is an island in the stormy sea, in front of Egypt; they call it Pharos. He rose at once and went to

Pharos, which at that time was still an island a little above the Canobic mouth of the Nile, but which has now been joined to the mainland by a causeway. When he saw that the site was eminently suitable (it is a strip of land similar to a fairly broad isthmus, running between a large lagoon and the sea which terminates in a great harbor), he exclaimed that Homer was admirable in other respects and was also an excellent architect, and ordered the plan of the city to be drawn in conformity with the terrain. Since there was no chalk available, they used barley-meal to describe a rounded area on the dark soil, to whose inner arc straight lines succeeded, starting from what might be called the skirts of the area and narrowing the breadth uniformly, so as to produce the figure of a mantle. The king was delighted with the plan, when suddenly a vast multitude of birds of every kind and size flew from the river and the lagoon on to the site like clouds; nothing was left of the barley-meal and even Alexander was much troubled by the omen. But his seers advised him there was nothing to fear (in their view the city he was founding would abound in resources and would sustain men from every nation); he therefore instructed his overseers to press on with the work.

(Anonymous, 2020)

Regardless of the truth of these stories, the city that Alexander founded in Egypt came to be what many of the ancients considered to be the greatest in the world. Alexander had little to do with the actual building of the great city himself, as he left Egypt soon after his arrival to continue his conquests and died soon after. The work was begun admirably by his commander Cleomenes and was brought to its full magnificence by the first two Ptolemy rulers of Egypt. Alexandria 'built of white marble and spectacularly endowed with cannades, monuments and palaces, the city was to achieve a magnificence that survived for seven hundred years and a commercial prominence, especially in the spice trade, that would endure for seventeen hundred years' (Key, 2006, p. 47).

Many magnificent edifices and buildings were constructed in Alexandria (Key, 2006; Mark, 2018b). The most famous was a towering lighthouse built on the Island of Pharos, which guided ships safely through the turbulent seas into the great harbour of Alexandria. There was the extensive Serapis Temple complex, high on a man-made hill, of which the most glittering attraction was a massive library containing over 500,000 books from all corners of the literate world. The city attracted the world's most renowned scholars, scientists, philosophers, mathematicians, artists and historians including Eratosthenes ($c.276–194$ BCE), who was the first to accurately calculate the circumference of the earth; Euclid, who was the father of geometry; and Archimedes (287–212 BCE), the great mathematician and astronomer. The engineer Hero was born in Alexandria where he developed the first vending machine, the force-pump and even a theatre of automated figures that could dance.

Egypt Under the Ptolemies

When Ptolemy I assumed rule of Egypt in 305 BCE after Alexander's death, he continued to push hard to make Alexandria the worldwide centre of trade and he succeeded mightily. For the next two centuries under Ptolemy rule, Egypt became a major player in international maritime trade. At the core of their trade network was their grain, a commodity that they had in surplus from their rich Nile Valley. Nowhere else in the world was grain so plentiful and the Ptolemies turned Egypt back into the 'grain supplier par excellence of the eastern Mediterranean' (Kelder *et al.*, 2018). The city states of Greece, while they could rely on many of their far-flung colonies for much of their grain needs, still depended greatly on Egypt to feed their masses.

But grain was not the only trade commodity that the Ptolemies had to offer the world. Egypt was the sole source of paper made from papyrus, the standard writing material of the ancient world. Only on the banks of the Nile did this plant grow. And then there were the spices that were flowing into Egypt from Africa, Arabia and India! The Greeks had their home-based saffron to taste, smell and use as a pigment, but lusted mightily for the other exotic aromas and flavours they could get only through international trade. These precious cargoes arrived in Egypt through their Red Sea ports, from which they were hauled across the desert to the Nile and then sailed downriver to Alexandria and across the Mediterranean. The Nabataeans played a central role in these trade activities, soon abandoning Gaza for Alexandria as their Mediterranean outlet (Anonymous, 2021). Many of them settled in Alexandria and immersed themselves into the Greek culture, even taking on Greek names.

Rome and Carthage Rise and Fight for Mediterranean Supremacy

According to legend, ancient Rome was founded in 753 BCE by the two brothers, Romulus and Remus. The story goes that the two brothers fought incessantly over who would rule the city until Romulus killed Remus and named the city after himself. Rome rapidly grew in size and power through trade, positioned perfectly on an easily navigable waterway to move its goods. Greek culture and civilization served as the model on which the Romans built their own culture, borrowing Greek literacy, religion and architectural foundations. The Etruscans to the north also provided a model 'for trade and urban luxury' (Cartwright, 2018).

While trade made Rome prosperous in its early years, its rising military power elevated it to domination in the ancient world. Its wars with the North African city of Carthage consolidated Rome's power and played a key role in the city's gaining wealth and prestige. Once Carthage was defeated, Rome held almost complete dominance over the Mediterranean region.

Carthage was founded before Rome in North Africa by Queen Dido in 814 BCE. According to Mark (2020):

> Dido was allegedly fleeing the tyranny of her brother Pygmalion of Lebanon, landed on the coast of North Africa, and established the city on the high hill later known as the Byrsa. The legend claims that the Berber chieftain who controlled

the region told her she could have as much land as an ox hide would cover; Dido cut a single ox hide into thin strips and lay them end-to-end around the hill, successfully claiming it for her people.

Carthage remained a small city for a long time until, in 322 BCE, the conquests of Alexander the Great greatly swelled its population with refugees from the city of Tyre. It continued to grow and ultimately developed a huge maritime trade with ports all over the Mediterranean. Tribute and tariffs contributed much to the city's wealth in addition to its expansive trade network. It also supported an impressive, extremely productive agriculture. The writer Mago of Carthage wrote a tome of 28 volumes on agriculture and veterinarian science that was considered so important it was spared by the Romans after Carthage's final defeat in 146 BCE (Mahaffy, 1889).

Carthage prospered greatly for centuries until 264 BCE when it began warring with Rome for control of Sicily. As Mark tells it:

> Control of Sicily was divided between Rome and Carthage who supported opposing factions on the island which quickly brought both parties into conflict directly with each other. These conflicts would be known as the Punic Wars from the Phoenician word for the citizens of Carthage (given in Greek as Phoinix and in Latin as Punicus). When Rome was weaker than Carthage, they posed no threat. The Carthaginian navy had long been able to enforce the treaty which kept the Roman Republic from trading in the western Mediterranean.
>
> When the First Punic War (264–241 BCE) began, however, Rome proved far more resourceful than Carthage could have imagined. Though they had no navy and knew nothing of fighting on the sea, Rome quickly built 330 ships which they equipped with clever ramps and gangways (the corvus) which could be lowered onto an enemy ship and secured; thus turning a sea battle into a land battle. After an initial struggle with military tactics, Rome won a series of victories and finally defeated Carthage in 241 BCE. Carthage was forced to cede Sicily to Rome and pay a heavy war indemnity.
>
> (Mark, 2020)

For the next several decades, Carthage seethed for revenge. The Second Punic War broke out in 218 BCE, when Hannibal rose to power and attacked Italy after crossing the Alps with his North African war elephants. Hannibal was able to occupy most of southern Italy for 15 years but was never able to pin down the Romans and win a decisive victory. A Roman invasion of North Africa eventually forced Hannibal to return to Carthage and he was defeated at the Battle of Zama in Tunisia in 202 BCE.

Carthage was again forced to pay reparations and even when the debt was paid the Romans wanted to keep the Carthaginians weak, ordering them to dismantle their city and move it further inland. Unsurprisingly, the Carthaginians refused, and the Third Punic War began (149–146 BCE). Rome laid siege to the city and after three years captured it and burned it to the ground. The great city lay in ruins for years until the Romans rebuilt it and redeveloped its agriculture. Carthage became Rome's breadbasket for centuries until it was conquered by the Germanic Vandals in 439 CE.

And Then the Romans Controlled Egypt

About a century into the Greek rule of Egypt, the Ptolemies began forming alliances with Rome. Over time, these alliances grew stronger and stronger in Rome's favour until Egypt was finally fully annexed in 30 BCE and made into the Roman province of Aegyptus. Full Roman control of Egypt came when Octavian (the future emperor Augustus) defeated his arch-rival Mark Antony and deposed his lover Queen Cleopatra VII, the last Ptolemy ruler of Egypt. The main Roman interest in Egypt was the regular and reliable delivery of grain to feed the city of Rome.

The Romans not only took over the commerce of the Ptolemies, but expanded it greatly, joining it with their expansive trade network stretching across the Mediterranean from Spain to Syria. As Casson described it:

> The Mediterranean was now one Roman world, with trade routes crisscrossing it in every direction. The major ones all led to Rome itself, the capital of this mighty state, whose population, perhaps a million or even more, had its needs met almost totally from sources overseas. The emperors, we are told, kept the city's mob contented with 'bread and circuses'; the bread was baked from fifteen million bushels of wheat imported annually, chiefly from North Africa and Egypt, and the games required shiploads of wild animals from Asia and Africa ... To handle its shipments, particularly the huge cargoes of grain, the city of Rome boasted a fleet of sailing ships which was the largest and heaviest as a class that the world was to know until the eighteenth century. To handle what these brought in, the emperors built from scratch just north of the Tiber's mouth a huge port with long quays backed by lines of warehouses. A thriving town, Portus 'The Port,' sprang up about it to house and feed the army of required personnel – clerks, shipwrights, bargemen, ferrymen, stevedores, and so on.
>
> (Casson, 1981, p. 40)

The Romans built massive trading vessels to move goods around the Mediterranean. According to Casson's description:

> The average seagoing Roman merchantman had a capacity of some 300 tons, while those that hauled grain from Egypt were veritable super-freighters of their day, measuring 200 feet in length and having a capacity of over 1,000 tons ... They were exceptionally strong. This was because of the unique way in which they were constructed, a feature that has become known only in recent years, thanks to the discovery and investigation by underwater archaeologists of dozens of ancient wrecks. Indeed, an American diving team has raised and restored almost the complete hull of a small freighter measuring 47 feet or 14 meters in length of the fourth century BC that was found off the north coast of Cyprus. Their findings reveal that Greek and Roman shipwrights did not fasten a vessel's planks to a skeleton of ribs, as has been standard European practice for centuries, but to each other by means of close-set mortise and tenon joints, thereby forming a tight shell of wood. Then into this shell they inserted a complete set of ribs as stiffening. The result was a hull of formidable strength that needed little or no caulking.
>
> (Casson, 1981, p. 42)

The Roman vessels sailed directly between ports, on the shortest line across open water. Casson (1981, p. 42) tells the story about Emperor Caligula who advised the Jewish prince Agrippa when he was going to head home to Palestine, to go by way of Egypt: "'[They] are crack sailing craft," he explained, "and their skippers the most experienced there are; they drive their vessels like race horses on an unswerving course that goes straight as a die.'" Not only did they sail the shortest route, right over open water, but the ships' great size promised a fair degree of comfort to passengers.

References

Anonymous (2020) Plutarch on the foundation of Alexandria. Translated by M.M. Austin. *Livius.org: Articles in ancient history*. Available at: https://www.livius.org/sources/content/plutarch/plutarchs-alexander/the-foundation-of-alexandria/ (accessed 26 March 2021).

Anonymous (2021) Alexandria, center of trade. *Nabataea.net*. Available at: https://nabataea.net/explore/travel_and_trade/alexandria-center-of-trade/ (accessed 26 March 2021).

Bernstein, W.J. (2008) *A Splendid Exchange: How Trade Shaped the World*. Grove Press, New York.

Cartwright, M. (2013) Ancient Greece. *Ancient History Encyclopedia*. Available at: https://www.ancient.eu/greece/ (accessed 28 January 2021).

Cartwright, M. (2016) Trade in the Phoenician world. *Ancient History Encyclopedia*. Available at: https://www.ancient.eu/article/881/trade-in-the-phoenician-world/ (accessed 28 January 2021).

Cartwright, M. (2018) Trade in the Roman world. *Ancient History Encyclopedia*. Available at: https://www.ancient.eu/article/638/ (accessed 28 January 2021).

Cartwright, M. (2019) Mycenaean civilization. *Ancient History Encyclopedia*. Available at: https://www.ancient.eu/Mycenaean_Civilization/ (accessed 28 January 2021).

Casson, L. (1981) Maritime trade in antiquity. *Archaeology* 34(4), 37–43.

Casson, L. (1991) *The Ancient Mariners: Seafarers and Seafighters of the Mediterranean in Ancient Times*, 2nd edn. John Hopkins University Press, Baltimore, Maryland.

Dewan, R. (2015) Bronze Age flower power: the Minoan use and social significance of saffron and crocus flowers. *Chronika V*, 42–55.

Gilboa, A. and Namdar, D. (2015) On the beginnings of South Asian spice trade with the Mediterranean region: a review. *Radiocarbon* 57(2), 265–283.

Kelder, J.M., Cole, S.E. and Cline, E.H. (2018) Memphis, Minos, and Mycenae: Bronze Age contact between Egypt and the Aegean. In: Spier, J., Potts, T. and Cole, S.E. (eds) *Beyond the Nile: Egypt and the Classical World*. The J. Paul Getty Museum, Los Angeles, California, pp. 9–17.

Key, J. (2006) *The Spice Route*. University of California Press, Berkeley, California.

Knox, J.S. (2017) Solomon. *Ancient History Encyclopedia*. Available at: https://www.ancient.eu/solomon/ (accessed 28 January 2021).

Mahaffy, J. (1889) The work of Mago on agriculture. *Hermathena* 7(15), 29–35.

Mark, J.J. (2009) Byblos. *Ancient History Encyclopedia*. Available at: https://www.ancient.eu/Byblos/ (accessed 28 January 2021).

Mark, J.J. (2018a) Israel. *Ancient History Encyclopedia*. Available at: https://www.ancient.eu/israel/ (accessed 28 January 2021).

Mark, J.J. (2018b) Alexandria, Egypt. *Ancient History Encyclopedia*. Available at: https://www.ancient.eu/alexandria/ (accessed 28 January 2021).

Mark, J.J. (2020) Carthage. *World History Encyclopedia*. Available at: https://www.ancient.eu/carthage/ (accessed 1 April 2021).

Namdar, D., Gilboa, A., Neumann, R., Finkelstein, I. and Weiner, S. (2013) Cinnamaldehyde in early Iron Age Phoenician flasks raises the possibility of Levantine trade with Southeast Asia. *Mediterranean Archaeology and Archaeometry* 13(2), 1–19.

Paine, L. (2013) *The Sea & Civilization: A Maritime History of the World*. Vintage Books, New York.

Smith, R. (2009) *Premodern Trade in World History*. Routledge, London and New York.

Willard, P. (2002) *Secrets of Saffron: The Vagabond Life of The World's Most Seductive Spice*. Beacon Press, Boston, Massachusetts.

5

Land of Punt and The Incense Routes

Setting the Stage – Land of Punt

The ancient Egyptian demand for frankincense and myrrh led their pharaohs to send massive trade expeditions south to a place they called the 'Land of Punt' (Saggs, 1989; Tyson, 2009). Huge ships were built and launched down the Nile, disassembled and dragged across the desert to the Red Sea and then sailed to southern Arabia and Somalia in search of frankincense, myrrh and other exotic riches.

Many detailed accounts of these missions are recorded in texts, paintings and bas-reliefs dating from 2500 to 1000 BCE. While numerous descriptions of Punt exist, the Egyptians left no maps, and no one knows for sure where they were going. However, they must have been headed to southern Arabia and the Horn of Africa, as these are the only places on earth where frankincense and myrrh trees are native (Watt and Sellar, 1996; Singer, 2006).

The scale of the Egyptian trade missions was simply astonishing. By this time, the Egyptians were building huge wooden ships propelled by wind and rowers. From the reign of King Sahure in the twenty-fifth century BCE, the 'Palermo stone' describes the first expedition to Punt which returned with 80,000 measures of myrrh, along with 23,030 staves of wood and alloys of gold and silver. Other wall inscriptions dating from 2000 and 1500 BCE portray expedition forces of 3000 and 3700 men travelling to Punt. One record from the Sixth Dynasty (2345–2181 BCE) describes a single official travelling on 11 different expeditions. An ancient papyrus from the twelfth century BCE records that Rameses III 'constructed great transport vessels ... loaded with limitless goods from Egypt ... They reached the land of Punt, unaffected by (any) misfortune, safe and respected' (Tyson, 2009).

In the temple of Queen Hatshepsut (1480 BCE) there is a breathtaking bas-relief that shows a flotilla of ships departing and returning laden with the treasures of the distant land of Punt (Fig. 5.1) (Tyson, 2009). In one panel, five mammoth ships are shown heading out to sea with 30 rowers straining hard on their oars. A pilot looms over them standing on the yard of a square-rigged sail. Fish are illustrated swimming in the waters beneath the ship in such detail that they can be keyed out today. Scenes of Punt are depicted in another panel showing the local chief and his family, palm and myrrh trees, domed huts on stilts and livestock. Another large scene with hieroglyphics portrays sailors 'loading the ships very heavily with the marvels of the Land of Punt: with good herbs of God's Land and heaps of nodules of myrrh, with trees of fresh myrrh, with ebony and pure ivory' (Tyson, 2009). Other trade goods are shown including gold, wood, eye paint, live animals (baboons, monkeys and hounds), leopard skins and servants. The story ends with scenes of the flotilla heading back out to sea and arriving joyfully back home. A triumphant procession of Egyptians and Puntites is illustrated presenting their bounty to Queen Hatshepsut.

Travel to Punt via the Red Sea must have been challenging to say the least. The Egyptians first sailed down the Nile from their capital at Thebes to Coptos, the port closest to the Red Sea. This was the easiest part of the journey, with northerly winds propelling the ships. Once landed, the ships were disassembled and transported by donkey caravan across 100 miles of foreboding desert to the Red Sea port of Saww (today's Mersa Gawasis). Here the ships were reassembled and then floated for another thousand miles on the Red Sea to Punt. The Red Sea was likely a navigational nightmare for the Egyptians, with its fierce storms, chains of reefs and submerged islands. When the expedition had finally arrived in Punt and their trading was completed, they headed back, repeating the whole arduous journey in reverse with an additional challenge. The prevailing winds on the Nile and Red Sea were northerly, requiring the ships to be propelled by rowing. Thirty straining oarsmen are depicted in the ships of Queen Hatshepsut.

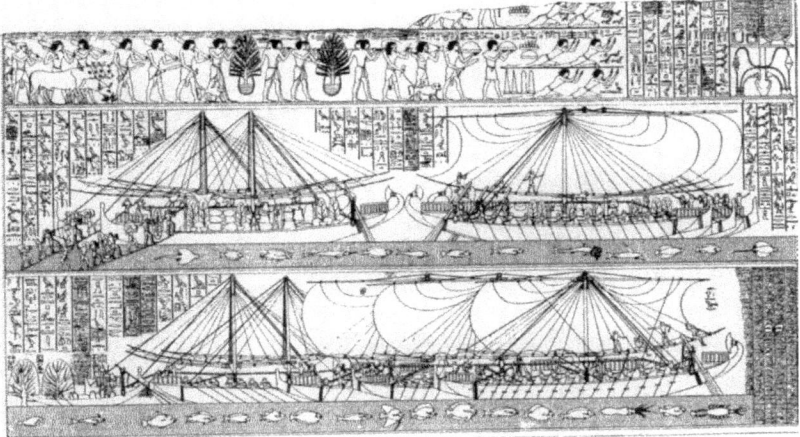

Fig. 5.1. Bas-relief at Deir el-Bahri, of a flotilla of Queen Hatshepsut's ships departing and returning from Punt (1480 BCE). (Source: https://maritimehistorypodcast.com/wp-content/uploads/2014/05/HatschepsutFlottePunt.jpg?x85308, accessed 28 January 2021).

Red Sea Trade after Rameses III

Egyptian power went into a steady decline after Rameses III. Egypt split into two kingdoms sometime after 1100 BCE and then fell completely under the control of the southern Nubians in 734 BCE. In 671 BCE, the Egyptian delta was conquered by the Assyrians, who made Egypt a vassal state until the locals regained control for a few decades. In 525 BCE, the Persians took over Egypt and held it until 332 BCE, when Alexander the Great made it part of his empire. When Alexander died, his general Ptolemy took over control and began a dynasty of Greek-speaking pharaohs. In 30 BCE Egypt became part of the Roman Empire after the death of Cleopatra.

From the death of Rameses III until Ptolemy, few expeditions were sent to Punt by the pharaohs of Egypt. It is likely that during this period Red Sea travel and trade became the purview of the Phoenicians based around Lebanon. The Book of Kings in the Bible describes a Phoenician expedition sent by King Solomon in the tenth century BCE to a place called Ophir to trade for gold, silver, ivory, gems, apes, peacocks and probably incense. The precise location of Ophir is not known, but it has been variously described as Sudan, Somalia and Yemen. It may even have been India, as the trip was said to take three years (Hourani, 1951).

Over the millennium of Egyptian decline before Ptolemy, it was only Pharaoh Necho II in the sixth century BCE who flexed much muscle in the Red Sea. He may even have commissioned an expedition by Phoenicians through the Red Sea and around Africa (Bernstein, 2008). Herodotus left us the following report in his *Histories*:

> Libya is washed on all sides by the sea except where it joins Asia, as was first demonstrated, so far as our knowledge goes, by the Egyptian king Necho, who, ... sent out a fleet manned by a Phoenician crew with orders to sail west about and return to Egypt and the Mediterranean by way of the Straits of Gibraltar. The Phoenicians sailed from the Arabian Gulf into the Southern Ocean, and every autumn put in at some convenient spot on the Libyan coast, sowed a patch of ground, and waited for next year's harvest. Then, having got in their grain, they put to sea again, and after two full years rounded the Pillars of Heracles in the course of the third, and returned to Egypt. These men made a statement which I do not myself believe, though others may, to the effect that as they sailed on a westerly course round the southern end of Libya, they had the sun on their right – to northward of them. This is how Libya was first discovered by sea.
>
> (Lendering, 2019)

This would make the first circumnavigation of Africa about 2000 years before Vasco de Gama, the explorer generally considered the first to sail round Africa!

Canal of the Pharaohs

It was not lost to the rulers of Egypt that a canal linking the Nile with the Red Sea would be preferable to dragging ships across the desert (Redmount, 1995).

Pharaoh Necho II in 600 BCE was probably the first to undertake such a venture, digging a deep-water canal about 130 kilometres (80 miles) long linking a tributary of the Nile called the Pelusiac with the Bitter Lakes, which at that time were connected to the Red Sea. Herodotus reports that over 120,000 slaves toiled and died in this venture. Necho's canal was not completed, however, when he became spooked that such a canal could subject his kingdom to invasion from the west.

When Persia's Darius the Great subjugated Egypt in 497 BCE, he revisited the dream of the canal across the Suez. He saw the canal as an opportunity to transport wheat to Persia more efficiently and bring back soldiers. He ordered that Necho's old canal be completed and dug wide enough for two triremes (war galley with three banks of oars) to move down it, side by side. This new attempt was also probably not completed, when the engineers of Darius realized that the water level of the Nile was lower than the Red Sea, meaning salty water would flow into the canal from the Bitter Lakes and ultimately poison irrigation waters.

Ptolemy II (282–246 BCE) may have been the first to complete the canal, by first dredging out the canal of Darius and then connecting it with the Red Sea through a series of locks that kept the salty waters at bay. This canal was maintained on and off through the subsequent Roman and Muslim empires but keeping it open was a continual struggle against shifting desert sands and silt deposits from the Nile.

The Rise of the Incense Kingdoms

The Arabian Peninsula was first settled between 6000 and 4000 BCE by Neolithic farmers originating from the Levant (Crassard and Drechsler, 2013). They brought with them sheep, goats and cattle, as well as the whole crop assemblage domesticated in the Near East. To support their agriculture in the desert climates of Arabia, these farmers learned to be masters of hydraulic engineering, building dams and water routes across the fertile valleys. These production methods proved so successful that they came to support magnificent cities, adorned with temples and royal palaces. When their agriculture was well established and producing dependable harvests, the people turned to the trade of frankincense and myrrh to enrich themselves, establishing the 4800-kilometre (3000-mile) long trading routes across the Arabian Peninsula to Egypt, Mesopotamia, the Levant and eventually Greece and Rome (Singer, 2006; Lawler, 2010).

Four kingdoms arose in the deserts of Yemen that came to control the Incense Route: Hadhramaut, Saba, Qataban and Ma'in (Singer, 2006). Each of these kingdoms spanned an alluvial valley with rich soils and a *wadi* (Arabic for 'valley') which brought them floodwaters during the monsoons. Tens of thousands of acres of farmland were maintained by these kingdoms. Each had its own language, a unique style of art and architecture, and its own gods.

Agriculture was the bedrock of their societies, but the trade of incense made them wealthy. Trees of frankincense and myrrh were located in only two of these kingdoms, but all had a hand in the early overland trade. These four states of ancient Yemen came to be called in Latin *Arabia Felix*, meaning 'happy Arabia'.

Of the four ancient Yemen kingdoms, Saba (called Sheba in the West) was the first to control the Incense Route and became fabulously wealthy (Stewart, 1978). Both the Qu'ran and the Bible refer to the legendary Queen of Sheba who ruled around 900 BCE. The Qu'ran observes 'that the country is ruled by a Queen who has been given everything and she has a magnificent throne' (*An-Naml* 27:23). The First Book of Kings in the Bible talks about her visiting King Solomon 'with a very great train, with camels that bare spices, and very much gold, and precious stones' (1 Kgs 10:2). She apparently was fascinated by this other powerful ruler, wanted to see the splendour of his court at first hand and question him about his god Yahweh. Myths abound about this visit, including stories that the queen had a club foot and a hairy leg that repulsed the king and the two had a round of lovemaking that produced a son who became a king of Ethiopia (Wood, 2011).

When the Queen of Sheba visited King Solomon in the ninth century BCE, Hadhramaut, Ma'in and Qataban were all essentially territories in the Sabaean Kingdom. Around 400 BCE, these territories wrestled free of Sheba and gained almost equal control of the incense trade routes. Over the next couple of centuries, the power of these kingdoms shifted back and forth until they were all defeated by an upstart Himyarite Kingdom from the south-western corner of Arabia in the last century BCE and first century AD. The Himyarite Kingdom dominated the incense trade for the next 500 years, raking in tremendous profits.

Domestication of the Camel

It was the domestication of the camel some 4000 years ago in Arabia that made the long-distance trade of incense possible. In fact, one of the greatest technological advances of humankind was the domestication of the dromedary, the one-humped camel *Camelus dromedarius* (Bernstein, 2008; Orlando, 2016). Donkeys had long been used to transport goods from one point to another, but they were poorly adapted to the harsh deserts of Arabia. They needed a regular source of water and food and wilted under the pervasive burning sun. The camel was much less limited.

The camel's ability to survive in desert conditions without water for long periods of time is rivalled by no other animal. Camels can endure losing 30–40% of their body weight in water and survive almost a week without drinking. In their hump, they can store as much as 35 kilograms of fat, providing them with a long-range energy source. Dromedaries can obtain much of their water reserves from desert vegetation, and when water is available, can drink almost 60 litres at a time. They also have a flexible 'thermostat' and do not sweat until their body temperature approaches $c.43°C$ (110°F).

The camel was likely first domesticated for its milk in the Horn of Africa about 3000 BCE and for transport in Arabia about 2000 BCE (Orlando, 2016). Camels are among that rare group of animals that have the set of characteristics necessary for domestication including docility, lack of fear of humans, ease of herding and the ability to breed in captivity (Diamond, 1999). Humans may have actually saved the camel from extinction, as the evidence is strong that they disappeared from the wild soon after they were domesticated. Camels were adapted well to surviving harsh desert conditions, but their slow loping pace made them an easy target for predators.

Caravan Routes

The great caravans that carried frankincense and myrrh from Yemen to the Mediterranean had to weave their way across vast stretches of largely inhospitable terrain. The traders had to travel around towering mountains, find water and shelter in desert infernos, and fend off bands of thieves. There was essentially only one trail in southern Arabia that met these necessities – a route from Shabwa to Timna, north-east to Ma'rib then Najran, where the trail separated (Fig. 5.2). All these cities were centred where important valleys entered the plain.

The frankincense after harvest was first transported by camels to Shabwa (Shabwah today), where Pliny tells us:

> a single gate is left open for its admission. To deviate from the high road while conveying it, the laws have made a capital offence. At this place the priests take by measure, and not by weight, a tenth part in honour of their god, whom they call Sabis; indeed, it is not allowable to dispose of it before this has been done: out of this tenth the public expenses are defrayed, for the divinity generously entertains all those strangers who have made a certain number of days' journey in coming thither.
>
> (Bostock and Riley, 1855–57, 12.32)

When the taxes had been paid, the merchants purchased their supplies for the northern journey. In the beginning, the merchants were not from Shabwa, but from the Kingdom of Ma'in. These traders were well known all across the civilized world and were responsible for the shipment of incense to Syria, Egypt and Assyria, and then Greece and Rome. Settlements of Minaean traders from the Mediterranean were also established at key points along the Incense Route, as well as in Egypt.

The caravan routes were lined by a series of way stations along the way, each collecting its own taxes (Singer, 2006). As Pliny further relates, 'all along the route, there is at one place water to pay for, at another fodder, lodging at the stations, and various taxes and imposts besides' (Bostock and Riley, 1855–57, 12.32). There would be local guides to pay, along with armed guards who travelled ahead of the convoy to root out bandits.

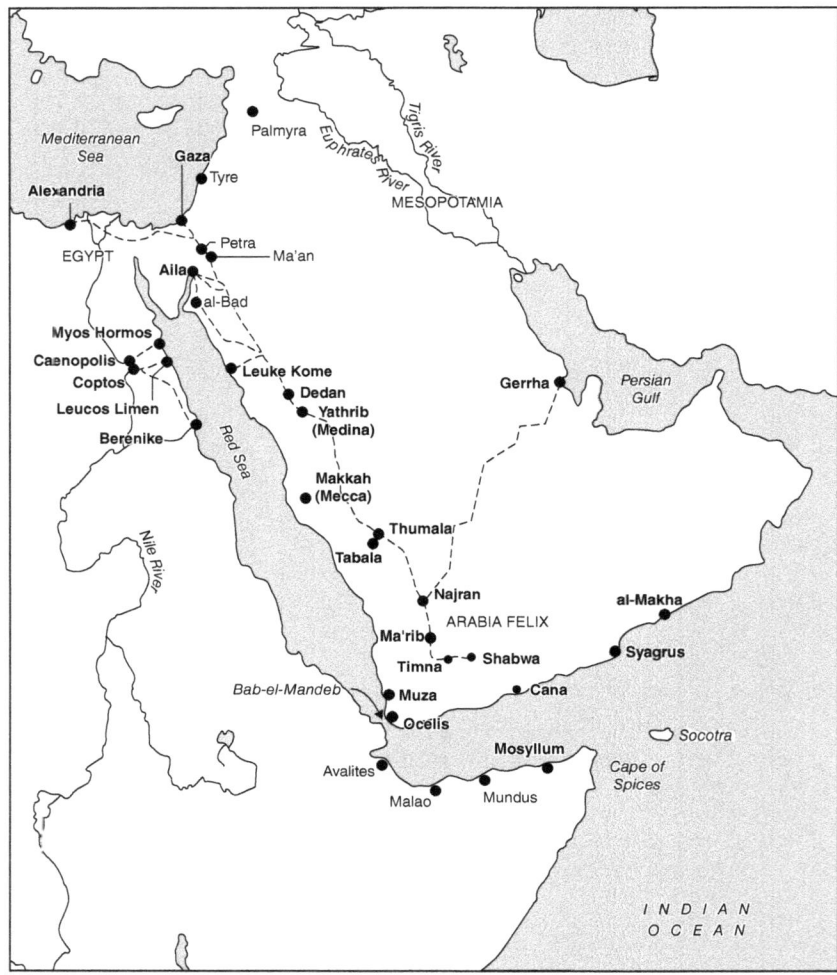

Fig. 5.2. Incense routes through ancient Arabia. (Redrawn from Singer, C. (2006) The incense kingdoms of Yemen: an outline history of the South Arabian incense trade. In: Peacock, D.P.S. and Williams, D.F. (eds) *Food for the Gods: New Light on the Ancient Incense Trade.* EBSCO Publishing, Ipswich, Massachusetts, Figure 2.6.)

The caravan leaders were well versed in the overland routes, knowing every oasis, well, shelter, settlement and taxation point. By 500 BCE, the caravans were huge, encompassing 200 camels or more. These large convoys provided mutual security against roaming bandits and protection against the harsh, unforgiving desert. A fully loaded camel could predictably travel at a rate of two-and-a-half miles an hour, so the caravan leaders could decide where and when the caravan would rest each day.

From Shabwa, the caravans journeyed south-east through the desert to the capital of the Qatabanian Kingdom, Timna. Here exotic goods from India, Ceylon and South East Asia could be added to their cargo for transport north. These goods were brought here from the South Arabian ports, principally Aden. The Greeks referred to Aden as *Arabia Emporion*, or 'Arabia's emporium'.

From Timna, the caravans headed north-east to Ma'rib, about 90 miles away in the kingdom of the Sabaeans. It could be reached directly through a desert with towering, dry sand dunes, but most merchants probably took the longer route through the mountains between the two cities. The Qatabanians had gone to great pains to build special passes through the most difficult points which allowed travellers to pass single file on levelled surfaces. Of course, there were numerous religious shrines and Qatabanian tax collectors along the way.

Ma'rib Dam

In Ma'rib, the travellers passed by a massive dam which has been called by many the engineering marvel of the age (Romey, 2015). As Stewart describes it:

> Faced with meticulously cut stone blocks, the dam spanned an 1,800-foot gap cut through the Balaq Hills by the Wadi Adhanah, and rose 15 feet above the watercourse, according to modern estimates. Awed ancient writers gave greater dimensions – 100 feet high and over 5,000 feet long – but they apparently confused the dam proper with its extensive associated irrigation system. This branched out from two massive stone and mortar abutments at the dam's north and south ends – they still stand today – that were connected to the sides of the Balaq Canyon by 25-foot thick walls. From the abutments, water was distributed to the extensive cultivated areas along the downstream banks of the wadi by a system of branching canals – the main one on the northern end more than a mile long – equipped with gates and sluices.
>
> (Stewart, 1978, p. 26)

The Ma'rib dam was twice as long as the Hoover Dam built in the Black Canyon of the Colorado River, USA in the 1930s. This Ma'rib dam was built in the seventh century BCE by a ruler named Sumhu' Alay Yanuf and his son Yatha'-Amar Bayyin. Its reservoir was used to irrigate nearly 10,000 hectares that supported large acreages of millet, barley, grapes, flax and sesame. The Ma'rib dam was maintained for over 1000 years by the Sabaeans and then the Himyarites, the civilization that replaced the Sabaeans.

The Sayhad Desert and Further Points North

It must have been a great relief for the travellers to have come to the green oasis surrounding the great Ma'rib dam and it is likely that the travellers lingered there for a few days, purchasing supplies and resting. Next facing them was the Sayhad desert, of which there was no way around. The Romans called it *Arabia*

Deserta ('desert Arabia') in comparison to *Arabia Felix* ('happy Arabia') from where they came. They dove deep into the desert until finally arriving at the oasis of Najran. Here the trail bifurcated, with one route heading north-east to the port city of Gerrha on the Persian Gulf and the other heading directly north to Yathrib (present-day Medina). Gerrha was a thriving metropolis of middlemen who bought aromatics and spices from the South Arabians and sold them goods from all over the known world including 'purple-dyed wools from Phoenicia, multicolored textiles from Persia and embroideries from Anatolia' (Singer, 2006, pp. 32–33). Yathrib served not only the caravans but also pilgrims heading to the holy city of Makkah (Mecca), a few days away. Yathrib also achieved fame as the place the Prophet Mohammed sought sanctuary in 622 CE 'after fleeing persecution in Mecca'.

From Yathrib, the merchants had the choice of heading north-east to Mesopotamia (Babylonia or Assyria) along the edge of the great Nafud Desert, or north-west to the Mediterranean. Those merchants heading to the Mediterranean would have continued trudging seven more days to Dedan, the capital of the Lihyanite Kingdom, where a thriving colony of Minaean merchants lived 2000 years ago. Hegra replaced Dedan late in the first century BCE when the Nabataean Kingdom took power in the region.

Merchants heading north to the Levant and Anatolia would have then travelled through Petra, the capital city of the Nabataeans. Petra served as an important hub of trade, linking the Red Sea ports of Egypt with southern Arabia, East Africa and the coast of India. It also came to serve as a terminus of the Silk Route reaching all the way to north-eastern China. The Nabataean port of Leuke Kome ('white port') on the Red Sea coast was the entry point of maritime trade into the Nabataean Kingdom. This activity will be described more fully in Chapter 6 (this book).

The last stop on the northern route was Gaza, the 'main spice entrepôt of the ancient Graeco-Roman world' (Singer, 2006, p. 33). This city sat on the southern edge of the Mediterranean, midway between Jerusalem and Alexandria, and was the terminus of all the overland trade routes carrying not only incense, but also spices and silk. From Gaza, the trade goods were shipped to Alexandria, which became the most important processing centre for goods destined for the Greek and Roman empires.

When the incense arrived at Alexandria, security reached fever pitch. Pliny relates:

> But, by Hercules! at Alexandria, where the incense is dressed for sale, the workshops can never be guarded with sufficient care; a seal is even placed upon the workmen's aprons, and a mask put upon the head, or else a net with very close meshes, while the people are stripped naked before they are allowed to leave work.
>
> (Bostock and Riley, 1855–57, 12.32)

At journey's end, the incense from southern Arabia had travelled a prodigious distance. Pliny the Elder estimated the distance between Timna, at the start of the Incense Route, and Gaza, at the northern terminus, was 2,437,500 steps or about 1500 miles (2400 kilometres). The journey would have taken about 65 days by camel. The incense that found its way through Gaza to Puteoli (today's Pozzuoli) on its way to Rome travelled another 1500 miles by sea for a total journey of about 3000 miles (4800 kilometres).

Profits Along the Way

Frankincense and myrrh generated income all along the supply chain. The wealth was shared by the families collecting the sap, the series of middlemen who procured the incense and serviced the caravans, the camel drivers trudging along the desert trails and the shippers moving cargo across the Mediterranean. Taxes were collected at multiple points along the way by all forms of authority, from priests, city authorities and national customs officials. The final import duty collected by the Roman government was a flat rate of a quarter of the cargo's final value (McLaughlin, 2014).

When the incense finally arrived in Rome it brought a rich profit (Bernstein, 2008). The best quality frankincense sold for 5–6 denarii a pound which equalled about a week of wages for a skilled labourer; myrrh received twice as much. The overall transport costs were 2 denarii, so the profits for a typical camel-load of 500 pounds of frankincense was 1500 to 2000 denarii. Each camel driver could handle six animals, which meant he was responsible for a total cargo worth as much as 12,000 denarii – a sum likely far more than the camel driver would earn in a lifetime.

The Great Intermediaries: the Nabataeans

By the end of the first millennium BCE, the mighty Nabataean Kingdom in northern Arabia had become the key intermediary in the trade of incense from southern Arabia to Rome. At its peak, the Kingdom of Nabataea spanned the Sinai Peninsula, the north-west corner of Arabia and across the Levant, between the Euphrates River and the Red Sea. It grew rich by controlling the end of the caravan trade from southern Arabia to the Mediterranean coast.

The Nabataeans were originally a nomadic people who roamed with their camel herds across the resource-poor deserts and were ignored by the powerful regimes that surrounded them. The Greek historian Diodorus (90 to 30 BCE) tells of a Nabataean who claimed:

> We live in the desert, in a land that has neither water, nor grain, nor wine, nor anything else needed by you. But we are not willing to be and so we have taken refuge in a land that lacks all the things that are valued amongst other people. We have chosen to live in the desert like harmless wild animals.
>
> (McLaughlin, 2014, p. 50)

In their travels across the desert, the Nabataeans had come across other nomadic peoples trading in incense. They realized that if they could get this incense to Mediterranean markets, they would be richly rewarded by foreign merchants. Using their camel herds to carry this precious cargo the final leg to Mediterranean ports, they did just this, and in the process became a wealthy and successful kingdom.

The Nabataeans managed to thrive as traders even though they were located in the middle of the western and eastern forces battling over the remains of the Seleucid Empire. They had numerous bloody tussles with the Greeks and Herod but kept their territorial independence for centuries as an ally of the Roman Empire (McLaughlin, 2014). The Nabataeans were an independent kingdom from the fourth to the first century BCE, and then a client state of Rome for another 100 years. They were not fully annexed by the Romans until 106 CE as *Arabia Petraea*.

Petra – Jewel of the Nabataeans

In 150 BCE, the Nabataeans built a great caravan city called Petra (Greek for 'rock') in the valley of the Dead Sea in Jordan. Called 'Wadi Moussa' (Arabic for 'valley of Moses'), it was located ideally between the southern incense routes of Arabia, the Red Sea ports of Aila and Leuke Kome and the western silk roads that wound their way through India and China.

Petra became an important:

> processing and repackaging center for many of the raw products acquired by the Nabataeans in their trading enterprises. Frankincense and myrrh were converted into medicines and cosmetics before they were sold to the Nabataeans international customers ... They diversified their merchandise into a wide array of exotic and luxury products – spices and incenses of many kinds, ivory, sugar, a glittering array of precious and semi-precious gems, and a curious assortment of strange creatures from far-off lands. There was a never-ending supply of buyers for everything on offer, especially among the wealthy clientele in the Greek and Roman as well as the Near Eastern worlds.
>
> (Bryce, 2014, p. 242)

In and around Petra, the Nabataeans built an astonishing network of massive shrines and tomb complexes. As described elegantly by Liu in his book *The Silk Road in World History* (Liu, 2010, pp. 24–26): 'Approaching the city from the peninsula's southern coast, the caravan trader would see a huge tomb network, marked by four monolithic obelisks glowing red under the Arabian sun.' Here the Nabataeans had also built an ingenious series of dams and dikes to protect them from flooding. 'The caravans continued through a gorge that the Arabs called the Siq. Flanked by high cliffs, the gorge was so narrow that only a thread of light could penetrate to the ground'. At a spring in this gorge, there was a larger-than-life group of statues representing robed men leading camels that pointed the way through the darkness to the 'shining city of Petra'.

Liu continues:

> Once through the dark passage, the sudden splendor of the façade of the huge temple of Khasneh would have dazzled the eyes of the traders. At the first level of the façade, six Corinthian columns supported the roof. Above the roof were three niches, also flanked by obviously a female deity flanked by two males.
>
> After Khasneh, the valley widened. To the left of the road, the Nabataean's sacred place of sacrifice, at the top of the highest mountain in the region, rose high above its summit ... Following the road westward, very close to the road, the caravan trader would have seen an amphitheater carved out of rose-colored sandstone ... After the traveler passed this theater, the valley opened into a basin. On the right one could then see, far away, a façade of Corinthian columns, adorned with uniquely Petraean conical tops, along the red sandstone cliffs. On the left, where the stream that was diverted above now joined the Wadi Moussa ... one finally reached the urban area of Petra ... [Here] a typical entrance in cities built in the Hellenistic style of the time greeted the traveler first entering the city: a shrine to the left and a nymphaeum, a structure with a fountain and a statue, to the right. Along the main street was arrayed a colonnade of freestanding red sandstone columns that extended for about three hundred yards along the river flowing through the Wadi Moussa. Shops and hotels with colonnaded porticos lined the main street; behind them, rows of houses climbed into the foothills. On the main street, monumental buildings dominated the scene. Three elevated platforms served as markets for the traders. On the other side of the main street, opposite the market, stood the royal place. Two major temples flanked the west end of the colonnade, which was marked by a huge gate decorated with bas-relief. From there, the traveler could see the largest shrine of the city to the west.
>
> (Liu, 2010, pp. 24–26)

Petra remained an important caravan city for centuries. From the second century BCE until the early second century CE, its leading citizens were well known to the Greeks and Romans (Liu, 2010). Petraean merchants supplied incense, spices and textiles to the Romans from Arabia, Africa and India and even settled in those far-off places. It wasn't until the second century AD that their trading status began to fade as silk and other goods began to flow into rival Palmyra in Syria via the Silk Route from China.

Maritime Incense Trade

For the first 400 years of their existence, the Nabataeans depended on overland routes to move frankincense and myrrh, but by the first century CE they began to shift their attention to the Red Sea. They loaded ships at island ports near today's Yemen, shipped the goods north by sea, and then unloaded them at the Arabian ports of Aila (today's Aqaba) and Leuke Kome. From there they were carried overland by camel caravan either north to Gaza or west to Alexandria in Egypt.

The Nabataeans began their Red Sea trade clandestinely with tiny Himyar, which had emerged in 115 BCE in the south-eastern corner of Arabia under the long shadow of the Sabaeans. The Himyarites supported themselves by secretly trading with the Nabataeans. They would float frankincense and myrrh to an island near their coast on rafts made of animal skins. There, out of sight from any mainland observers, the Nabataeans would load the incense on to their dhows and carry it up the Red Sea to their markets in the north.

The Nabataeans developed two Red Sea ports: Aila on the Gulf of Aqaba on the extreme north-eastern corner of the Red Sea near Gaza; and Leuke Kome, several hundred miles further south. The ancient geographer Strabo talked of the Nabataean sailors and suggested that it took about ten weeks to travel from the southern Arabian ports to Aila (McLaughlin, 2014). There were strong prevailing northern winds in the upper reaches of the Red Sea, encouraging many ships to dock and unload at Leuke Kome, from which camel caravans would carry the cargoes another 300 miles to Petra. In the Roman period, Leuke Kome became the premier trade port of the Nabataean Kingdom. Strabo, *Geography* (1903, 16.4.23) tells us that from there 'traders travel safely and easily on the route to and from Petra, and they move in such numbers of men and camels that they resemble an army.' The Romans maintained a garrison of soldiers there and a customs station that collected the standard 25% tax on all trade goods.

Roman Invasion of the Incense Route

Slowly but surely, the Himyarites grew in strength and power through their trade with the Nabataeans, and they turned their eyes towards the inland incense kingdoms. The Himyarite ambitions were aided in 24 BCE, when Nabataea was forced to support a Roman expedition to find and conquer the source of incense (Anonymous, 2021). The Romans had become greatly distressed at the negative impact of the incense trade on their economy, and Aelius Gallus, the Roman prefect of Egypt, decided they must cut out the middleman and find their own source of incense. The Nabataeans were at first dismayed at the prospect of helping the Romans but came to realize that a Roman invasion might weaken the Kingdoms of Saba, Ma'in and Hadhramaut, and aid their trading partners the Himyarites. Syllaeus, the brother of the king, was designated to help the Romans.

After an exploratory fleet deemed a maritime attack on South Arabia was not possible, a great force of 10,000 Romans, 500 Jewish soldiers sent by Judean King Herod and 1000 Nabataean soldiers led by Syllaeus headed south to conquer the incense states. It was Leuke Kome that received the Roman fleet. In a disastrous journey across the Red Sea and march through central and southern Arabia, thousands of the soldiers died of exposure and disease. It took the Roman general Aelius Gallus 80 days to get to the borders of the Sabaean Kingdom, crossing a hot and forbidding desert from oasis to oasis. They did manage to destroy key cities in Ma'in and Hadhramaut, but were forced by thirst, exhaustion and disease to abandon their quest outside the walls of

Ma'rib. By then many of the soldiers were suffering from scurvy-induced paralysis of their limbs, whose origin was unknown to the Romans. They assumed it must be the local water causing the symptoms (McLaughlin, 2014).

The 1000-mile journey back was an unmitigated disaster for the Romans, and Syllaeus was later executed for failing to aid them in their goal. However, the balance of power in the region had been shifted inadvertently to the Himyarites, who subsequently defeated the other weakened South Arabian kingdoms. They gradually wrestled control of the southern portion of the Incense Route and became almost the exclusive source of the world's frankincense and myrrh; Saba fell in 25 BCE, Qataban in 50 CE and Hadhramaut in 100 CE. With all but the Himyarites out of the picture, the Nabataeans gained a monopoly on the vastly lucrative incense trade between Arabia Felix and the Mediterranean via the Red Sea.

References

Anonymous (2021) History of Nabataea. *Nabataea.net*. Available at: https://www.nabataea.net/explore/history/history/ (accessed 31 March 2021).

Bernstein, W.J. (2008) *A Splendid Exchange: How Trade Shaped the World*. Grove Press, New York.

Bostock, J. and Riley, H.T. (1855–57) *The Natural History of Pliny. Translated with Copious Notes and Illustrations*, Vol. 3. Harry G. Bohn, London.

Bryce, T. (2015) *Ancient Syria: A Three Thousand Year History*. Oxford University Press, Oxford.

Crassard, R. and Drechsler, P. (2013) Towards new paradigms: multiple pathways for the Arabian Neolithic. *Arabian Archeology and Epigraphy* 24, 3–8.

Diamond, J. (1999) *Guns, Germs and Steel*. Norton, New York.

Hourani, G.F. (1951) *Arab Seafaring in the Indian Ocean in Ancient and Early Medieval Times*. Princeton University Press, Princeton, New Jersey.

Lawler, A. (2010) A forgotten corridor rediscovered. *Science* 328, 1092–1097.

Lendering, J. (2019) Herodotus on the first circumnavigation of Africa. *Livius.org: Articles in ancient history*. Available at: https://www.livius.org/sources/content/herodotus/herodotus-on-the-first-circumnavigation-of-africa/ (accessed 31 March 2021).

Liu, X. (2010) *The Silk Road in World History*. Oxford University Press, New York.

McLaughlin, R. (2014) *The Roman Empire and the Indian Ocean: The Ancient World Economy and the Kingdoms of Africa, Arabia and India*. Pen & Sword Books Ltd, Barnsley, UK.

Orlando, L. (2016) Back to the roots and routes of dromedary domestication. *Proceedings of the National Academy of Sciences USA* 113, 6588–6590.

Redmount, C.A. (1995) The Wadi Tumilat and the 'Canal of the Pharaohs'. *Journal of Near Eastern Studies* 54, 127–135.

Romey, K. (2015) 'Engineering marvel' of Queen of Sheba's city damaged in airstrike. *National Geographic*. Available at: https://www.nationalgeographic.com/science/article/150603-Yemen-ancient-Sheba-dam-heritage-destruction-Middle-East-archaeology (accessed 27 March 2021).

Saggs, H.W.F. (1989) *Civilization before Greece and Rome.* Yale University Press, New Haven, Connecticut.

Singer, C. (2006) The incense kingdoms of Yemen: an outline history of the South Arabian incense trade. In: Peacock, D.P.S. and Williams, D.F. (eds) *Food for the Gods: New Light on the Ancient Incense Trade.* EBSCO Publishing, Ipswich, Massachusetts, pp. 4–27.

Stewart, R.T. (1978) A dam at Marib. *Aramaco World: Arab and Islamic Cultures and Connections* 29(2). Available at: http://archive.aramcoworld.com/issue/197802/a.dam.at.marib.htm (accessed 28 January 2021).

Strabo, *Geography* (1903) H.C. Hamilton, Esq., W. Falconer, M.A., Ed. *The Geography of Strabo. Literally translated, with notes, in three volumes.* George Bell & Sons, London.

Tyson, P. (2009) Where is Punt? *NOVA.* Available at: http://www.pbs.org/wgbh/nova/ancient/egypt- punt.html (accessed 27 March 2021).

Watt, M. and Sellar, W. (1996) *Frankincense and Myrrh: Through the Ages, and A Complete Guide to Their Use in Herbalism and Aromatherapy Today.* C.W. Daniel Company Limited, Saffron Walden, UK.

Wood, M. (2011) In search of myths and heroes: the Queen of Sheba. *PBS.* Available at: http://www.pbs.org/mythsandheroes/myths_four_sheba.html (accessed 28 January 2021).

Origins of the Spice Trade in the Indian Ocean

Setting the Stage – Central Role of Rivers

The ancient civilizations of Egypt, Mesopotamia and Harappa all evolved along mighty rivers that provided rich alluvial soils, irrigation waters and transportation. The Tigris and Euphrates Rivers were the backbone of ancient Sumer and the Mesopotamian kingdoms that followed. The ancient Harappa civilization relied on the Indus River. The Nile served as the lifeblood of the Egyptian kingdoms. These rivers fed the agricultural productivity of these great states and served as a vehicle to move grain, ores, stone and luxury products.

Ultimately, these great societies became connected by their seafarers following their rivers through linking seas and gulfs to the Indian Ocean. Once linked, they became united through maritime trade. This trade was driven by the human desire for the exotic – manufactured goods, precious stones, metal ores, unusual animals and spices. In particular, the spices cinnamon, ginger and pepper came to generate the lion's share of profits to the merchants willing to tackle the treacherous sea journeys. The movement of these spices came to be known as the 'Spice Route' and eventually spanned 15,000 kilometres (9300 miles) from the west coast of Japan, through the islands of Indonesia, around India to the Middle East and across the Mediterranean to Europe.

Persian Gulf Routes

Water transport down the Tigris and Euphrates was likely born about 6000 years ago to transfer copper from the Ergani mines in Anatolia to the southern Sumerian settlements around Uruk in southern Iraq. Travel down the turbulent rivers would have been rapid with prevailing north to south winds; upstream travel would have been very difficult. The earliest vessels of the Mesopotamians were made of reeds or skins to prevent their hitting the bottoms of shallows (Paine, 2013). Many were round like huge reed baskets. They also used rafts made buoyant with inflated animal skins or airtight ceramic pots that could be disassembled upon arrival and their pieces sold or transported back to their source.

The world's first literate people, the Sumerians, settled in Mesopotamia around 3200 BCE, several hundred years before the start of Dynastic Egypt. They built their cities along the Tigris and Euphrates Rivers and actively traded grain for metal and building materials. The 'Fertile Crescent' of Mesopotamia supported a bounteous breadbasket but had no stone, metal or hardwood timber. These had to be brought from distant points north and south to build the cities of the Sumerians.

By 3000 BCE, Mesopotamia had maritime links through the Persian Gulf all the way to the Indus Valley, a network covering over 1850 kilometres (1150 miles) (McIntosh, 2008; Harappa.com, 1995–2021). The furthest reach of Mesopotamian trade was what the Sumerians called Meluhha, where the Harappan civilization thrived. Meluhha covered an area of over a million square kilometres, spanning from the highlands of Afghanistan to western India.

The Sumerians did not travel all the way to the Indus Valley, but the Meluhhans did travel and dock in Mesopotamian ports, and some of them even settled in Sumer. The distinctive Harappan clay and porcelain tags used to seal the rope around bundles of goods have been found throughout the Persian Gulf region. Mesopotamian ships sailed only as far as the western coast of the Gulf to modern-day Oman, which in antiquity was called Magan. The main trading port of the Sumerians was Dilmun, a great city that arose on an island off the central coast of the Persian Gulf in today's Bahrain. Here, the Sumerians exchanged shipments of grain for ore. Harappan ships also travelled to Dilmun and by the second millennium BCE this city was the centre of all Persian Gulf trading.

The evidence suggests that many more trade goods were moving from the Indus Valley to Mesopotamia than vice versa. The Indus Valley exported carnelian stones and stone beads, copper and bronze fishhooks, cotton textiles, live chickens, ivory seals and boxes, gold jewellery, shell and bone inlays, timber and water buffaloes, while Mesopotamia exported only bitumen, silver, tin and copper ingots, and woollen textiles.

The spices cinnamon and ginger grew wild in proximity to the Indus Valley and could also have found their way from there to Mesopotamia, but there is no evidence of their use by either the Harrapans or Mesopotamians. The Assyrians and Babylonians are also thought to have carried on an extensive trade with the Indian subcontinent in cardamom and cinnamon, but no physical evidence of this remains (Ravindran, 2017). The earliest evidence of spice trade from India to Mesopotamia comes from the traces of cinnamaldehyde found in Iron Age Phoenician clay flasks (eleventh to tenth century BCE) used to ship liquids.

To the Red Sea and Beyond

The Red Sea and India–Gulf trading spheres were probably fully connected by the end of the third millennium BCE (Boivin *et al.*, 2009; Fuller *et al.*, 2011). Several coastal societies became involved from southern Arabia, Africa and in the Indus Valley. It was during this period that the first African crops reached India and became important staples in the drier regions, including sorghum, pearl millet and finger millet. Moving in the reverse direction was Asian broomcorn millet, which likely began its journey from China. Zebu cattle were also introduced to Yemen and East Africa from India about this time. These zebu hybridized with local cattle to produce hybrids that were critical 'in the long-term success and southernmost spread of cattle pastoralists in eastern Africa' (Fuller *et al.*, 2011, p. 547).

By the second millennium BCE, ships were sailing regularly between India and Arabia to conduct trade. Their travel was greatly facilitated by the seasonal monsoon winds that blow from the south-west during the summer and from the north-east in winter. Most of the goods from India were carried to Aden in south-eastern Arabia and then delivered to Egypt on the overland Incense Route or across the Red Sea.

For materials to get from the Red Sea to Egypt they had to travel overland to the Nile via the 240-kilometre- (150-mile)-long desert corridor called the Wadi Hammamat. We have already discussed how dissembled boats were dragged to the coast by the Egyptian state during the Dynastic Period to travel down the Red Sea in search of Punt to acquire incense and other exotic goods. We also discussed how the Nabataeans began shipping most of their incense to Egypt through the Red Sea, loading ships at their island port in the south-west corner of Arabia. They also carried spices, textiles and pearls brought to their ports by Indian traders. These goods were shipped to Leuke Kome on the Red Sea and then moved by camel caravan either north to Petra or west to Alexandria and Gaza.

The trans-Arabian Sea transport provided the Arabians with a wide range of items besides incense to trade with Egypt and the Mediterranean. To their own frankincense and myrrh were added such luxury items as cinnamon, cassia, gemstones and pearls from South East Asia and ivory from Africa. Of course, the Arabs led their northern trading partners to believe that they were the origin of all these goods, discouraging Egyptian and Mediterranean sailors from establishing direct trade routes to India.

Early Indonesian Seafarers

The Indians were not the only peoples trading spices with the West. By the end of the first millennium BCE, large quantities of cinnamon, cassia and perhaps other spices were also being carried to the West from South East Asia by Indonesian mariners (Loewe, 1971; Brierley, 1994). The cassia was first shipped to Indonesia from China in exchange for cloves. Large outrigger cargoes of spices were then piloted to Madagascar by native Indonesians and then on to Rhapta in the Rufiji Delta, south of today's Dar es Salaam. From there the spices were transported to Cape Guardafui in Somalia, known to the ancient Egyptians as Punt. From the east coast of Africa, the cassia was finally delivered to North Africa and the Mediterranean via the Red Sea.

The Indonesians established a large trading colony in Madagascar. Even today, there is strong evidence of Indonesian culture in Madagascar, evinced by 'the ceremonies associated with the cultivation of wet and dry rice, the cattle that work the plough, the types of fish traps used on the island, the houses built on stilts, and the many words in the Malagasy language which have Indonesian roots' (Brierley, 1994, p. 12).

The trip from Indonesia to Madagascar and back must have been epic. The outrigger canoes loaded with spices were almost totally dependent on the whims of the sea. The Roman Pliny recorded in his encyclopaedia that the 'Troglodytes' of Africa obtained cinnamon from others who:

> carry it over vast tracts of sea, upon rafts, which are neither steered by rudder, nor drawn or impelled by oars or sails. Nor yet are they aided by any of the resources of art, man alone, and his daring boldness, standing in place of all these; in addition to which, they choose the winter season, about the time of the equinox, for their voyage, for then a south easterly wind is blowing; these winds guide them in a straight course from gulf to gulf, and after they have doubled the promontory of Arabia, the north east wind carries them to a port of the Gebanitæ, known by the name of Ocilia. Hence it is that they steer for this port in preference; and they say that it is almost five years before the merchants are able to effect their return, while many perish on the voyage.
>
> (Bostock and Riley, 1855–57, 42.19)

Royal Road of the First Persian Empire

In the sixth century BCE, the Persian Achaemenid Empire (530–330 BCE) took control of the Middle East and spread across a wide area encompassing Egypt in the west to Anatolia in the north, and through Mesopotamia to the Indus River in the east. On land, their trade centred on the Royal Road (Fig. 6.1), a communication and trade route built by Darius the Great which extended for nearly 2400 kilometres (1500 miles) from Sardis in Asia Minor through Mesopotamia and all the way down the Tigris to Susa (Hirst, 2018). Because the Royal Road follows a rather tortuous route between the important cities of the Persian Empire, it is likely that Darius's Royal Road was cobbled together from more ancient routes. Many archaeologists believe the western sections of the road may have originally been built by the Assyrian kings. Its other precursors were a road built by Cyrus the Great between Gordion and the coast during his conquest of Anatolia and a series of ancient roads built by the Hittites in the tenth century BCE to facilitate trade.

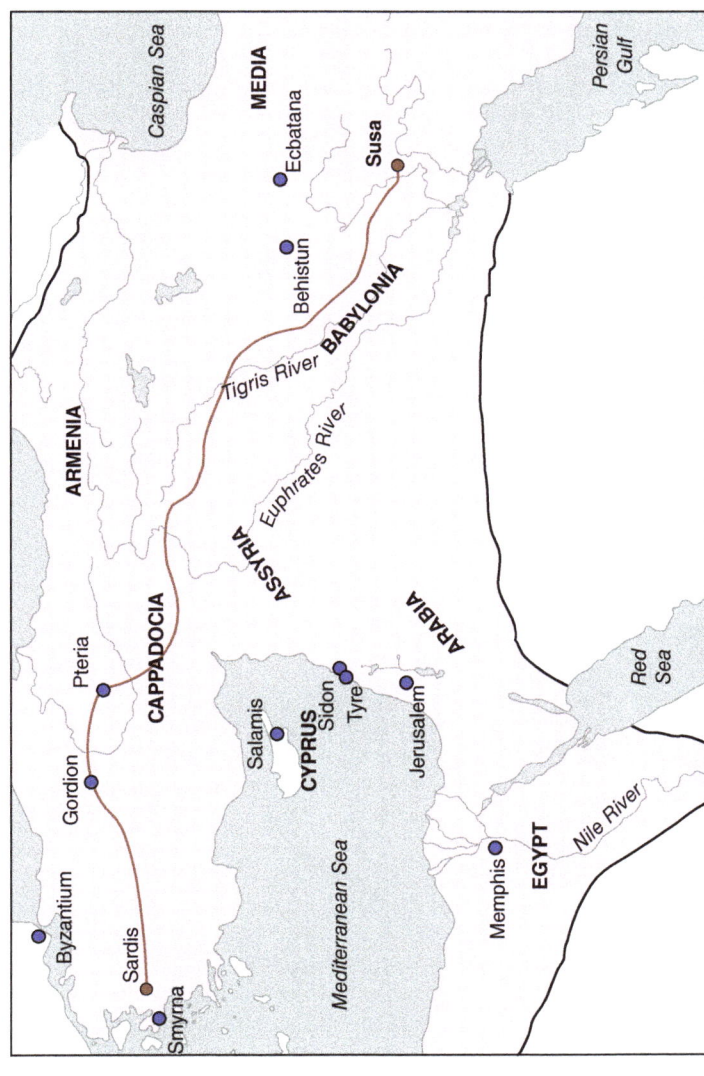

Fig. 6.1. Persian Royal Road in the sixth century BCE. (Redrawn from MossmapsCorrections according to *Oxford Atlas of World History 2002*, *The Times Atlas of World History* (1989), *Philip's Atlas of World History* (1999) by पालनीहार, CC BY-SA 4.0 <https://creativecommons.org/licenses/by-sa/4.0>, via Wikimedia Commons. File: Achaemenid Empire at its greatest extent according to Oxford Atlas of World History 2002.jpg)

Most of what is known about the Royal Road comes from descriptions of Herodotus about the Achaemenid postal system. He wrote:

> There is nothing that travels faster, and yet is mortal, than these couriers; the Persians invented this system, which works as follows. It is said that there are as many horses and men posted at intervals as there are days required for the entire journey, so that one horse and one man are assigned to each day. And neither snow nor rain nor heat nor dark of night keeps them from completing their appointed course as swiftly as possible. The first courier passes on the instructions to the second, the second to the third, and from there they are transmitted from one to another all the way through, just as the torchbearing relay is celebrated by the Hellenes in honor of Hephaistos. The Persians call this horse-posting system the angareion.
>
> (Strassler, 2009, p. 642)

On the Royal Road, couriers riding in relays could travel the distance in seven to nine days, while normal traders on foot would take perhaps three months (Hirst, 2018). Most of the road was unpaved, but a few intact sections of paved road still exist in the cities of Gordion and Sardis. These are built with cobblestones, are 4.8 to 7.6 metres (16 to 25 feet) wide and have a curbing of dressed stone. Along the route were way stations for postal couriers and ordinary travellers. One hundred and eleven of these rest stops have been identified between Susa and Sardis, where fresh horses were kept for the couriers and lodging could be found for the simple travellers. Royal outposts also dotted the route manned with soldiers to protect travellers.

Persian and Greek Explorations

The Persians also maintained robust maritime trade networks extending across the eastern Mediterranean Sea and south through the Persian Gulf to India; however, there is little evidence they went in the other direction from the Gulf along southern Arabia to the Red Sea. This lucrative trade route was left to sailors from the Indus Valley and southern India, and perhaps a few Arabians who ventured that far east.

The great Persian ruler Darius I (521–485 BCE) recognized the value of making such a link and commissioned the Greek Scylax of Caryanda to explore the Indus River and chart a way to the Red Sea. In a 30-month journey, Scylax followed the coastline of Pakistan and Iran, crossed the Gulf of Oman to the Arabian Peninsula, entered the Red Sea and travelled all the way to the Suez. Scylax wrote his own periplus about this trip that is quoted by Herodotus, as well as Aristotle, Strabo and Avienus, but no manuscript has survived.

Despite these groundbreaking discoveries, none of Darius's successors shared his enthusiasm for this potential trade route, and nothing more happened until Alexander the Great picked up the quest and initiated his own missions of discovery. When his eastward march of conquest stalled in 325 BCE at the Indus River (bowing to the wishes of his army), he commissioned his naval officer Nearchus to explore the Indus River and find his way to the Persian Gulf. Alexander then led his army across the Arabian Peninsula intending to link up with Nearchus on the other side near the Strait of Hormuz. The two did eventually find each other, although Alexander had a disastrous march across the oppressive Gedrosian Desert, losing a third of his men and most of his horses to the blazing sun and burning sand.

Alexander's last great goal was to possess all of Arabia and establish a trade route linking India and Egypt (Ashley, 1998). He knew little about Arabia and sent four naval expeditions to reconnoitre its coast and nearby islands. Three of these journeyed from the Indian Ocean and explored the eastern side of Arabia. The other voyage led by Anaxicrates sailed west from the Gulf of Suez as far as Yemen, where he discovered the source of frankincense and myrrh before being forced to turn back due to diminishing water supplies. Alexander now had all the intelligence needed to begin a trade network from Egypt to India, but his untimely death delayed its implementation for a few more years. It wasn't until his former general Ptolemy I took over as ruler of Egypt that the first Greeks forged a trade route to India.

Arab Stranglehold on Egyptian Trade

For almost 2000 years the Arabs had a stranglehold on the spice trade to Egypt. They dominated Red Sea trade and kept the Egyptians completely in the dark as to where spices originated. In the first century CE, the Egyptians' myopic view of the world was dramatically altered as an accidental consequence of war.

Schoff described very succinctly the central role the Arabians played in worldwide trade:

> With the spread of culture in both directions, Egypt and the nations of Ancient India came into being, and a commercial system was developed for the interchange of products within those limits, having its center of exchanges near the head of the Persian Gulf ... The growth of civilization in India created an active merchant marine, trading to the Euphrates and Africa, and eastward we know not whither. The Arab merchants, apparently, tolerated the presence of Indian traders in Africa, but reserved for themselves the commerce within the Red Sea; that lucrative commerce which supplied precious stones and spices and incense to the ever-increasing service of the gods of Egypt. This was their prerogative, jealously guarded, and upon this they lived and prospered according to the prosperity of the Pharaohs. The muslins and spices of India they fetched themselves or received from the Indian traders in their ports on either side of the Gulf of Aden; carrying them in turn over the highlands to the upper Nile, or through the Red Sea and across the desert to Thebes or Memphis.
>
> One Arab kingdom after another retained the great eastern coast of Africa, with its trade in gold and ivory, ostrich feathers and oil; the shores of the Arabian Gulf produced an ever-rising value in frankincense and myrrh; while the cloths and

precious stones, the timbers and spices – particularly cinnamon – brought from India largely by Indian vessels, were redistributed at Socotra or Guardafui, and carried to the Nile and the Mediterranean. Gerrha and Obollah, Palmyra and Petra, Sabbatha and Mariaba were all partners in this commercial system. The Egyptian nation in its later struggles made no effort to oppose or control it. The trade came and the price was paid.

(Schoff, 1912, pp. 3–4)

The Arabs did everything in their power to protect their Indian trade monopoly. As Schoff further relates:

No information was allowed to reach the merchants in Egypt, and every device the imagination could create was directed toward discouraging the least disturbance of the channels of trade that had existed since human memory began. And in an unknown ocean, with only the vaguest ideas of the sources of the products they sought, and the routes that led to them, it might have been many years before a Roman vessel, coasting along hostile shores, could reach the goal. But accidents favored Roman ambition.

(Schoff, 1912, pp. 3–4)

War Elephants and Red Sea Travel

In the post-Alexander wars between the Ptolemaic Dynasty in Egypt and the Seleucid Empire in Syria, elephants came to play a key battlefield role as essentially tanks (Bernstein, 2008; Paine, 2013). To obtain these elephants, the Egyptians had no other option but to send missions down the Red Sea to Africa to get them. The pachyderms were captured in Sudan and Eritrea, transported about 480 kilometres (300 miles) on specifically designed ships called *elephantegoi* and then marched across the desert from Red Sea ports to the Nile, where they ultimately found their way to the battle zones. They began their land journey from the fortress city of Berenike on the Red Sea.

The sea travel with elephants must have been particularly harrowing. As the Greek historian and geographer Agatharchides describes in his *On the Erythraean Sea* (113 BCE):

But the elephant transports, which ride deep in the water because of their weight and are burdened with their gear, encounter great and terrible dangers in these areas. For running with sails set and often continuing through the night, because of strong winds, they are wrecked when they run aground on the rocks and submerged bars. The sailors are unable to disembark because generally the water is deeper than the height of a man. When they do not succeed in saving their ship with their poles, they throw everything overboard except the food. If they do not escape in this way, they fall into great despair because there is neither island nor headland nor another ship to be seen in the vicinity. For these places are completely inhospitable and rarely do people sail through them in ships.

(Bernstein, 2008, pp. 141–142)

On to India

Once the Greeks had established regular Red Sea travel, the obvious next step was to discover the way to India and the Far East. In his *Geography* (2.3), the Greek historian Strabo writes that around 120 BCE, a shipwrecked Indian sailor was discovered washed up on the Red Sea coast of Egypt and was taken to Alexandria, where he learned Greek. When he was sufficiently fluent, he told his benefactors that he was a Tamil from southern India and had sailed on a vessel across the Indian Ocean to Arabia, and then north up the Red Sea. As a reward for his salvation, he taught his hosts how to make the trip themselves.

Strabo relates in his *Geography* (2.3) that the first journey from Egypt to India was made by a Greek sea captain named Eudoxus of Cyzicus in 118 or 116 BCE, guided by the Indian sailor. He likely 'hugged the southern and then the eastern Arabian shores to the Strait of Hormuz at the mouth of the Persian Gulf and finally navigated the coasts of what are now Iran and Pakistan to the southern Indian trading centers, a total distance of 5,000 miles' (Bernstein, 2008, p. 38).

Along the way, Eudoxus must have discovered the Indian Ocean monsoon winds. The Indian Ocean maintains its temperature; while the Asian landmass heats up in summer, producing low atmospheric pressure, and cools down in winter, producing high pressure. The changes in atmospheric pressure fuel seasonal winds that blow from the north-east between November and April and from the south-west between May and October. This allowed Greek traders to cross the Arabian Sea in a few weeks, trade in Indian ports while they waited for the winds to shift and then return again in a matter of weeks.

It is likely that very few Greek ships sailed all the way to India from Red Sea ports. The trip was long and dangerous, subject to piracy and massive storms, and the customs rates were brutal: one-quarter in Egypt and one-fifth in India (McLaughlin, 2014). The Greeks preferred to meet the Indians in the middle, first on Socotra Island near the Horn of Africa and later at Aden on the coast of today's Yemen.

The Roman Sea

The amount of trade passing through Red Sea ports was upped dramatically when Emperor Augustus brought Egypt under Roman control in 30 BCE. As described by Singer (2006, p. 42), this event 'ushered in a period of great prosperity and peace. With Egypt and its Red Sea ports now part of the Roman empire, the volume of sea-traffic increased at an incredible rate.' Strabo, *Geography* (1903, 16.4) reports that 'in his day 120 vessels sailed regularly from Egypt to India, whereas previously very few made the journey.'

Goods were first shipped to Alexandria from the Roman ports of Ostia or Puteoli and then to Coptos on the Nile. It took about 20 days for the ships to reach Alexandria from Italy and another 11–12 days to move the goods down the Nile to Coptos. A cornucopia of Roman goods flowed into Coptos including sacks of gold and silver coins, fabricated tin, copper, iron, locally produced barley, wheat and sesame oil, Alexandrian glass vessels, grape juice and wine from Italy and Syria, and purple cloth from Phoenicia. Coptos was a beehive of commercial trading and transport, and a menagerie of merchants and financiers collected there from Rome, Egypt, Arabia and India.

The merchandise arriving at Coptos was carried overland by camel caravans to the Red Sea ports of Berenike and Myos Hormos. These caravans passed through a rugged desert under a hot and unrelenting sun, with bandits hiding in the surrounding hills ready to pounce on them at any time. Sidebotham relates that:

> The traveler [hoping to] journey between the Nile and Berenice had to deal with an unforgiving hyper-arid environment, which was often the haunt of bandits seeking refuge from authorities on Nile, of tax evaders and other malcontents, and of local 'nomads' who when possible robbed wayfarers or extorted protection money from them.
>
> (Sidebotham, 2011, p. 3)

The pressure from bandits encouraged the Romans to build garrisons along the desert trails. These stations, called *phrouroi*, could be quite large, housing hundreds of soldiers (McLaughlin, 2014). Watchtowers were set up in the hills surrounding the trails and when sentries observed suspicious characters, a chain of signals was passed between the watchtowers to alert the nearest garrison. Armed patrols also accompanied some of the larger and more important caravans for the entire journey.

The Romans also built fortified way stations along the route called *hydreumata*. Some of these were huge affairs providing hundreds of people with respite from the unrelenting heat of day. At these stations, Roman engineers had either dug deep wells where groundwater was available or built cisterns to collect and store rainwater. In the time of Pliny, there were eight of these stations between Coptos and Berenike and this number rose to at least 18 over the ensuing decades (McLaughlin, 2014).

It took about seven days for the caravans to make their way to Myos Hormos (177 kilometres, 110 miles) and 12 days to more southerly Berenike (370 kilometres, 230 miles) (Cappers, 2006; Sidebotham, 2011). Myos Hormos was closer to Coptos than Berenike, but the strong northerly winds of the upper Red Sea made voyages to southern African ports slower and more difficult from there; Berenike was sheltered from the high northern winds by the Ras Banas Peninsula and ultimately became the premier trade emporium on the Red Sea. It remained a major trade centre for almost 800 years, linking the Mediterranean basin, Near East and Egypt with the African coast, India, China and South East Asia.

Two major trade routes emerged under the Romans that spanned about 4800 kilometres (3000 miles) (Cappers, 2006; McGrail, 2015). There was a southern African route that went down the Red Sea coast, through the

Bab-el-Mandeb Strait and then along the eastern coast of Africa to Rhapta, close to present-day Dar es Salaam. The other route also went through the Red Sea and Bab-el-Mandeb Strait but then went east across the Indian Ocean to ports in India. The full journey down the coast of Africa from Egypt took about two years to complete, while that to India and back was closer to a year.

Periplus of the Erythraean Sea

Incredibly, a detailed eyewitness description of ancient travel to Africa and India has been left to us in the *Periplus of the Erythraean Sea*, written by an unknown, Greek-speaking Egyptian author in the first century CE (Schoff, 1912). The Erythraean Sea was the ancient name for the body of water located between the Horn of Africa and the Arabian Peninsula. In this remarkable account, the conditions of the routes are described, along with the emporiums and anchorages along the way (Fig. 6.2), the demeanour of the locals and the major imports and exports. Eudoxus' navigator Hippalus is credited in this account with discovering the monsoon winds, although we now know that Indians had been travelling to Arabia and back for thousands of years.

Ships carrying goods bound for both Africa and India left the Red Sea ports between July and September to catch the favourable northerly winds. The ships were steered south down the middle of the Red Sea as much as possible to avoid the dangerous coastlines. The ships bound for African ports then headed to Cape Guardafui on the Horn of Africa and then south to Rhapta, hugging the African coast. They did not venture at all to the Arabian coast. The ships bound for India sailed to the ports of Aden and Cana on the southern coast of Arabia, and then grabbed the monsoon winds across the open waters of the Indian Ocean to south-west India. There they visited ports along the Indian coast from Barbarikon, on the Indus River, Muziris (Kodungallur) on the south-western Malabar Coast and then Ceylon. Important trading partners were the Tamil dynasties of the Pandyas, Cholas and Cheras in southern India. From India they headed home, rarely travelling deeper into South East Asia.

At each stop along the way, different local commodities were offered for trade, sometimes for gold and coins and other times for barter including goods like cloth, silver and gold statues, cereals, wine and olive oil. Frankincense and myrrh from South Arabia were very popular in India along with gold and silver specie, for which the Indians traded their locally produced pepper, cotton and pearls, along with silks they had obtained from Chinese traders. In the trip down the African coast, Egyptian linens, glass, wine and metal products would be traded for African ivory, tortoiseshell, myrrh and frankincense, along with cinnamon, Indian cloth, sashes and fine muslins obtained from their trade with Indian merchants.

The first major spice trade centre in the world became Muziris, located in the present-day Indian state of Kerala on the south-western coast of India (Perur, 2016). The exact location is not known. Probably established by 3000 BCE,

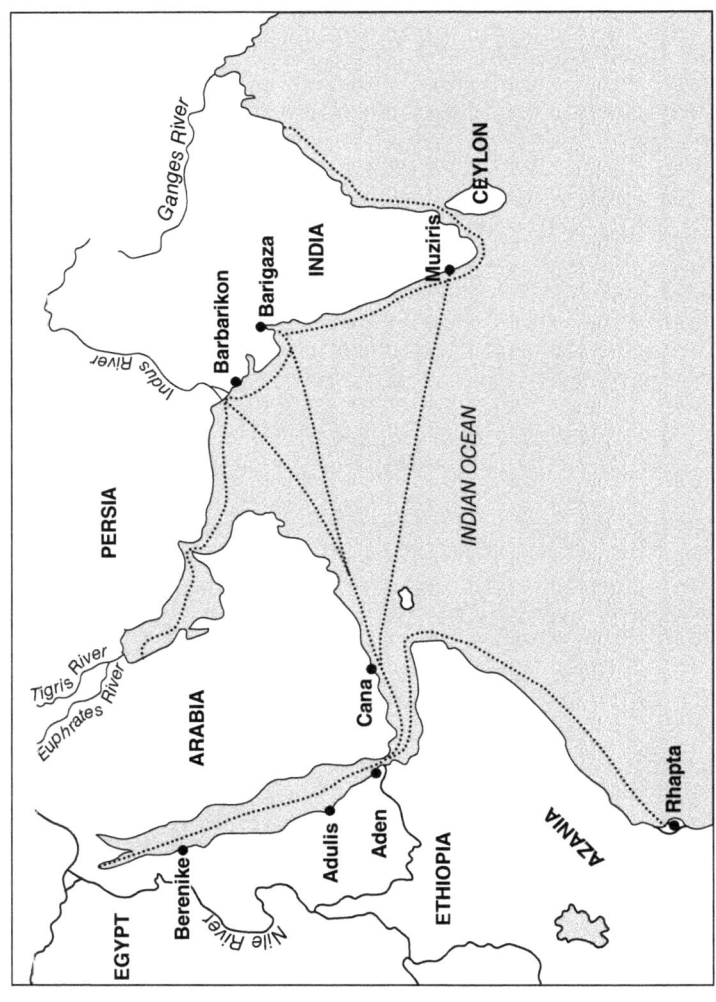

Fig. 6.2. Main routes described in *Periplus of the Erythraean Sea*. (Routes redrawn from Young, G.K. (2003) *Rome's Eastern Trade: International Commerce and Imperial Policy 31 BC–AD 305*. Routledge, London, Map 2.1.)

it remained one of India's most important trading ports through the Roman period. In the *Akananuru*, a collection of ancient Tamil poetry, it was described as 'the city where the beautiful vessels, the masterpieces of the Yavanas (westerners), stir white foam on the Periyar, river of Kerala, arriving with gold and departing with pepper' (Perur, 2016). Black pepper was the major export of this great emporium, but other trade items included locally gathered ivory and pearls, along with semi-precious stones and silks from the Gangetic valley and east Himalayan regions.

Once their ships were filled, the traders would head back to the Egyptian ports of Myos Hormos and Berenike. There, consignments of their treasures were sent overland on camel caravans and then shipped to the commercial hub of the Egyptian Roman empire, the city of Alexandria. The diversity of the goods sent across the desert would have been simply breathtaking – Arabian frankincense, Sri Lankan and Chinese cinnamon, Indian pepper, pearls and precious stones, Chinese silks and porcelain, African myrrh, ivory, rhinoceros horn and tortoiseshell.

Rome's Breathtaking International Trade Network

Under Greek and then Roman control Alexandria became the epicentre of international trade, solidly tapped into three age-old world trade networks. The land-based Incense Route winding from southern Arabia supplied the merchants of Alexandria with frankincense, myrrh and cinnamon. The two Red Sea maritime routes, that led down the African coast or through the Arabian Sea to the Indian Ocean, also provided them with cinnamon, along with ginger and pepper.

The length and breadth of these ancient trade networks were simply phenomenal. The Incense Route across Arabia passed over 1500 miles from Timna at its start to Alexandria, and another 1500 miles across the Mediterranean Sea to Rome. The incense travelled for at least four months to get to its final destination. The spices that were carried by camel along the early land-based route from Timna to Gaza would have taken over two months. The Nabataean ships carrying incense across the Red Sea to Egypt travelled about two-and-a-half months depending on the strength of the prevailing northern winds. The trip between Alexandria and Berenike on the Red Sea down the Nile and across deserts lasted about a month. Once the goods had found their way to Alexandria, it took another 20 to 30 days to reach Rome by ship.

The trade routes through the Red Sea down Africa and across to India also spanned more than 1500 miles. The full journey from Red Sea ports to Dar es Salaam and back took about two years to complete, while the round trip to India was closer to a year. Once the aromatics and spices from these far-flung corners of the world arrived in Alexandria, they then travelled another 1500 miles by sea to Rome. In total, all the spices that Romans had become so dependent upon had travelled a prodigious distance of 3000 to 4500 miles (4800 to 7240 kilometres) to get to Rome!

The everyday life of wealthy Romans was filled with the bounty of products flowing in from these distant, exotic locations. As Singer describes:

> Wealthy Romans looked to the east for their fashions. Clothes were brighter, dyed with a range of costly pigments and stitched from Chinese brocades and textiles imported via India. Houses were filled with ornamental furniture made from teak, tortoiseshell, ivory and ebony. Menageries of unusual pets became popular; monkeys, tigers and leopards were regularly shipped from Africa, and played an important role in public games and festivals. Precious gems like turquoise and lapis lazuli regularly hung from Roman wrists.
>
> (Singer, 2006, p. 35)

Cosmetics and perfumes were made up of among the most expensive ingredients money could buy, including cinnamon and myrrh. Singer continues:

> Roman cuisine was infused with exotic flavours, including ginger from China and pepper from India ... Indian pepper was particularly popular, and extremely expensive. It was used in fish and meat sauces, in medicines and in stimulating tonics which were believed to cure impotence. Romans also mixed pepper and other aromatics into their wine: ingredients such as frankincense, myrrh, cinnamon flowers, ginger and cardamom were added, and the wine was heated over a slow fire.
>
> (Singer, 2006, p. 35)

References

Ashley, J.R. (1998) *The Macedonian Empire: The Era of Warfare under Philip II and Alexander the Great, 359–323 BC*. McFarland and Company, Jefferson, North Carolina.

Bernstein, W.J. (2008) *A Splendid Exchange: How Trade Shaped the World*. Grove Press, New York.

Boivin, N., Blench, R. and Fuller, D.Q. (2009) Archaeological, linguistic and historical sources on ancient seafaring: a multidisciplinary approach to the study of early maritime contact and exchange in the Arabian Peninsula. In: Petraglia, M.D. and Rose, J.I. (eds) *The Evolution of Human Populations in Arabia*. Springer, Dordrecht, the Netherlands, pp. 251–278.

Bostock, J. and Riley, H.T. (1855–57) *The Natural History of Pliny. Translated with Copious Notes and Illustrations*, Vol. 3. Harry G. Bohn, London.

Brierley, J.H. (1994) *Spices: The Story of Indonesia's Spice Trade*. Oxford University Press, Kuala Lumpur and New York.

Cappers, R.T.J. (2006) *Roman Footprints at Berenike: Archaeobotanical Evidence of Subsistence and Trade in the Eastern Desert of Egypt*. Monograph No. 55. Cotsen Institute of Archaeology, University of California, Los Angeles, California.

Fuller, D.Q., Boivin, N., Hoogervorst, T. and Allaby, R. (2011) Across the Indian Ocean: the prehistoric movement of plants and animals. *Antiquity* 85, 544–558.

Harappa.com (1995–2021) Kenoye, M. Meluhha: the Indus civilization and its contacts with Mesopotamia. Available at: https://www.harappa.com/video/meluhha-indus-civilization-and-its-contacts-mesopotamia (accessed 1 April 2021).

Hirst, K.K. (2018) The Royal Road of the Achaemenids: international highway of Darius the Great. *ThoughtCo*. Available at: https://www.thoughtco.com/royal-road-of-the-achaemenids-172590 (accessed 28 January 2021).

Loewe, M. (1971) Spices and silk: aspects of world trade in the first seven centuries of the Christian era. *Journal of the Royal Asiatic Society* 2, 166–179.

McGrail, S. (2015) *Early Ships and Seafaring: Water Transport beyond Europe*. Pen & Sword Books Ltd, Barnsley, UK.

McIntosh, J. (2008) *The Ancient Indus Valley: New Perspectives*. ABC-CLIO, Inc., Santa Barbara, California.

McLaughlin, R. (2014) *The Roman Empire and the Indian Ocean: The Ancient World Economy and the Kingdoms of Africa, Arabia and India*. Pen & Sword Books Ltd, Barnsley, UK.

Paine, L. (2013) *The Sea & Civilization: A Maritime History of the World*. Vintage Books, New York.

Perur, S. (2016) Lost cities #3 – Muziris: did black pepper cause the demise of India's ancient port? *The Guardian*. Available at: https://www.theguardian.com/cities/2016/aug/10/lost-cities-3-muziris-india-kerala-ancient-port-black-pepper (accessed 28 January 2021).

Ravindran, P.N. (2017) *The Encyclopedia of Herbs and Spices*, Vol. 1. CAB International, Wallingford, UK.

Schoff, W.H. (1912) *The Periplus of the Erythraean Sea: Travel and Trade in the Indian Ocean by a Merchant of the First Century*. Longmans, Green, and Co., New York.

Sidebotham, S. (2011) *Berenice and the Ancient Maritime Spice Route*. University of California Press, Berkeley, California.

Singer, C. (2006) The incense kingdoms of Yemen: an outline history of the South Arabian incense trade. In: Peacock, D.P.S. and Williams, D.F. (eds) *Food for the Gods: New Light on the Ancient Incense Trade*. EBSCO Publishing, Ipswich, Massachusetts, pp. 4–27.

Strabo, *Geography* (1903) H.C. Hamilton, Esq., W. Falconer, M.A., Ed. *The Geography of Strabo. Literally translated, with notes, in three volumes*. George Bell & Sons, London.

Strassler, R.B. (ed.) (2009) *The Landmark Herodotus: The Histories*. Knopf Doubleday Publishing Group, New York.

7

Silk Route Beginnings

Setting the Stage – Ancient Steppe Routes

What came to be called the Silk Route was really the gradual amalgamation of numerous shorter routes spanning the wide geography covering China, Central Asia and the Middle East. While the Chinese were inventing silk and began discovering the civilizations that surrounded them, trade routes had been operating for millennia across the vast Eurasian Steppe and through Mesopotamia to the Mediterranean Sea. Widespread systems of trade in the steppes of inner Eurasia likely trace to at least 3000 BCE. On what has come to be called the 'Steppe Roads' (Fig. 7.1), goods were not transported by caravans or imperial envoys, but between the tribes of steppe people who populated the region (McLaughlin, 2016; Jones, 2018).

These steppe routes were closely tied to the rise of horse pastoralism. As Christian describes it:

> [the use of horses] allowed more intensive exploitation of livestock for their draft power, their furs, and their milk, as well as their meat. More intensive exploitation of domestic animals allowed whole communities to live mainly from livestock and that, in turn, allowed whole communities to settle the Inner Eurasian Steppe for the first time. Pastoralism always tended to mobile lifeway than agriculture, for the simplest way of feeding herds of livestock was to move them from pasture to pasture throughout the year. The mobility of Inner Eurasian pastoralists ensured that contacts and exchanges of ideas, technologies, goods, languages, would be extensive and vigorous throughout the Inner steppe lands, and would also flow across the ecological borders with neighboring agrarian societies. From their earliest appearance, exchanged their produce (livestock, meat, hides, wool) to neighboring sedentary communities, such as the agrarian Tripolye of Ukraine.
>
> (Christian, 2000, pp. 10–11)

By 2000 BCE, the pastoralist communities had spread from the Pontic–Caspian region of the Ukraine to eastern Kazakhstan, western and northern China and parts of Mongolia. Trade goods were being vigorously exchanged between the pastoralists and pockets of agrarian communities scattered

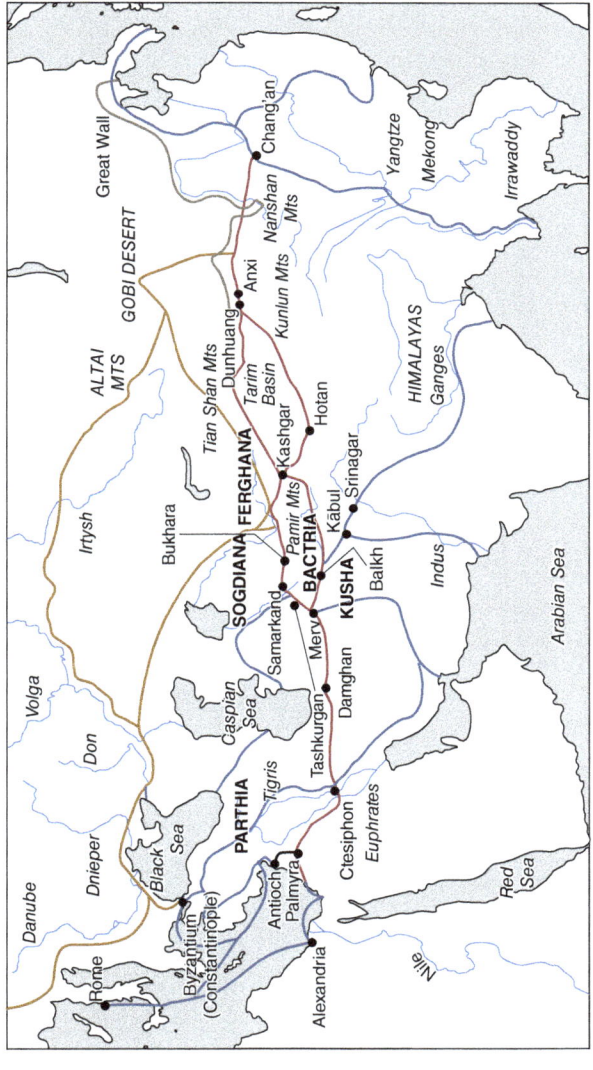

Fig. 7.1. Steppe and Silk Routes across Central Asia. (Redrawn from http://www.east-site.com/images/silk_road_map.gif)

throughout the region. Mesopotamian wheat and barley cultivation likely had spread by then through the Eurasian Steppe to China. A trading route composed of many links likely spanned a distance of about 8000 kilometres (5000 miles) between western Asia and China. It was not a continuous route travelled by one person, but a chain of many shorter links.

Christian asserts that:

> by 2000 BCE the network of exchanges we know as the 'Silk Roads' was already functioning as a system of vigorous and widespread exchanges within and sometimes beyond the Inner Eurasian steppes. And these early systems of exchange depended largely on the role of pastoralist communities.
> (Christian, 2000, p. 14)

Franck and Brownstone attest that:

> at a very early time, nomads were bringing to the cities copper, tin, and turquoise from Iran, gold from the Altai Mountains of Mongolia, lapis lazuli and rubies from Afghanistan, furs from Siberia, incense from Arabia, cottons from India, and their own products like wool, hides, and livestock. In the process, they carved out the main routes across Asia, among them the Silk Road.
> (Franck and Brownstone, 1986, p. 48)

One of the dominant forces in Central Asia became the Scythians who originated in southern Siberia and spread throughout the expansive area from the Ukraine to Mongolia. They reigned supreme in the area from about the ninth century BCE to the second century CE. The Scythians became feared and admired for their prowess in war and horsemanship. The Scythians acted as trading middlemen between the civilizations of Greece, Persia, Palestine and Egypt and the frontier people to the north and east. They traded grain, honey, furs, cattle and slaves for wine, pottery, textiles, olive oil, bronze mirrors, Egyptian beads, weapons and works of art. They became expert metallurgists, crafting bronze and gold designs featuring horses, eagles, griffins and bearded men. After Alexander's invasions in the third century BCE, they became strongly influenced by the Greeks, developing an art style that ultimately was adapted by Chinese jade carvers.

The Horse and the Balance of Power in Central Asia

Nomadic people on horseback dominated the steppes for millennia and terrorized settled people across the region through plunder and warfare (Downs, 1961; Creel, 1965). The possession of a great mounted cavalry became critical for a nation's self-defence, and Central Asia's strong and tall horses played a pivotal role in the development of the Silk Route linking East with West. As Creel (1965, p. 648) tells it: 'For some two thousand years China's foreign relations, military policy, economic well-being, and indeed its very existence as an independent state were importantly conditioned by the horse.'

It is thought that horses were first domesticated in the western part of the Eurasian Steppe around 4000 BCE. Horses were likely used first for pulling wagons carrying men of rank and as war chariots. The earliest evidence of horse domestication comes in the form of a harness excavated in the Ukraine dating to about 4000 BCE and the presence of bit-wear marks on horse teeth excavated in northern Kazakhstan from 3500 to 3000 BCE. The oldest domesticated horse remains found in China are 3500 years old and are buried next to chariots.

Horseback riding was probably invented about 1000 BCE and the practice spread rapidly across the steppes. Being on horseback had dramatic effects on the number of sheep, goats and cattle that could be herded and contributed greatly to the effectiveness of military forces. Mounted cavalry, raining arrows on opponents, became greatly feared by the settled people and to avert these attacks they were forced to pay tribute for their own safety. Having large herds of horses became an important symbol of power and nomadic horse people became immensely powerful, including the Scythians in the west and the Xiongnu in the east. Their culture and customs are richly described in the ancient records of both Western and Eastern civilizations including the *Histories* of the Greek Herodotus and Ban Biao and his son's *Book of Han*.

The native sources of horses, now extinct, were much smaller than those ridden today. The only living wild relative of the domesticated horse, the Mongolian wild horse or Przewalski's horse, is only 48–56 inches (122–142 centimetres) at its shoulders, while the tallest of the domesticated horses in Central Asia was almost 64 inches (163 centimetres) (Creel, 1965). It is likely that the ancient steppe peoples were conducting selective breeding by 600 BCE to develop horses that were larger, stronger and had greater endurance than the wild sources. Recent genomic studies suggest that horse domestication originated in the western part of the Eurasian Steppe and then spread across Central Asia. It was a cultural diffusion rather than a repeated invention. As knowledge of horse culture moved east, the herds were restocked with the local wild horses to produce regional distinctiveness (Gaunitz *et al.*, 2018).

The success of local breeders came to play a significant role in the balance of power in Central Asia and the development of the Silk Route. The Mongolian Xiongnu had horses far superior to the Chinese, which forced them to seek better horses from the west. They came to desire a landrace of tall, powerful horses from the people living in the Ferghana (Fergana) Valley of today's eastern Uzbekistan. This story will be told below.

China Stretches Its Muscles

Of the great early civilizations, China was initially the most isolated from the outside world. As Richard Smith describes China's situation:

> It was accessible by water from the Indian Ocean, but this required passage around Southeast Asia and across the dangerous South China Sea. Overland was

even a greater challenge since separating China from South Asia, West Asia and the Mediterranean were the highest mountains and some of the worst deserts in the world.

(Smith, 2009, p. 121)

China's communication and expansion westward were greatly restricted along its south-west by the Tibetan Plateau with the towering Himalayas and on its north-west by the Tarim Basin dominated by the Taklamakan Desert. That desert is surrounded by the Kunlun Mountains to the south, the Pamir Mountains and Tian Shan Mountains to the west and north, and the Gobi Desert to the east.

The earliest Chinese trade networks relied on the nomadic pastoralists who lived in the steppes surrounding China. The pastoral communities lived in areas that would not support agriculture and required grain and manufactured goods from settled people including metal, medicines and silk. Their livelihood relied heavily on horses, which came to be their most important commodity.

The Chinese traded with all the peoples around them, but their most important foreign trade was with the West, initially with their neighbours in Central Asia but soon extending to India, Persia and ultimately the Mediterranean. As Smith describes it:

> goods from China that ended up in India, Persia or Europe often had passed from group to group in informal relay networks as nomads made their seasonal rounds. Nomads also taxed the professional merchants who traveled across their lands in return for providing services like protection and sometimes guides and transport animals.

(Smith, 2009, p. 126)

The most important items of trade with their nomadic neighbours were silk and horses (Smith, 2009). Silk was used by the nomadic chiefs to recruit soldiers into their armies, while the horses, ironically, were used by the Chinese to fight those same nomadic chiefs.

Chinese Struggles with Mighty Xiongnu

The impetus for the Silk Route came from Han China's constant wars with the Xiongnu of Central Asia (modern Siberia and Mongolia) and the alliances China sought to combat them (McLaughlin, 2016). The Xiongnu were ferocious fighters who mustered hundreds of thousands of archers on horseback in repeated raids of China. They carried away with them grains, silk textiles, alcohol and anything else portable. The records of the early Han period battles and diplomacy are well represented in ancient Chinese documents such as 'Accounts of Ferghana' (in *Records of the Grand Historian*) and 'Geographical Accounts of the Western Area' (in *History of the Han Dynasty*).

China's wars with the Xiongnu began in the fifth century BCE and led to their building of the Great Wall to protect their agricultural communities along their northern and later western borders. The wall was built with gates where legitimate trading could be conducted with the other nomadic groups. One of those early trading partners were the Yuezhi who in the second century BCE were located north-east of China and west of the Xiongnu. They traded jade and horses for Chinese silks and grains through what came to be called the 'Jade Gate'.

The Great Wall slowed but by no means stopped the harassment from the Xiongnu. When the Han Dynasty rose to power in the late third century BCE, they tried to appease the Xiongnu by marrying their princesses to their leaders and offering large dowries of silks, grains and alcohol. This strategy had only limited success and the bloody raids continued. Finally, the Han emperor Wudi had enough and decided to conduct an all-out war against the Xiongnu.

Part of Wudi's strategy was to seek alliances with the surrounding nomadic peoples who had also been abused by the Xiongnu. In the second century BCE, he sent an envoy named Zhang Qian to look for their former trade partners, the Yuezhi, and ask them to help in his fight against the Xiongnu. Unknown to Han Chinese, to their west across the mountains were the remnants of the furthest extension of Alexander's Hellenistic world: Bactria, covering northern Afghanistan; and Sogdiana, encompassing today's Turkmenistan, Uzbekistan and Tajikistan. The Yuezhi had migrated into Sogdiana (Sogdia) at the end of the second century BCE and eventually settled down from their nomadic ways in Bactria.

Adventures of Zhang Qian

It took Qian a total of 13 years to find the Yuezhi and return. Along the way he was captured by the Xiongnu in Mongolia and spent ten years as a slave tending cattle and sheep. He eventually escaped and found the Yuezhi in Bactria, but they were no longer interested in fighting the Xiongnu. However, Qian was by now an expert on the western regions and had learned of other peoples in the valley of Ferghana that might help Wudi. These he called the 'Dayuan', the 'Great Ionians', who were located north of the Sogdian kingdom of Kangju. The Dayuan had strong, handsome horses of a quality previously unknown to the Chinese. Qian also learned of the distant Parthian Empire and its trade links with India.

Armed with this knowledge, Qian headed back, following a trail along the base of the Kunlun mountain range along the south of the Tarim Basin. There he discovered a series of distant oasis farming communities, fed by irrigation waters from the high mountains. As he approached the Chinese frontier, he was captured once more and imprisoned for another two years before finally escaping and finding his way home.

Wudi, who had long ago given up Qian for dead, was thrilled with the knowledge his envoy brought back and decided to set up embassies in Ferghana in hopes of gaining military support and getting a regular supply of what the

Chinese now called 'heavenly' horses that supposedly sweat blood. (The root of this phenomenon is unknown but could have been caused by a tiny parasite found in Central Asia and other areas of the Old World.) Qian was sent on a second diplomatic mission in 119 BCE west of the Tian Shan Mountains to treat with the Ferghanas, Sogdians and Bactrians, who received him favourably and sent emissaries to the Chinese capital. Qian returned with a gift of a few heavenly horses for the emperor.

Han Chinese Take Control of Their Borderlands

In 107 BCE, Wudi sent another group of envoys to the Dayuan of Ferghana carrying 1000 pieces of gold and a gold statue of a horse, to trade for more heavenly horses (Smith, 2009; McLaughlin, 2016). The Dayuan now rebuffed these offerings, feeling that the long, inhospitable distance between themselves and China offered protection against retribution. The frustrated envoys smashed the golden horse in response and were murdered as they attempted to head back home.

An infuriated Wudi decided to do the impossible and sent an expedition of tens of thousands of men 4000 kilometres (2500 miles) across some of the most formidable desert ranges and mountains on earth to bring the Dayuan to their knees. Most of the army died on route, not being able to live off the land, and the expedition failed even to reach the capital of Dayuan. When the survivors straggled back to China, Wudi locked them outside the Great Wall to die of hunger and thirst. Wudi then sent another expedition force of 100,000 soldiers carrying much more ample supplies. The bulk of this force survived the journey and they laid siege on the capital for 40 days. Finally, the Dayuan surrendered and offered the Han 30 heavenly horses and another 3000 of a lesser breed if they would end the assault.

This smashing victory resulted in the other nations of Central Asia sending emissaries and tributes to the Han including the Parthians. In the 'Accounts of the Western Area (Xiyu zhuan)' in the official dynastic history of the Han, the *Hansu*, the tribute flowing into China is described as follows:

> Reports of the heavenly horses and of grapes necessitated the opening of the roads to Dayuan and Anxi [Parthia]. From this, rare treasures such as brilliant jewels, turtle shell, rhinoceros horn, and kingfisher feathers filled the palace. The four splendid horses: Pusao, Longwen, Yumu, and the blood-sweating horses are kept within the palace gates; and elephants, lions, fierce dogs, ostriches kept in the outer gardens. Rare items of foreign lands arrive from all four directions.
>
> (Tanaka *et al.*, 2010)

The Han Chinese then turned their might back towards the Xiongnu. In multiple invasions involving hundreds of thousands of foot soldiers and supporting cavalry, they hurled themselves at the enemy. They killed thousands of men, captured huge herds of horses and sheep, and totally disrupted their traditional trade routes. Staggered by these relentless battles, the Xiongnu split into two confederations: an eastern one which submitted to the emperor and began trade relations; and a western one that did not surrender but moved further west out of reach.

China was now freed of harassment from the Xiongnu and had free reign to establish trade routes from the Ganzu Corridor through the Tarim Basin to Ferghana and beyond. In the heart of the Tarim Basin was the Taklamakan Desert where oasis city states ringed its borders as trading centres. The Chinese built protectorates and alliances with these city states and established new ones. A trade mission was sent to Persia where the emperor greeted them with open arms. By the last century BCE, Chinese power extended across all Central Asia as far as what is now Uzbekistan. Commercial relations were established across this corridor and what came to be called the Silk Route was born.

Silk Route Map

The long route was divided into several major areas of political influence: China, including the Hexi Corridor and the Tarim Basin; the Central Asian kingdoms of Greco-Bactrian origin; and the Middle Eastern Parthian and then Sasanian empires. The great independent caravan city of Palmyra played a final pivotal role in the trade with Rome.

The Han-Dynasty Silk Route began at the capital city of Chang'an (present-day Xi'an) and travelled west into Gansu Province through the Hexi Corridor to the Tarim Basin where the Chinese were confronted by the expansive Taklamakan Desert. This desert was greatly feared by travellers and essentially was unpassable, composed of ever-shifting dunes blown by fierce winds. The word *Taklamakan* in Turkic means 'go in and you will not come out'. The British consul-general at Kashgar in the 1920s, Sir Clarmont Skrine, described it in a book on Chinese Central Asia:

> To the north in the clear dawn the view is inexpressibly awe-inspiring and sinister. The yellow dunes of the Taklamakan, like the giant waves of a petrified ocean, extend in countless myriads to a far horizon with, here and there, an extra-large sand-hill, a king dune as it were, towering above his fellows. They seem to clamor silently, those dunes, for travelers to engulf, for whole caravans to swallow up as they have swallowed up so many in the past.
>
> (Skrine and Lamb, 1926, p. 111)

Chinese traders skirted the Tarim Basin by three routes to avoid the Taklamakan Desert: (i) a southern route that meandered west along the northern foot of the Kunlun Mountains; (ii) a central route that ran west along the southern base of the Tian Shan Mountains (called the 'Heavenly Mountains'); and (iii) a northern route that went west along the northern foot of the Tian Shan Mountains. All three met in Kashgar, the end of the Silk Route in China and the beginning of the route through the Kushan Empire.

Once the traders were at Kashgar, there were several choices of route. One wound westward through the Terek Pass in the Heavenly Mountains (Tian Shan) into Ferghana and Sogdiana and across the Oxus River to Merv. Others moved south through the towering Pamir Mountains to Tashkurgan and went along the Wakhan Corridor of Afghanistan to Balkh in Bactria. Still another route went south to Srinagar, passed over the Karakoram Pass and entered India.

The journey through the mountain passes must have been particularly arduous. McLaughlin describes the Karakorum Mountains as:

> a cold and arid region where caravans trekked along bleak rocky paths covered with rubble-like grey gravel. Some paths followed narrow precipices that led along cliff faces descending into steep rock-strewn valleys. Other routes crossed narrow gorges along suspension bridges made from knotted ropes and wood planks hung over steep chasms.
>
> (McLaughlin, 2016, p. 95)

From Merv the merchants continued west on an easier path to Damghan, the old capital of Parthia, and south-east to Hamadan, then on to Ctesiphon, south of today's Baghdad on the Tigris River. From here various routes led through Syria to Palmyra, then to Antioch and Tyre on the eastern shores of the Mediterranean Sea. Some travellers would have continued from Antioch through Anatolia to Sardis and west to Ephesus. Another route from Ctesiphon skirted the Euphrates River on the eastern side of Mesopotamia on the way to Sardis.

There was also an alternative northern route for those who didn't want to trade their goods with the Parthians and instead profit directly from commercial contacts. This route bypassed the Parthian Empire through the steppe lands north of the Caspian Sea (McLaughlin, 2016). This region was populated by a mounted pastoral people called the Sarmatians, who sent caravans from their capital Samarkand through Ferghana and along the Tarim Basin to Chinese cities in the Gansu Corridor.

The Engines of the Silk Routes

Two species of camel were domesticated by humans: the dromedary or single-humped camel (*Camelus dromedarius*) and the Bactrian or two-humped camel (*Camelus bactrianus*) (Potts, 2005). As has already been described in Chapter 5 (this book), the wild dromedary's distribution originated in East Arabia and it played the prominent role in the transport of goods along the Incense Route. The Bactrian camel's distribution was in Central Asia and it came to be the key carrier of goods along the Silk Routes.

Tracing the origin of the domesticated camel has proven difficult as the bones and teeth of Palaeolithic and Neolithic camels are little changed from the modern Asian species and there has been little change in the size and proportions of wild versus domesticated camels. In fact, the wild ancestor of the domesticated camel is now extinct based on DNA analysis (Orlando, 2016).

There is also little record of the objects used to control them, as the likely means of guiding camels in their early stages of domestication were wooden nose pins and rawhide control reins, which have not survived. As a result, the precise dates of camel domestication remain at best an educated guess, perhaps 5000 to 6000 years ago.

Regardless of when they were first domesticated, camels are well represented in the historical record of the Chinese (Olsen, 1988). Camels were commonly noted in the literature of the Zhou Dynasty (1046–771 BCE) and played an important role in Wudi's military exploits. According to the *Shiji* (*Historical Records*), his armies were supported by 100,000 oxen, 30,000 horses and tens of thousands of camels.

Over time, camels' use in caravans became much more important than as a baggage carrier for the Chinese armies. The camel was the primary beast of burden carrying goods along the western trade routes from China to the Black Sea via the Old Silk Route. Daniel Potts (2005) suggested that if the Silk Route is considered to be the bridge between the Eastern and Western cultures, the Bactrian camel was the principal means of locomotion across that bridge.

The Merchants of the Silk Routes

A people called the Sogdians came to be the dominant merchants on the western two-thirds of the Silk Routes. The Sogdians were an Iranian-speaking people who lived from the sixth century BCE to the eleventh century CE in the fertile valleys of today's Uzbekistan and Tajikistan (de la Vaissiere, 2003, 2004; McLaughlin, 2016). They never established a great kingdom themselves, but instead rose to power by embedding themselves as master merchants and translators inside the great kingdoms around them. The Sogdians were happy to acknowledge whatever military power was dominant in their homelands and were influenced by both Iranian and Greek culture. Once the Chinese had established diplomatic relations with the Tarim kingdoms, the Sogdians found themselves at the strategic centre of long-distance commerce connecting China with India and Iran. They became the master merchants responsible for much of the silk trade between China and the Persian Empires and had trading settlements scattered all the way from the Black Sea to Japan.

The Sogdians reached their greatest power between the fifth and seventh centuries to become the most important long-distance caravan merchants in Central Asia. Their main destination was China and many Sogdian merchants migrated there to establish and maintain their trade linkages. They also traded actively with the nomadic people in the steppes and to a large part controlled the trans-Asiatic trade routes to Persia. The Sogdian cities of Samarkand and Bukhara on the Parthian frontier became major centres of caravan trade.

Cultural Diffusion along the Silk Routes

The Silk Route served not only as a great portal of trade but also as a conduit of knowledge and beliefs. It played an important role in the arrival of Buddhism, and to a lesser extent Manicheanism and Nestorianism, into China. Buddhism began as a religious doctrine in the sixth century BCE in India and became their official religion in the third century BCE. Monks and missionaries travelled along the Silk Route from India to Central Asia and on to China, carrying Buddhist teachings, writings and paintings.

Converts and pilgrims from China also followed the Silk Route west in search of enlightenment (Sen, 2006). One was Faxian (337–442) who travelled by foot along the southern route through the Tarim Basin and over the Himalayas to India, later returning home by sea. He arrived back in China with a fabulous collection of Buddhist scriptures and statues, leaving us a travelogue of his journeys with the all-encompassing title: *A Record of Buddhist Kingdoms, Being an Account by the Chinese Monk Fa-Xian of his Travels in India and Ceylon in Search of the Buddhist Books of Discipline.*

The most famous Buddhist pilgrim was Xuanzang (600–664), who became one of the great Chinese translators of Buddhist writings. He journeyed for 14 years along the northern Silk Route to Bactria and over the Hindu Kush mountain range along today's Afghan–Pakistan border to the ancient Indian kingdom of Gandhara and ultimately to Ceylon. He became a renowned scholar winning debates against local authorities and returned to China by the northern Silk Route. He left an excellent (more pithily) entitled account of his journeys, *Records of the Western Travels*, which gives a thorough account of Buddhist thought in the seventh century.

As Buddhism spread along the Silk Route, monasteries and elaborately designed cave complexes were built near the oasis towns. The extent and artistic content of these shrines were simply breathtaking (Wood, 2002). These holy sites became a repository of hundreds of holy manuscripts and dramatic art in murals and statuary dating from the fifth to the tenth century (Wood, 2002; Onishi and Kitamoto, 2008). Large and small statues of Buddha in gold and bronze adorned the caves. Colourful cave frescoes portrayed great images of Buddha and wealthy cave benefactors in worship, to aid in their passage to nirvana. These massive complexes served as beacons of enlightenment, scattered across some of the most desolate, inhospitable terrain on earth.

Manichaeism and Nestorian Christianity also came to be widely assimilated along the Silk Route (Foltz, 2010). Manichaeism originated in Persia in the third century CE, reached Rome in the beginning of the fourth century and achieved its greatest popularity in China during the Sui (581–618) and Tang dynasties. Manichaeism views humans as being caught between dark and light, and arrival to Paradise upon death depends on achieving spiritual truth. Those who do not are condemned to rebirth in another body. The followers of Manichaeism were persecuted by the Christians in the fifth century and essentially disappeared in the tenth century, replaced by Islam in the West and Buddhism in the East.

Nestorianism emerged as a distinct sect of Christianity in the 430s in Constantinople. It is the Christian doctrine that holds that Jesus existed as two persons, the man Jesus and the Son of God. It was named after Nestorius (386–451), called the patriarch of Constantinople. It was soon outlawed by other Western theologians, forcing its disciples to flee to the Sasanian Empire and then China. It was banned in China in 845 and was almost wiped out by Islam in the eleventh century, although Marco Polo stumbled on some of the still faithful in Kashgar and Khotan in the thirteenth century.

Postscript – Discovery of the Buddhist Cave Complexes

The full majesty of the Buddhist cave complexes did not come to Western attention until the early 1900s through the efforts of European explorers/adventurers. These men came to be despised by the Chinese as 'foreign devils', for removing artefacts from the caves and shipping them to museums in the Western world (Wenjie, 1993; Onishi and Kitamoto, 2008). The most infamous was Sir Aurel Stein, a Hungarian-British archaeologist and geographer who was based at a couple of Indian universities. In the course of four expeditions between 1900 and 1930, he collected from grave sites and holy places countless documents and artefacts, from Neolithic stone tools to eighth-century-CE manuscripts (Wang, 2012).

Stein made his first major discovery at Niya, an ancient oasis on the southern edge of the Tarim Basin, where he found over 100 wooden tablets and letters written in Indian script, representing official orders and communications. These discoveries set off an international race of archaeologists from seven nations, all out to rob the ancient Buddhist treasures of the Taklamakan and Gobi deserts. This untethered competition didn't end until the mid-1920s when the Chinese finally forbade further exploration. The excavated artefacts of Stein and others wound up in more than 30 museums across Europe, America, Russia and East Asia.

In his second expedition in 1907, Stein uncovered his signature discovery near Dunhuang where the north and south roads of the Silk Route converged: the Mogao Grottoes or Cave of the Thousand Buddhas (Jiqing, 2012), Here, Stein bribed Wang Yuan, the leader of the monastic group that had taken charge of the caves, so he could carry away thousands of manuscripts written in Chinese, Sanskrit, Sogdian, Tibetan, Kök Turki and Uighur (Uyghur), as well as exquisite Buddhist paintings. Wang Yuan had reportedly tried to interest the government in the artefacts with no luck. Many of the documents had been carried there from India by Xuanzang himself for translation. Included in the cache was the world's oldest printed document, *The Diamond Sutra*, from 863 CE.

In Stein's final full expedition, he returned to Dunhuang to get the remaining documents in the cave temples. While there, he discovered a cemetery where the people of the Turfan region were buried and although the tombs had already been robbed, Stein was able to recover the ancient and beautiful silks encasing the corpses. Stein sent these along with hundreds of artefacts and manuscripts collected through the years to the British Museum.

The second most infamous of the 'foreign devils' was Stein's arch-rival, Albert von Le Coq, a German who had made his fortune as a brewer and a wine merchant before he began to study archaeology at the age of 40. In 1904, he uncovered the 'Lost Murals of Bezeklik Thousand Buddha Caves' mostly buried under shifting sands near Turfan on the northern branch of the Silk Route (Whitford, 2013). When he and his team began digging, they were astonished to find an extensive cave complex filled with manuscripts, statues and exquisite murals dating from the fifth to fourteenth century. These Le Coq removed from the caves, crated and shipped to Europe to the Berlin Ethnological Museum. Tragically, this transfer led to the ultimate destruction of many of them, as they were lost in air raids during World War II.

References

Christian, D. (2000) Silk Roads or Steppe Roads? The Silk Roads in world history. *Journal of World History* 11(1), 1–26.

Creel, H.G. (1965) The role of the horse in Chinese history. *The American Historical Review* 70, 647–672.

de la Vaissiere, E. (2003) Sogdians in China: a short history and some new discoveries. In: Waugh, D. (ed.) *The Silk Road Newsletter* 1(2). Silkroad Foundation, Saratoga, California. Available at: http://www.silkroadfoundation.org/newsletter/december/new_discoveries.htm (accessed 12 February 2021).

de la Vaissiere, E. (2004) Sogdian trade. *Encyclopaedia Iranica*. Available at: http://www.iranicaonline.org/articles/sogdian-trade (accessed 28 January 2021).

Downs, J.F. (1961) The origin and spread of riding in the Near East and Central Asia. *American Anthropologist* 63, 1193–1203.

Foltz, R. (2010) *Religions of the Silk Road*. Palgrave Macmillan, London.

Franck, I.M. and Brownstone, D.M. (1986) *The Silk Road: A History*. Facts on File Publications, New York.

Gaunitz, C., Fages, A., Hanghøj. K., Albrechtsen, A., Khan, N. et al. (2018) Ancient genomes revisit the ancestry of domestic and Przewalski's horses. *Science* 360, 111–114.

Jiqing, W. (2012) Aurel Stein's dealing with Wang Yuanlu and Chinese officials in Dunhuang in 1907. In: Wang, H. (ed.) *Sir Aurel Stein, Colleagues and Collections*. British Museum Research Publication No. 184. British Museum, London. Available at: https://www.britishmuseum.org/pdf/3_Wang_Jiqing.pdf (accessed 28 January 2021).

Jones, E.J. (2018) Long distance trade and the Parthian Empire: reclaiming Parthian agency from an Orientalist historiography. Master's thesis, Western Washington University, Bellingham, Washington. Western Washington University Graduate School Collection No. 692. Available at: https://cedar.wwu.edu/wwuet/692 (accessed 29 March 2021).

McLaughlin, R. (2016) *The Roman Empire and The Silk Roads: The Ancient World Economy and the Empires of Parthia, Central Asia and Han China*. Pen & Sword Books Ltd, Barnsley, UK.

Olsen, S.J. (1988) The camel in ancient China and an osteology of the camel. *Proceedings of the Academy of Natural Sciences of Philadelphia* 140, 18–58.

Onishi, M. and Kitamoto, A. (2008) *Overview of the Silk Road: time, space and themes. Silk Road in Rare Books*. National Institute of Informatics, Japan. Available at: http://dsr.nii.ac.jp/rarebook/10 (accessed 12 February 2021).

Orlando, L. (2016) Back to the roots and routes of dromedary domestication. *Proceedings of the National Academy of Sciences USA* 113, 6588–6590.

Potts, D. (2005) Bactrian camels and Bactrian–dromedary hybrids. In: Waugh, D. (ed.) *The Silk Road Newsletter* 3(1). Silkroad Foundation, Saratoga, California. Available at: http://www.silkroadfoundation.org/newsletter/vol3num1/7_bactrian.php (accessed 29 March 2021).

Sen, T. (2006) The travel records of Chinese pilgrims Faxian, Xuanzang, and Yijing. *Education about Asia* 11, 24–33.

Skrine, C.P. and Lamb, A. (1926) *Chinese Central Asia: An Account of Travels in Northern Kashmir and Chinese Turkestan*. Oxford University Press, Oxford.

Smith, R.L. (2009) *Premodern Trade in World History*. Routledge, London and New York.

Tanaka, Y., Sato, S. and Onishi, M. (2010) *The horses of the steppe: the Mongolian horse and the blood-sweating stallions. Silk Road in Rare Books*. Digital Silk Road Project, National Institute of Informatics, Japan. Available at: http://dsr.nii.ac.jp/rarebook/02/index.html.en (accessed 28 January 2021).

Wang, H. (ed.) (2012) *Sir Aurel Stein, Colleagues and Collections*. British Museum Research Publication No. 184. British Museum, London.

Wenjie, D. (1993) Dunhuang art: the treasure of the Silk Road. In: Agnew, N. (ed.) *Conservation of Ancient Sites on the Silk Road: Proceedings of the First International Conference on the Conservation of Grotto Sites*. Getty Conservation Institute, Los Angeles, California, pp. 1–3.

Whitford, S. (2013) A place of safekeeping? The Vicissitudes of the Bezeklik murals. In: Agnew, N. (ed.) *Conservation of Ancient Sites on the Silk Road: Proceedings of the Second International Conference on the Conservation of Grotto Sites*. Getty Conservation Institute, Los Angeles, California, pp. 95–106.

Wood, F. (2002) *The Silk Road*. University of California Press, Berkeley, California.

8 Silk Route Connections

Setting the Stage – Empires of the Middle East

The Achaemenid Persian Empire of Darius controlled the Middle Eastern terminus of the Silk Route for over 200 years and developed trade on a scale previously unknown. Caravan routes were established along the Royal Road which stretched from Sardis in Asia Minor through Mesopotamia along the Tigris to the ancient city of Susa. Other roads connected it with Bactria, the Indus Valley, the Mediterranean coast of Asia Minor, the northern shore of the Black Sea and on to the Russian steppes. Persian Gulf ports also connected the empire with the West, mostly in the hands of the Phoenicians and then the Greeks.

The Persian Empire fell to Alexander the Great in the third century BCE when he romped through the region in his astonishing rapid power grab. He had little time before he died to fully reap the benefits of the enormous trade network he had inherited, but when his general Seleucus I took control of the Iranian Plateau and Persian Gulf, he moved quickly to strengthen the trade networks of the Achaemenid Persians. Seleucus established strong diplomatic links with the Mauryan courts of Chandragupta and Bindusara in India and built garrisons on the islands of Failaka (near today's Kuwait) and Tylos (today's Bahrain). This provided the Seleucids with control of the enormous, profitable trade network in silk, frankincense, myrrh, cassia and cinnamon.

It did not take long for another native Iranian kingdom, Parthia, to rise to power in the Middle East and threaten the control of the Seleucids. In the mid-third century BCE, Parthia became a player when Arsaces I of the Parni tribe in Iran's north-east corner began a rebellion against the Seleucid Empire. This burgeoning empire was greatly expanded in the second century CE by Mithridates I who seized Media and Mesopotamia from the Seleucids. The Parthian Empire ultimately grew to stretch from central-eastern Turkey to eastern Iran. It came to be the centre of trade between the Mediterranean and the Han Empire of China, dominating the western reaches of the Silk Route.

Roman Intrusions into the Middle East

In the last century BCE, Rome began to vigorously thrust at the western border of Parthia in a war that raged on and off for centuries with few long-lasting consequences until the very end. The first large-scale Roman invasion of Parthia came in 53 BCE and ended when the Roman Crassus was routed by the Persians in the Battle of Carrhae (Barnett, 2007). Thirty thousand men were captured or killed, Crassus was beheaded, and the eagle standard of the Romans called the *aquila* was taken. The Roman general Ventidius avenged the loss of Carrhae in 40 BCE by taking control of Armenia and Syria, only to have Antony later lose ground. A period of relative calm then ensued for a few decades and the Parthians even returned the *aquila* (Barnett, 2007). In a bid for peace, Augustus of Rome gave the Parthian emperor (called a *shahanshan*) an Italian concubine, Musa, who became his favourite wife and gave him a son. That son became emperor when his mother poisoned the *shahanshan* and then married her son.

As the last century BCE waned, the hostilities between the Romans and Parthians heated back up, and for centuries they battled for control of Armenia and Mesopotamia. The Romans managed to take over the Parthian capital of Ctesiphon three times in the second century CE, but the decentralized Persians were able to absorb each thrust and recoup their territory. The Romans never did fully defeat the Parthians, but they weakened them sufficiently that a warlord named Ardashir the Unifier was able to lead a coup in 224 that established what became the Persian Sasanian Empire (224–621).

The long struggle for control of the Near East between the native Iranians and the Romans lasted for over 600 years. In a seemingly never-ending struggle, border towns and provinces in the Near East repeatedly passed back and forth between the warring parties. Smack in the middle of these conflicts were the independent city states of Petra and Palmyra that flourished on international trade. Astonishingly, battles raged on for centuries in the Middle East, but through it all, lucrative trade in spices, silks and fragrances continued mostly unabated between the invading Romans and the local Syrians and Iranians.

Palmyra Emerges as the Greatest of the Middle Eastern Trading Centres

In spite of the wars periodically raging in the Middle East, another great caravan city, Palmyra, emerged in the second century CE in Syria. Popularly known as the 'Bride of the Desert' (Bryce, 2015), it was founded in an oasis about halfway between the Mediterranean Sea and the Euphrates River. In its heyday, much of the silk that reached Rome eventually passed through this great city. As trade along the Silk Route grew in intensity, so did the power of Palmyra and it came to eclipse Petra's role as Rome's major Middle Eastern trading partner:

> Palmyra became the eastern hub of international trading operations. Inscriptional and archeological evidence from the first decades of the 2nd century CE informs us of the enormous range of goods that passed through the city – slaves, salt, dried foods, purple cloth, perfumes, prostitutes (from an inscription on a seal of 137 CE), silk, jade, muslin, spices, ebony, incense, ivory, precious stones, and glass. Palmyra's centrality in the Near Eastern trading network, and the oft-quoted insatiable demand in the Roman world for the goods that passed through the city as they were conveyed east to west contributed substantially to the enormous wealth of the city and its merchant class.
>
> (Bryce, 2015, p. 279)

The Palmyraeans were very 'hands on' in their relationships with their trading partners. They avoided using middlemen in their trade relationships and instead established colonies at critical junctures along their extensive trade routes (Bryce, 2015; Celentano, 2016). There were enclaves of Palmyraean traders scattered across the far-flung corners of the ancient world from Babylon in Mesopotamia to Coptos in Egypt and Merv on the Parthian border. Palmyraeans even sailed with their merchandise on the Red Sea.

Palmyra rose to prominence in a large part through 'the competition and compromise' between Rome and the Persian Empires (Liu, 2010, p. 28). Before the arrival of these two powers, Palmyra (then named Tadmore) had been under the control of the Hellenistic Seleucid Empire, which left a strong Greek imprint on the city. It was named Palmyra ('city of palm trees') by the Romans. The Palmyraean traders kept politically neutral and were able to tap into the caravan trade routes linking the eastern Mediterranean cities with the harbours of the Persian Gulf and western coast of India. They must have been masters at finding agreement as they had to deal with a diverse array of political authorities representing Rome, Parthia, Kusha and the nomadic tribes of the desert.

When the Romans began their push into the Middle East, Palmyra was only a minor sheikhdom. It was originally left largely independent and didn't receive the attention of Rome until 41 BCE when it had become wealthy enough for the Roman general (and Cleopatra's lover) Mark Antony to send a force to sack it (Hekster and Kaizer, 2004). The Palmyraeans are said to have abandoned the city before the Romans arrived, who left it unmolested when they found it empty.

Palmyra was made an official part of the Roman province of Syria by Emperor Tiberius in about 14 CE and Emperor Hadrian declared it a free city in 129 CE. During the reign of Septimius Severus (193–211 CE) the city was elevated to the status of a Roman *colonia*, the highest civic status that could be accorded a city of the empire; in effect, its inhabitants now enjoyed full Roman citizenship status (Bryce, 2015, p. 280). It held its privileged status in the Roman imperial period until 273 CE, when it was destroyed by Emperor Aurelian after its upstart leader Zenobia took possession of all of Syria and Egypt, during Rome's 'Crisis of the Third Century'. This story is told below.

Palmyra was a magnificent, marbled city overshadowing even the grandeur of Petra. As described by Liu:

> The colonnade of Palmyra was more magnificent, and its temples grander than those of Petra or any other Middle Eastern caravan city ... Palmyra's Grand Colonnade, which was on the main thoroughfare, was at the center of its urban area. More than 375 Corinthian columns lined each side of this avenue. At its eastern end, the colonnade was marked by a great gate with three arches. On the south side of the grand avenue stood three huge public water fountains, the agora, the Senate house, and the theater. Outside Palmyra was a large necropolis, containing sculptures as elaborate and artistic as those designed for the living in other parts of the city.
>
> (Liu, 2010, p. 29)

Many of these were removed and displayed in museums across the world.

Zenobia Grabs Power

A remarkable power play was made in 269 CE by the Palmyraean queen Zenobia to take over all the Roman-controlled Middle East. Gauging that Rome was too occupied with internal problems to respond, she sent the head of her army to Egypt to claim it for her own and through negotiations added the Levant and Asia Minor to her Palmyrene Empire. This gave her control of an empire covering modern-day Iraq, Turkey and Egypt. She had access to the rich breadbasket of Egypt along with all the major trading ports lining the Mediterranean and Red Seas where Palmyraean merchants were plying their lucrative trade (Bryce, 2015).

By all accounts she was a remarkable woman – stunningly beautiful, yet having the pluck to march in battle and lead her soldiers by steely example (Mark, 2014; Lewis, 2018). The great historian Edward Gibbon described her thus:

> Zenobia is perhaps the only female whose superior genius broke through the servile indolence imposed on her sex by the climate and manners of Asia. She claimed her descent from the Macedonian king of Egypt, equalled in beauty her ancestor Cleopatra, and far surpassed that princess in chastity and valour. Zenobia was esteemed the most lovely as well as the most heroic of her sex. She was of a dark complexion. Her teeth were of a pearly whiteness, and her large black eyes sparkled with uncommon fire, tempered by the most attractive sweetness. Her voice was strong and harmonious. Her manly understanding was strengthened and adorned by study. She was not ignorant of the Latin tongue, but possessed in equal perfection the Greek, the Syriac, and the Egyptian languages. She had drawn up for her own use an epitome of oriental history, and familiarly compared the beauties of Homer and Plato under the tuition of the sublime Longinus.
>
> (Gibbon, 1821, p. 355)

It took Rome a little while to respond to Zenobia's audacious takeover but in 272 CE Emperor Aurelian went after her fragile empire after dealing with a western incursion by the Vandals. He marched easily through Asia Minor destroying cities faithful to Zenobia and sparing the others. When he arrived in Syria, he overwhelmed Zenobia's Palmyrene forces in the Battle of Immae and pursued her and her remaining forces to the city of Immae, where he crushed her army once again. She escaped to Palmyra where, as Edward Gibbon relates (1821, p. 347): 'She retired within the walls of her capital, made every preparation for a vigorous resistance, and declared, with the intrepidity of a heroine, that the last moment of her reign and of her life should be the same.' However, when it became clear that Palmyra would fall, she fled to Persia with her son on camelback but was captured as she tried to cross the Euphrates River. Her haughty, arrogant rule of the western part of the Roman Empire was over.

What happened to her next has been mostly lost in the mist of history. As Mark tells it:

> She was brought back to Aurelian in chains where she protested her innocence and blamed her actions on the bad advice given her by her advisors, chiefly Cassius Longinus, who was promptly executed. Zenobia was then brought back to Rome. What happened to her next varies with the account one reads. According to Zosimus, she and her son drowned in the Bosphorus while being transported back to Rome, but he also claims she arrived in Rome, without her son, was put on trial, and acquitted; after which she lived in a villa and eventually married a Roman. The *Historia Augusta* relates the story of her being paraded through the streets of Rome in gold chains and heavily-laden with jewelry during Aurelian's triumph parade, after which she was released and given a palace near Rome where she 'spent her last days in peace and luxury'. Zonaras claims she was taken back to Rome, never was paraded through the streets in chains, and married a wealthy Roman husband, while Aurelian married one of her daughters.
>
> (Mark, 2014)

Kushans Take the Centre of the Silk Routes

After being forced out of their homeland by the Xiongnu in 165 BCE, the Yuezhi moved westward and then south into Sogdiana and Bactria, kept on the run by constant warfare with the local peoples they encountered along the way (McLaughlin, 2016; Benjamin, 2017). When the peripatetic Yuezhi arrived in Sogdiana, the area was ruled by the Kangju/Sogdians. It is likely that the rulers of Kangju gave the Yuezhi safe passage through their territory, by accepting a subservient relationship to avoid military conflict. In Bactria, the Yuezhi finally settled down after defeating the remnants of the Greco-Bactrian states and formed what came to become the Kushan Empire.

The Kushans were an eclectic people borrowing aspects of their culture from both the Greeks and Persians (Szczepanski, 2017). They borrowed much of the Greek alphabet and minted coins honouring their leaders in the Greek fashion. For their religion they initially adopted Zoroaster from the Persians but later turned to Buddhism. They used their fabulous wealth to lavishly support Buddhist monasteries and statues and sent Buddhist missionaries across Central Asia to China and Tibet.

At the height of its power from the first to the third century CE, the Kushan Empire sat squarely at the centre of the Silk Route controlling Bactria, the western Tarim Basin and northern India. They grew fabulously wealthy by exploiting trade. The acquisition of the Indian territories added greatly to the resources of the Kushans, providing them with elephants, rhinoceroses, turtle shell, gold, silver, copper, lead and tin. Their trade resources were greatly expanded by the procurement of cotton cloth, wool carpets, salt, sugar, ginger and pepper. They also took control of the maritime trade routes connecting India and Egypt, which provided them with goods from Rome.

Kushan Connections

Not only did the Kushan Empire play a pivotal role in the movement of silk from China to Persia, but it also served as an important crossroads with other trade routes to India and beyond (Fig. 8.1). In the early centuries of the first millennium CE, the Kushan Empire had Indian trade connections that reached all the way to the Arabian Sea. This made it an important meeting point of goods from China, India, Persia and Rome.

The Kushan capital at Begram (or Bagram) in today's Afghanistan was at the centre of all the major trading routes connecting the first-century world! Archaeologists have found extensive evidence of Begram's international trade connections (Seland, 2013). There are two ancient ruins in Begram that represent two fortified enclosures, the Old Royal City in the north and the New Royal City in the south. In excavations at these locations, objects have been found from all over the world. As Sanjyot Mehendale relates:

> among them, is an Indianesque piece of earthenware referred to as the 'Kinnari' pot, Graeco-Roman objects such as a bronze satyr head, painted glass beakers with analogies to Roman Alexandria, pillar-moulded bowls found also in several sites of the Arabian peninsula and in Arikamedu in India, and plaster medallions. Also found were fragments of Chinese lacquer objects the decoration of which is similar to ones found in Noinula, Mongolia and in Lo-lang, Korea, as well as numerous carved ivory and bone objects generally thought to originate either from India either north-central or southern.
>
> (Mehendale, 1996, p. 50)

According to Mairs (2012, p. 6), at Begram archaeologists 'uncovered two storerooms full of luxury goods from such far-flung regions of the ancient world as the Graeco-Roman Mediterranean, India and China'. Of greatest note 'are cut-glass vessels, mould-blown glass, glass with faceted decorations, coloured enamelled vessels and vessels with applied moulded relief decoration, even bowls of millefiori or mosaic glass'. This glassware originated from Egypt and was blown using technologies first developed on the Syro-Palestinian coast.

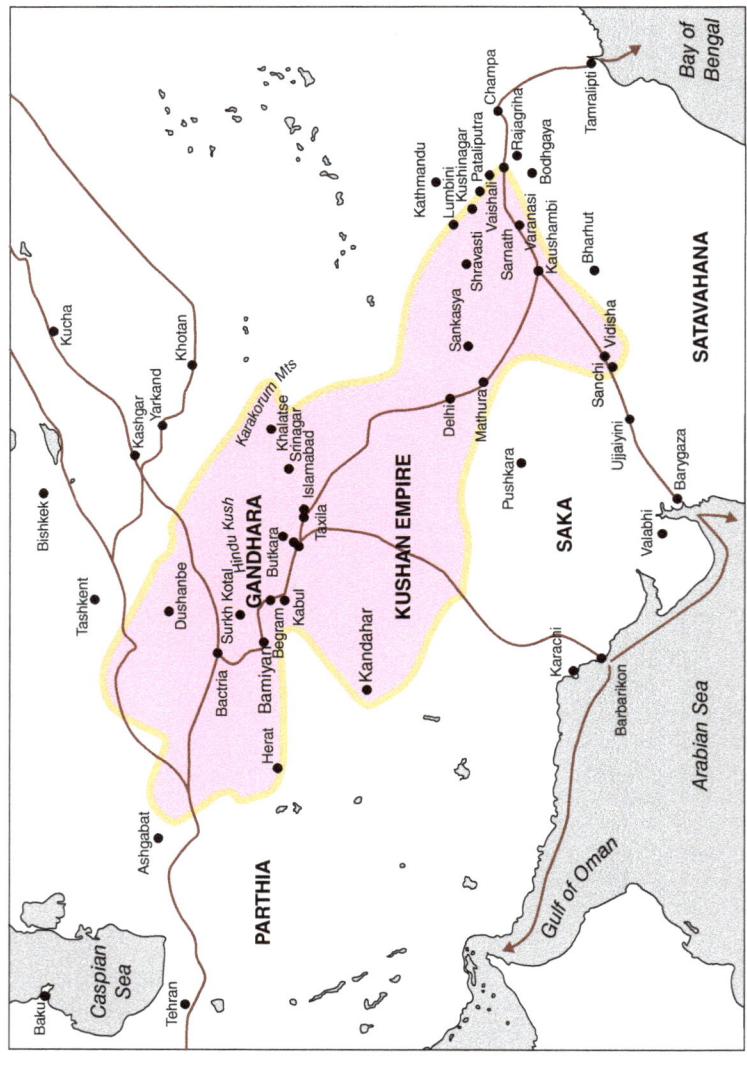

Fig. 8.1. Silk Routes through Kushan Empire (100–240 CE). (Redrawn from *Ancient History Encyclopedia* by Muhammad Bin Naveed, published on 23 June 2015 under the following license: Creative Commons Attribution-NonCommercial-ShareAlike. File: Kushan Empire (Illustration) - Ancient History Encyclopedia)

To get to Begram, the glassware must have been shipped down the Nile to Coptos, carried overland by camel caravans across the desert corridor to the Red Sea port of either Berenike or Myos Hormos, put on a ship bound for India which first sailed to Aden and Cana on the southern coast of Arabia and then grabbed the monsoon winds across the open waters of the Indian Ocean to north-west India and likely Barbarikon (Ghosh, 2014). From there it was transported north along the Indus River to Taxila and across the Khyber Pass through the Kush Mountains and then across the Kohdaman Plains to Begram. A truly remarkable journey!

Parthia Controls the Terminus

The Parthians served as the middlemen between Kusha and Rome by preventing the Chinese merchants from travelling through their territory. The Chinese were allowed to pass through Kusha by paying a tariff but were not allowed by the Persians to go any further than Merv. When the traders were finally stopped there, they carried not only silk from China but also a plethora of goods from Central Asia and Kusha, which had their extensive trade links stretching all the way to northern India.

Parthians produced little of interest themselves for trade but prospered mightily by trading with the eastern merchants for goods that had originated from the far-off Roman Empire. A full account of the nature of the trade between the Parthian and Chinese merchants was recorded in Fan Ye's *Hou Hanshu*, the official history of the Eastern Han Dynasty from 25 to 221 CE (Thorley, 1971). From Roman traders, the Parthians got gold and silver which were the principal objects desired by the Chinese. They also traded coral from the Mediterranean and Red Seas, high-quality glass and gold-embroidered rugs made in Alexandria and Syria, and asbestos cloth made in Syria from the attachment hairs of Mediterranean shellfish. Their trade goods also included the medicine storax obtained from a shrub growing in Syria and Asia Minor, amber from Scandinavia, cinnabar from the island of Socotra off the Horn of Africa, and even jade from the Khotan region of the Tarim Basin (a primary jade source of the Chinese themselves).

Perhaps most surprising, the Persians also traded silk cloth of a delicate, much more translucent form than that of the Chinese. This was the fabric that the Romans had come to adore. It was produced in Syria by unravelling and reweaving the heavy silk brocades that the Persians had originally obtained from China. In essence, the Chinese were buying back their own silk from the clever Parthian merchants.

Sasanians Take Over

As the third century AD began, the long dominant Parthian Empire had grown weak, and power was divided among a tapestry of small kingdoms. In 215, the ruler of one of those kingdoms, Ardashir, decided to rebel against the Parthian king and began invading territories around his kingdom. By the time the Parthian king Artabanus IV responded, the balance of power had shifted and when Artabanus was killed in battle, Ardashir became the dominant Persian ruler. Ardashir went on to found what came to be called the 'Sasanian Empire'.

As the Sasanian Empire grew in strength, it turned its military might east and invaded the Kushan Empire, taking control of the Kushans' territory west of the Tarim Basin and pushing them all the way back into north-east India. This gave the Sasanian Empire direct contact with the Chinese and allowed them to completely control the silk trade with Rome. By now, the Sasanians also had control of most of the spice trade from India, making them Rome's primary supplier of luxury goods. In the sixth century CE, the Sasanians completed their stranglehold on the spice trade by invading Southern Arabia and taking control of the entrance to the Red Sea. The Sasanians wound up dominating the world's luxury trade for 300 years until their empire came crashing down in 651 when their armies were crushed by the invading Muslims.

The Sasanian dynasty's takeover of the Parthian and Kushan empires was an immensely profitable one. They now controlled the geographic terminus of the Silk Routes. They also developed their own state-run weaving industry to further enhance their silk profits (Feltham, 2010). The Sasanians imported raw silk from China, then dyed and wove it into twills and brocades. These textiles were then exported back to China and the Far East or shipped west to Byzantium and Europe.

The Sasanian weavers:

> developed a compound weft silk twill with elaborate repeating motifs such as winged lions, hunting scenes, tree of life patterns, and opposing birds, each motif enclosed within a pearl-like roundel, and each group of roundels separated by scrolling, geometric plant forms. Both the heraldic animal and human elements and the interlocking plant motifs would inspire Eastern and Western design for centuries to come.
>
> (Feltham, 2010, p. 16)

The lion became the dominant decorative design as a symbol of power and prestige.

Ebbs and Flows of the Silk Route

By the Christian era the Silk Routes were fully formed, and silk was moving continuously from China to the eastern Mediterranean across Central Asia. The total length of the historical Silk Route was about 10,000 kilometres (6214 miles), of which approximately 3000 kilometres (1864 miles) were inside China's territory (Onishi and Kitamoto, 2008).

Over the following centuries, the amount of movement along the Silk Routes ebbed and flowed depending on the politics of the kingdoms through which they passed (Onishi and Kitamoto, 2008). As Christian relates succinctly:

the importance of the Silk Roads waxed and waned, partly as a function of the degree of stability to be found in the borderlands between the steppes and the agrarian civilizations of China, India, Iran, and Mesopotamia, and partly as a result of economic and political conditions in the major regions of agrarian civilization. When agrarian civilizations or pastoralist empires dominated large sections of the Silk Roads, merchants traveled more freely, protection costs were lower, and traffic was brisk.

(Christian, 2000, p. 6)

For a period of about 40 years from AD 90 to 130, the Silk Route was peaceful across its entire route from north China to Rome. Conditions for trade were as favourable as they would be before or after. China and Rome were in periods of great prosperity and only two commercially minded states lay between them, Parthia and Kusha (Thorley, 1971).

China was no longer being harassed by the Xiongnu and had political control all the way across the Ganzu Corridor through the Tarim Basin to Ferghana. To get to Rome, most silk moved through only three polities – China, the Kushan Empire and Parthia. The remaining silk was traded to Rome through Nabataean Petra, Rome's ally and then client state.

Ten state-sponsored caravans left China headed to the Tarim Silk Routes each year that were composed of as many as 600 camels. These were loaded with up to 10,000 silk rolls (4 tons of fabric) and managed by 200 handlers. These caravans did not go the whole route – their goods were generally dropped off at weigh stations where they were picked up by other caravans and taken to the next city. Thus, the silk hopscotched from city to city until it reached its destination. Each transaction added to the cost of the product, rising dramatically until the goods arrived in the hands of wealthy Romans. It likely took a roll of silk more than a year to be carried from China to a Mediterranean port. The trip from its origin in China across the Tarim Basin probably took eight months. From the western edge of China to the Syrian frontier would have taken at least five more months.

The journey that silk made across the whole of Central Asia was truly remarkable. As described by Feng:

> The early trade in silk was carried on against incredible odds by great caravans of merchants and animals travelling at a snail's pace over some of the most inhospitable territory on the face of the earth – searing, waterless deserts and snowbound mountain passes. [...] Blinding sandstorms forced both merchants and animals to the ground for days on end [...] and altitude sickness and snow blindness affected both man and beast along cliff-hanging and boulder-strewn tracks. Death followed on the heels of every caravan.
>
> (Feng, 2005, p. 7)

A Plague Slows Trade

In the middle of the second century AD, the world suffered under what came to be called the 'Antonine Plague' (Gilliam, 1961). It led to a dramatic decline in the maritime trade all along the Spice Route (McLaughlin, 2016). The long-active maritime trade of the Romans down the Red Sea and through the Indian Ocean slowed to a trickle, leaving the Kushans with an excess in luxury goods and no one with whom to trade. The world suffered through a major economic decline.

The Antonine Plague originated in China and raced along the Silk Route, passing to Central Asia, India, Parthia and ultimately Rome. It was first recorded in Parthia in 165 during the Roman invasion, where it quickly spread among the army and then through the densely populated Roman East. It likely contributed mightily to the decline and fall of the Roman Empire itself, as too few people survived to adequately support the massive Roman infrastructure.

The German historian Niebuhr wrote in his *Lectures on the History of Rome*:

> This pestilence must have raged with incredible speed and it carried off innumerable victims. As the reign of Aurelius forms a turning point in so many things, and above all in literature and art, I have no doubt that this crisis was brought about by that plague ... The ancient world never recovered from the blow inflicted upon it by the plague which visited the reign of M. Aurelius.
>
> (Niebuhr, 1844, p. 134)

The plague was named after the family name of Emperor Aurelius.

The most popular modern diagnosis of the disease is smallpox, although insufficient details remain to pinpoint its cause (Littman and Littman, 1973). The great Greek physician Galen described the disease as a slow killer, resulting in fever, diarrhoea and pharyngitis accompanied with dry and pustular skin eruptions on the ninth day. At least 25% of those who contracted the disease died and this plague killed upwards of 10% of the entire Roman population (probably five million people). At its peak, 2000 people died in Rome daily.

Silk Trade after 400 CE

China became divided during the Southern and Northern Dynasties period (420–589) and several Chinese and non-Chinese steppe people took over control of parts of the Central Asian Silk Routes until the reunification of China under the Sui Dynasty in the late sixth century, who reinstalled the rule of the ethnic Han Chinese. This became a period of great prosperity for East–West trade across Central Asia, as the Chinese again subjugated all the countries throughout the region. Chinese power held sway over Central Asia for another 200 years until its defeat against the Arab Abbasid Caliphate at the Battle of Talas in 751.

By the end of the eighth century, sea routes became active from the southern coastal city of Guangzhou (Canton) in China to the Middle East. By then the art of making silk had been mastered by the Persians and the heyday of the Silk Route was over. The Tang Dynasty also went into free fall about that time and its economy became less able to support foreign imports of luxury goods. Simultaneously, many thriving oasis towns along the Silk Route, supporting monasteries and grottoes, began to disappear as the glacier-fed streams that had supported them since the end of the Ice Age began running dry or changed course. The spread of Islam from the Middle East also played a key role in the disappearance of the Buddhist communities along the Silk Route, as the followers of Allah defaced and destroyed the Buddhist relics or left them to crumble. Only those previously buried in the sands survived for later discovery by the 'foreign devils'.

References

Barnett, G. (2007) Mark Antony's Persian campaign. *HISTORYNET*. Available at: http://www.historynet.com/mark-antonys-persian-campaign.htm (accessed 12 February 2021).

Benjamin, C. (2017) The Yuezhi. *Oxford Research Encyclopedias: Asian History*. Available at: http://asianhistory.oxfordre.com/view/10.1093/acrefore/9780190277727.001.0001/acrefore-9780190277727-e-49 (accessed 28 January 2021).

Bryce, T. (2015) *Ancient Syria: A Three Thousand Year History*. Oxford University Press, Oxford.

Celentano, C. (2016) Palmyrenes abroad: traders and patrons in Arsacid Mesopotamia. *Phasis. Greek and Roman Studies* 19, 30–57.

Christian, D. (2000) Silk Roads or Steppe Roads? The Silk Roads in world history. *Journal of World History* 11(1), 1–26.

Feltham, H.B. (2010) *Lions, Silks and Silver: The Influence of Sasanian Persia*. Sino-Platonic Papers No. 206. University of Pennsylvania, Philadelphia, Pennsylvania. Available at: http://sino-platonic.org/complete/spp206_sasanian_persia.pdf (accessed 28 January 2021).

Feng, J. (2005) UNESCO's efforts in identifying the World Heritage significance of the Silk Road. In: *15th ICOMOS General Assembly and International Symposium: 'Monuments and sites in their setting – conserving cultural heritage in changing townscapes and landscapes', 17–21 October 2005, Xi'an, China*. World Publishing Corporation, Xi'an, China. Available at: http://openarchive.icomos.org/id/eprint/428/1/4-20.pdf (accessed 30 March 2021).

Gibbon, E. (1821) *The History of the Decline and Fall of the Roman Empire*, Vol. 1. R. Priestley, London.

Gilliam, J.F. (1961) The plague under Marcus Aurelius. *American Journal of Philology* 82, 225–251.

Ghosh, S. (2014) Barbarikon in the maritime trade network of early India. In: Mukherjee, R. (ed.) *Vanguards of Globalization: Port Cities from the Classical to the Modern*. Primus Publications, New Delhi, pp. 59–74.

Hekster, O. and Kaizer, T. (2004) Mark Antony and the raid on Palmyra: reflections on Appian, Bella Civilia V 9. *Latomus* 63(1), 70–80.

Littman, R.J. and Littman, M.L. (1973) Galen and the Antonine Plague. *American Journal of Philology* 94, 243–255.

Lewis, J.J. (2018) Zenobia: warrior queen of Palmyra. *ThoughtCo*. Available at: https://www.thoughtco.com/queen-zenobia-biography-3528385 (accessed 28 January 2021).

Liu, X. (2010) *The Silk Road in World History*. Oxford University Press, New York.

McLaughlin, R. (2016) *The Roman Empire and the Silk Roads: The Ancient World Economy and the Empires of Parthia, Central Asia and Han China*. Pen & Sword Books Ltd, Barnsley, UK.

Mairs, R. (2012) Glassware from Roman Egypt at Begram (Afghanistan) and the Red Sea trade. *British Museum Studies in Ancient Egypt and Sudan* 18, 1–14. Available at: https://www.academia.edu/1489668/_Glassware_from_Roman_Egypt_at_Begram_Afghanistan_and_the_Red_Sea_Trade_ (accessed 3 April 2021).

Mark, J.J. (2014) Zenobia. *Ancient History Encyclopedia*. Available at: https://www.ancient.eu/zenobia/ (accessed 28 January 2021).

Mehendale, S. (1996) Begram: along ancient Central Asian and Indian trade routes. *Cahiers d'Asie centrale* 1, 47–64.

Niebuhr, B.G. (1844) *Lectures on the History of Rome, From the Earliest Times to the Fall of the Western Empire*. Taylor, Walton, and Maberly, London.

Onishi, M. and Kitamoto, A. (2008) Overview of the Silk Road: time, space and themes. *Silk Road in Rare Books*. National Institute of Informatics, Japan. Available at: http://dsr.nii.ac.jp/rarebook/10 (accessed 12 February 2021).

Seland, E.H. (2013) Ancient Afghanistan and the Indian Ocean: maritime links of the Kushan Empire ca 50–200 CE. *Journal of Indian Ocean Archaeology* 9, 66–74.

Szczepanski, K. (2017) The Kushan Empire. *ThoughtCo*. Available at: https://www.thoughtco.com/the-kushan-empire-195198 (accessed 28 January 2021).

Thorley, J. (1971) The silk trade between China and the Roman Empire at its height, circa AD 90–130. *Greece & Rome* 18, 71–80.

9

Ancient South East Asian Maritime Trade

Setting the Stage – Indonesia Awakes

South East Asian trade began in about 500 BCE, when sailors from south-east India and Ceylon began reaching out to the islands of Indonesia. As the traders moved into the Indonesian archipelago, their religions travelled with them. First, Hindu kingdoms began to spring up as early as 300 BCE and then Buddhist ones in the early centuries CE. An almost symbiotic relationship developed between the traders and the developing religious communities.

The resources of the Indonesian archipelago and its central location between China and India ultimately made it an important crossroads in international trade. The different regions and islands of Indonesia produced a number of commodities that stimulated internal and external trade. The cycle of monsoons on the Indonesian archipelago's waters facilitated trade among islands and brought the archipelago into an open system of world shipping. Traders rode the north-western monsoon winds from the Strait of Malacca to Batavia (on Java Island) and then to Makassar (on Sulawesi Island) and on to the Spice Islands (the Moluccas).

Origin of Trade between India and South East Asia

Widespread commercial activity in India began in the Gangetic plains in the fifth century BCE and fully coalesced with the establishment of the Mauryan Empire in about 324 BCE. It came to dominate most of the Indian subcontinent, except for the Tamil-speaking south. The Mauryan Empire grew wealthy from its extensive trade networks by land to China through the Silk Route and by sea to Sumatra to the east, Ceylon to the south and to Persia and Africa to the west. By the third century BCE, Indian sailors were regularly travelling long distances in all directions, riding the monsoon winds in their triangle-sailed dhows.

The Mauryan dynasty arose after Alexander left India in 322 BCE, when a non-royal named Chandragupta overthrew the small Nanda dynasty then ruling the Ganges Valley and began toppling the hodgepodge of Greek governors remaining in Alexander's wake. Even the powerful Seleucids, who had inherited most of Alexander's eastern empire, retreated to Persia rather than battle Chandragupta, giving up a large tract of India in exchange for 500 war elephants and a marriage relationship (Hirst, 2018).

During the glory years of the Maurya Empire, several other kingdoms came to rule large stretches of the southern reaches of India including the Pandyas (fourth century BCE to sixteenth century CE) in the west, the Cheras (third century BCE to twelfth century CE) in the east and the Anuradhapura Kingdom (377 BCE to 1017 CE) in Ceylon. These kingdoms, like the Mauryans, grew wealthy from international trade across the Bay of Bengal to Indonesia and South East Asia.

A wide range of artefactual evidence scattered across the eastern part of the Indian subcontinent and the western part of South East Asia signals that Indo-South East Asian networks were in full operation in the late centuries BCE (Bellina, 2003; Gupta, 2005). One of the most prominent indicators of widespread trade is Northern Black Polished Ware, a special ceramic of the Gangetic Valley that has been recovered in large amounts on the island of Bali and Vietnam. Origin-specific stone beads (etched agates, carnelian), glass beads, pendants and bronze vessels are also found all across Burma (present-day Myanmar), coastal Thailand, Peninsular Malaysia and the Indonesian islands of Sumatra and Java. Stamped pottery sherds of South East Asian origin have been found at several archaeological sites across south-eastern India (Rao, 2001).

Some of the richest evidence of early trade between India and South East Asia has been found on the island of Bali, off the eastern coast of Java (Ardika, 2018). Excavations at Sembiran and Pacung in north-eastern Bali have uncovered hundreds of indented sherds of Indian 'rouletted wares' dating between 150 BCE and 200 CE, along with glass and carnelian beads similar to south Indian samples and gold-foil eye covers similar to those found in graves on the south-east India coast. Incredibly, a tooth has been recovered from one of the excavations that dates from 2500 BCE and its DNA reveals it came from a trader of likely Indian extraction (Lansing *et al.*, 2004).

Maritime Trade of the Anuradhapura Kingdom

Ceylon's central position in the Indian Ocean also made it a particularly active port-of-call in the early days of Indian Ocean trade, both with South Asia and the rest of the world (Kingwell-Banham *et al.*, 2018). Goods from the East including India, Indonesia and China were traded for those from the West coming from as far away as Europe, Arabia and Africa.

Perhaps the most extensive evidence of the central position of Ceylon in Indian Ocean trade has been found in Manthai, on the island's north-western coast (Kingwell-Banham *et al.*, 2018). This site was the main seaport of the Anuradhapura Kingdom for nearly two millennia. By *c*.200 BCE, 7% of the ceramic assemblage found in archaeological sites was of Middle Eastern origin and 2% Chinese. Of the grain fragments recovered from *c*.100–250 CE, 57% could be identified as locally produced rice but 2.5% were wheat. Since wheat is not native to Ceylon, this suggests that Roman traders had settled into the region by that period. The wheat fragments were without chaff, suggesting that the crop was imported after thrashing as a clean grain from either West Asia or the Mediterranean.

The ancient archaeobotanical assemblage found at Manthai has also provided direct evidence of trade in spices. Black pepper has been found in Manthai dating to around 500 CE and a clove dating from *c*.900 to 1100 CE. Black pepper was native to the forest of the western Ghats and was likely first shipped from ports in south-west India to Ceylon. The clove had to have come 7000 kilometres (4350 miles) from the Molucca Islands in Indonesia, probably traded by intermediaries across the Bay of Bengal from South East Asian ports.

Indianization of Indonesia

Indonesia is the largest archipelago in the world, with more than 13,000 islands spread across 5000 kilometres (3100 miles) of sea. The largest islands are Java, Sumatra, Borneo, Sulawesi and New Guinea. Strategically located between China and India, it came to be a vital link in the East–West trade of spices.

There is little written record of Indonesia until the late centuries BCE, leaving the early development of the region cloudy; however, it is known that Hindu kingdoms began to spring up in Java and Sumatra as early as 300 BCE and by the early centuries CE, Buddhists came to rule large areas of these islands. These kingdoms were greatly influenced by Indian culture and the local rulers freely combined components of Indian government with their own perceptions of how to rule.

The Buddhists were sent by the Mauryan Empire's greatest leader, Ashoka (265–238 BCE), whose wealthy kingdom was fuelled by the trade of silk, spices, incense, glass, precious stones, beads and lumber. Ashoka sent Buddhist emissaries throughout all of South East Asia to facilitate that trade and the spread of Buddhism became closely associated with the growth of commercial networks into Malaysia, Indonesia and beyond to China (Donkin, 2003).

These Buddhist monastic institutions developed a mutual support network with the merchant communities. As Sen describes it:

> the transmission of Buddhist doctrines [across Indonesia] to China was the result of an interdependent and reciprocal relationship between Buddhist monks and merchants travelling between India and China. Merchants regularly assisted the growing number of Buddhist monks journeying across the overland and maritime routes to China, met the increasing demand for ritual items and actively financed monastic institutions and proselytizing activities. Buddhist monks and

monasteries, in turn, fulfilled the spiritual needs of the itinerant merchants and helped introduce new items into the stream of commodities traded between India and China ... offerings of precious objects to Buddhist monastic institutions, created and sustained the demand for commodities such as pearls, lapis lazuli and coral exported from India.

(Sen, 2019, p. 17)

China's Slow Entry into the South East Asia Trade Network

The first records of Chinese maritime trade don't appear until the fifth century BCE in the middle of the Zhou Dynasty (1027–256 BCE) (Gungwu, 1958; Schottenhammer, 2013). By this time the north-eastern Chinese had begun trading silks and other manufactured goods with the Yue in south China for cassia, ivory, pearls, tortoise shells, kingfisher and peacock feathers, rhinoceros horn and scented woods. Demand for these goods likely helped drive the dynastic expansions into the South China Sea in the second century BCE.

By the start of the Han Dynasty (206 BCE to 220 CE), the first Chinese junks were following the coast of China into the Gulf of Thailand and down the Malay Peninsula, in what was called by the Chinese 'Nanhai' trade. However, until the Song Dynasty (960–1269 CE), the Chinese themselves did little shipping across the South China Sea. In the early centuries of the first millennium CE trade across the region was a monopoly held by South East Asians, of whom the people of what is now Indonesia formed the majority. The people of Sumatra and Java were the pre-eminent ancient sailors of the world (Wang, 1958; Flecker, 2017). In the later centuries of the first millennium CE, Malay, Cham (Vietnamese) and Khmer (Cambodians) joined the party, followed by the Arabs and Persians after 750 CE. Rome never had significant direct trade with China.

Java Becomes the Nucleus of Indonesia

The early Indian and Chinese seafarers and merchants rarely travelled further than Malaya or Java and were supplied with Moluccan spices and other products primarily by Indonesians. The island of Java became a 'nucleus of the Indonesian archipelago' (Sulistiyono and Rochwulaningsih, 2013, p. 116).

Java was the 'rice basket' of the Indonesian archipelago in the precolonial era, supplying itself and many of the outer islands. The outer islands produced spices and forest products (camphor and sandalwood) that were traded to the Javanese for re-export to Western countries. The outer islands also bartered for textiles and various finished metal goods such as jewellery. Thus:

> Java acted as a warehouse of imported commodities before they were distributed to the surrounding regions of the Outer Islands such as Palembang, Lampung, Banjarmasin,

Bali and Lombok, and the Maluku islands. Ports along the northern coast of Java became a rendezvous for traders of the Outer Islands as well as their foreign fellows.
(Sulistiyono and Rochwulaningsih, 2013, p. 119)

The city of Ko-ying which probably lay in western Java was by the third century the eastern terminus for Indian shipping and was an entrepôt for spices and other luxury goods from the entire archipelago.

During the early centuries CE, the arrival of Hinduism and Buddhism from India to Sumatra and Java stimulated the formation of many small indigenous states (Frederick and Worden, 1993; Donkin, 2003). The first great empire to emerge in Indonesia was Srivijaya, which came to rule large areas of Sumatra, western Java and much of the Malay Peninsula. By dominating the Malacca and Sunda Straits, Srivijaya became rich and powerful through trade and taxes with passing ships of China, India and eventually Arabia. It controlled the shortest sea passage between the West and China and was on the route from the Spice Islands to the Strait of Malacca. By the fifth century, its main port of Palembang, accessible to the enterer by way of a river, became wealthy as an entrepôt for Chinese, Indonesian and Indian markets. It dominated trade between India and China until the tenth century when improvements in nautical technology allowed Indian and Chinese merchants to sail directly between their countries (Fig. 9.1).

Srivijaya had cultural and trade links first with the Buddhist Pala dynasty of Bengal, and later with the Islamic caliphates in the Middle East. Between the seventh and eleventh centuries, it was a hegemony in South East Asia linking the East with the West. It had lucrative trade links with China which lasted from the Tang to Song dynasties. Srivijaya dominated the Strait of Malacca until the thirteenth century when it succumbed to the rival Javanese Singhasari and Majapahit empires.

For centuries, Srivijaya was a magnet for pilgrims and scholars from all across Asia. These included the Chinese monk Yijing, who visited Sumatra for lengthy stays on his way to India between 671 and 695, and the eleventh-century Buddhist scholar Atisha, who was a major player in the development of Tibetan Buddhism in the eleventh century (Frederick and Worden, 1993; Sen, 2006).

The strongest rival to Srivijayan rule in Indonesia was the Mataram Kingdom, which controlled central Java from 760 to 830 (Frederick and Worden, 1993). The rule of Mataram was complicated by a rivalry and intermarriage of two lines of rulers, one supporting Hinduism (the Sanjaya) and the other supporting Buddhism (the Sailendra, who had family connections with the Srivijaya). The rulers of Mataram constructed the massive Buddhist Borobudur temple complex, considered one of the greatest monuments of world religious art, and the Prambanan temple complex, the Hindus' answer to the Buddhists' Borobudur.

The Chinese Pilgrims – Chroniclers of the Ancient Spice and Silk Routes

Hundreds of Chinese Buddhist monks made pilgrimages to India from 400 to 600 CE, travelling by land and sea along the Spice Routes. Three of the most

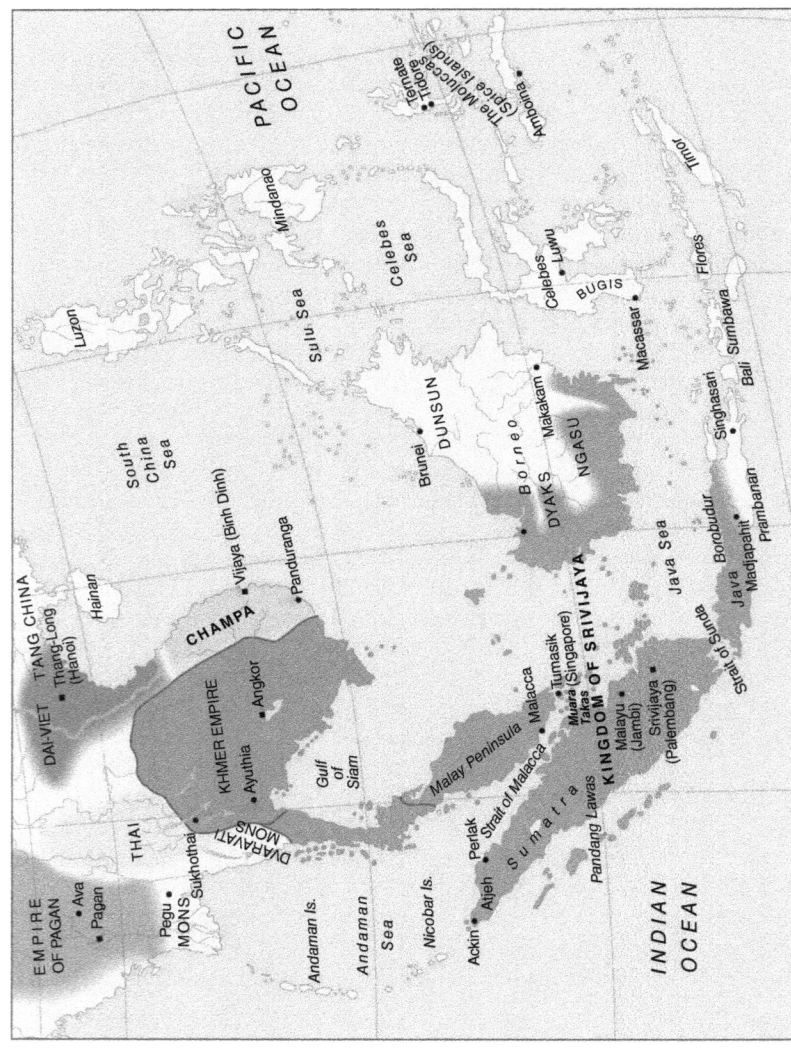

Fig. 9.1. South East Asia and the Srivijaya Kingdom, 1200 CE. (Redrawn from Gunawan Kartapranata under the Creative Commons Attribution-Share Alike 3.0 Unported license. https://commons.wikimedia.org/w/index.php?curid=7662386)

famous of these pilgrims were Faxian, Xuanzang and Yijing, who left behind detailed accounts of Buddhist practices, shrines in Central and South Asia and travel conditions along the Spice Routes. They translated hundreds of Buddhist texts into Chinese, described the social and political conditions of India, Ceylon and India, and 'offered remarkable insights into cross-cultural perceptions and interactions' (Sen, 2006, p. 24).

Faxian embarked on his trip at the age of 60 in 399 CE; he travelled for ten years across the deserts and mountains of the Silk Route to India where he visited major Buddhist pilgrimage sites. He then went to Ceylon for two years before heading home to China in a treacherous sea voyage passing through the Strait of Malacca. He arrived home in 413 and spent the last years of his life translating the scores of Buddhist writings he had brought back. He died aged 88.

Xuanzang followed the Silk Route to Kashmir and then headed down to the Ganges plains. He studied India's sutras in the revered Buddhist monastery Nalanda for five years and visited many sacred Buddhist sites in what are now Pakistan, India, Nepal and Bangladesh. Born sometime around 600 CE, he started his long and arduous journey in 629 and returned to China back across the Silk Route some 16 years later.

Yijing left for India in 671 and returned in 695, travelling mostly by sea from Yangzhou to Guangzhou and then through the Malay Archipelago to India and back. During his trek across India he was robbed by bandits, stripped naked and had to hide under leaves in the mud to escape capture. In his journeys, he visited Srivijaya for six months, studied the Malay peoples, learned Sanskrit and travelled to the Nalanda Monastery where he stayed for 11 years.

Faxian left us one of the best depictions of the nature of sea travel in antiquity. He wrote:

> On the seas (hereabouts) there are many pirates, to meet with whom is speedy death. The great ocean spreads out, a boundless expanse. There is no knowing east or west; only by observing the sun, moon, and stars was it possible to go forward. If the weather were dark and rainy, (the ship) went as she was carried by the wind, without any definite course. In the darkness of the night, only the great waves were to be seen, breaking on one another, and emitting a brightness like that of fire, with huge turtles and other monsters of the deep (all about). There merchants were full of terror, not knowing where they were going. The sea was deep and bottomless, and there was no place where they could drop anchor and stop. But when the sky became clear, they could tell east and west, and (the ship) again went forward in the right direction. If she had come on any hidden rock, there would have been no way of escape.
> (Sen, 2019, p. 33)

Early Trade in the Outer Reaches of Indonesia

The Moluccas or Spice Islands, with their nutmeg, mace and cloves, are found well outside the perimeter of the Indian and Chinese trading spheres, located in the Malay Archipelago between the islands of Celebes to the west and New Guinea to the east. The entry of these spices into world trade was dependent on contacts made by Malay and Indonesian sailors, with the Javanese being the primary traders.

It was not until the late centuries BCE that nutmeg, mace and cloves became important to the palates and medicines of the Indians, Chinese and Romans. But for thousands of years before, there was vigorous inter-island trade among the Spice Islands and the other islands of Halmahera, Seram, Kai and Aru. This trade was centred on the sago palm, *Metroxylon sagu*, which was the primary food source of the small, volcanic Moluccan and Banda Islands, where nothing else grew but coconut and spices (Ellen, 1979). The Bandanese became the undisputed leaders in the inter-island trade of sago for spices, travelling in fleets of *kora-kora* canoes, propelled by rowers on platforms of bamboo lashed five feet away on either side of the canoe proper.

The sago palm was the staple food of hundreds of thousands of people, but it received little attention from the outside world until 1869, when the great Victorian naturalist and Darwin's contemporary, Alfred Russel Wallace, described its characteristics at length in his epic *The Malay Archipelago*. About its taste, Wallace wrote:

> The hot cakes are very nice with butter, and when made with the addition of a little sugar and grated cocoa-nut are quite a delicacy. They are soft, and something like corn-flour cakes, but have a slight characteristic flavor which is lost in the refined sago we use in this country. When not wanted for immediate use, they are dried for several days in the sun, and tied up in bundles of twenty. They will then keep for years; they are very hard, and very rough and dry; but the people are used to them from infancy, and little children may be seen gnawing at them as contentedly as ours with their bread and butter. If dipped in water and then toasted, they become almost as good as when fresh baked; and thus treated, they were my daily substitute for bread with my coffee.
>
> (van Wyhe, 2015, p. 514)

The Golden Peninsula

The Chinese found Indian markets for their silk, spices and porcelain first through Malaysian intermediaries (Flecker, 2017). They began following the coast of China into the Gulf of Thailand and down the Malay Peninsula by 200 BCE, where they started trading with the Indians across the Malay Peninsula. The first Chinese junks arrived in the Strait of Malacca in about the first century CE.

The Malay Peninsula was settled by many Indian Hindus and Buddhists who established trading communities. Merchants from Indonesia, East India and China met there to trade gold, beads, precious stones and pearls for incense, pepper and porcelain (Sen, 2019). The peninsula became so famous for its gold that Ptolemy in his early world map named it the *Golden Chersonese* or Golden Peninsula (Wheatley, 1961).

Among the earliest trading centres to emerge in South East Asia was one at the Isthmus of Kra, the narrowest point on the Malay Peninsula spanning only 40 kilometres (25 miles). In the early stages of trade between India and China, goods would be offloaded on either side of the isthmus and land portaged in both directions. Riding the monsoon winds, the Indians would drop off their goods on the western side and the Chinese theirs on the east.

As many as 30 kingdoms arose and fell between the third century BCE and second century CE on the Malay Peninsula. Probably the greatest of the ancient kingdoms was Langkasuka (second to fifteenth century CE) (O'Reilly, 2006). This Buddhist kingdom was a walled, inland city state connected to the coast by canals where it conducted considerable international trade. It also became a popular resting place for Chinese Buddhist pilgrims headed to India. Langkasuka's glory years were in the sixth century CE when it was independent and regularly sent embassies to China. In the early part of its history, it was closely aligned with the kingdom of Funan and in the eighth century came under the control of the Kingdom of Srivijaya.

The First Great Trading Empire: Funan

The first important trading empire to emerge in South East Asia was what the Chinese called *Funan* (Sen, 2019). It arose in the first century AD and covered parts of modern Vietnam, Cambodia and Thailand. There is considerable debate over whether it was a large country with two major cities or a series of small, competing principalities that were scattered along the coast and throughout the Mekong Delta (Smith, 2009). It is most likely the latter. Funan is known to have been a bounteous rice belt where visiting sailors and merchants could 'hold over' while waiting for the monsoon winds to shift. It is also known to have been heavily influenced by Indian culture and was a place where Hindus and Buddhists coexisted peacefully for centuries.

Historical records concerning Funan have been written in Sanskrit and Chinese. Perhaps the fullest account was left by a pair of third-century Wu-Dynasty visitors from 245–250 CE, Kang Dai (K'ang T'ai) and Zhu Ying (Chu Ying). 'They described Funan as a sophisticated country of people living in houses raised on stilts and ruled by a king in a walled palace, who controlled trade and managed a successful taxation system' (Hirst, 2018). According to a myth reported in their account and others, Funan was formed when a female ruler named Liu-ye attempted to raid a visiting merchant ship and wound up marrying one of the ship's travellers, a Brahman from India. This story may or may not be true, but it is known that the Funan rulers brought in many Indians of the Brahmin caste to help in the administration of their courts. It is not known whether the establishment of the Funan empire was driven indigenously or through Indian émigrés, but both were likely important.

Funan's greatest port city was Oc Eo, located in the Mekong Delta on the Gulf of Thailand in today's Vietnam (Hirst, 2018). Between the second and fifth centuries CE, Oc Eo was the most important stopover between China and Malaya and came to be a critical node in worldwide trade. It flourished until the early sixth century until the Chinese shifted their trade from the Isthmus of Kra to the Strait of Malacca, thus bypassing Funan.

Oc Eo served as a major trading centre for Indian, Chinese and Indonesian goods. From India came pepper, precious stones, jewellery and cotton fabrics. These ships eventually carried goods from Rome including gold coins, gold and silver jewellery, and glassware. From China came silk and fine manufactured products such as mirrors (Ardika, 2018). Cinnamon came in from Ceylon and China. From Indonesia arrived tin from the Malay Peninsula, copper from Siam (Thailand), as well as balsamic and camphor resins from Sumatra. Cloves, nutmeg and mace also found their way into Funan from the tiny Spice Islands.

South East Asian Trading Spheres in the Early First Century CE

As the first century CE dawned, three separate trading spheres were operating in the Indian Ocean and South China Sea (Miksic, 1997). In the first, sailors from India and Ceylon were travelling to and from Bali, Java and Sumatra across the Bay of Bengal, while in the second, Indonesian seafarers were conducting trade across the vast archipelago itself and in a third they reached into South East Asia. Trade emporiums arose in Java and Sumatra, where Indian and Ceylonese sailors could access all the spices and commodities of South East Asia.

Indian ships typically sailed only as far east as the Strait of Malacca, and Indonesian ships made the other two journeys to eastern Indonesia and China. Indian sailors travelled to the Strait of Malacca between April and August, returning in December or January. Indonesian ships bound for the Moluccas left in October and came back in May or June. The Indonesian ships headed to China between June and August and returned in January and February.

No ships sailed the entire route and commodities leapfrogged in relays. As Miksic describes it:

> So just as the ships from China would be returning to Sumatra or Java with their cargoes of silks and metal, the other ships would be leaving those islands on their way to India. Merchants from the two legs of the network would not meet unless they remained in the Straits for nearly a year. Ships did not usually stay over, but traders could. They changed from ship to ship in relays. The third leg of the triangle involved what is now eastern Indonesia, the sources of some of the rarest spices, specifically cloves and nutmeg.
>
> (Miksic, 1997, p. 14)

European Connections

As the first century CE dawned, there was a worldwide maritime trading route that stretched all the way from Rome, across the Mediterranean to northern Africa, through the Indian Ocean to Indonesia and on to China, with India at its centre. The commodities shipped over these routes 'were high in value and low in bulk', including spices, incense, precious metals, textiles, and specialty items such as rhinoceros horns and kingfisher feathers (Miksic, 1997).

When the first Roman ships arrived in Indian ports, they could trade for not only the gems and spices of India, but also silk, porcelain and spices from Indonesia and China, brought in by the Indian sailors. By 100 CE, trade between South Asia and Europe was booming. As John Miksic at the National University of Singapore describes it:

> In the year AD 100, a distant observer of the Earth would have seen humankind on the verge of an explosion of commercial activity. Between the Mediterranean, where the Roman Empire under the Caesars was at its height, and China, stable and prosperous under the Han Dynasty, the kingdoms of South Asia were encouraging foreign traders to come to their shores. Ivory and incense from India were used in Roman palaces. Traders from the Graeco-Roman regions established outposts along the southern coasts of India, where Roman coins and pottery have since been found. The British archaeologist Sir Mortimer Wheeler, who excavated several of these sites, concluded that 'it is fair to envisage Indo-European commerce of the first century AD pretty closely in terms of that of the seventeenth century' (Wheeler, 1954: 125). The Romans established trading companies with powers and procedures similar to those of the Dutch and British factory systems used in Asia over a thousand years later.
>
> (Miksic, 1997, p. 14)

The Two Ways to Rome

Goods from South East Asia found their way to Rome not only through the Red Sea, but also through the Persian Gulf (Seland, 2011). The Red Sea route to Rome was much longer than the one through the Persian Gulf, but goods were obtained more cheaply by the Romans through the Red Sea. They had almost full control of the Red Sea corridor to India from 30 BCE until about 250 CE; however, the Romans controlled the head of the Persian Gulf for only brief periods in the second and third centuries CE. This meant that the goods procured by the Romans through the Persian Gulf corridor were more expensive, subject to the tariffs charged by the Parthian middlemen (Fig. 9.2).

The two routes were composed of varying land and sea distances. The Red Sea corridor required a trip by ship of 4500 kilometres (2796 miles) from India to Egyptian Red Sea ports, followed by a caravan route of 380 kilometres (236 miles) across the Egyptian Desert and then another 760 kilometres (472 miles) by ship on the Nile to the Mediterranean, for a total distance of 5640

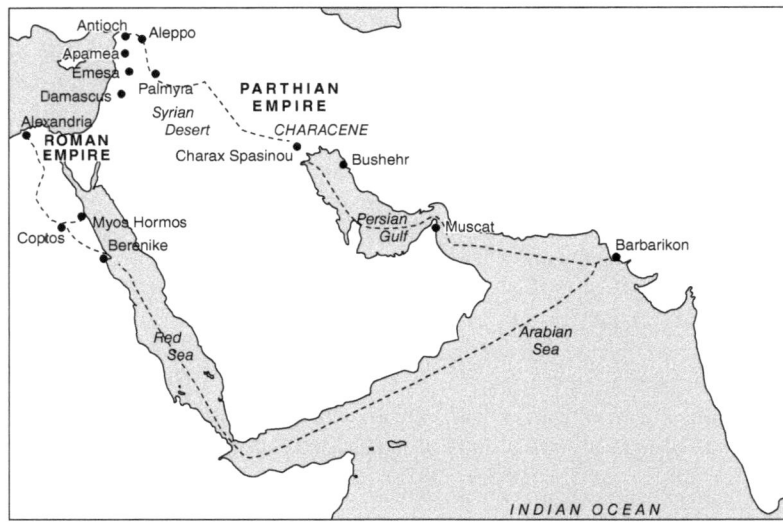

Fig. 9.2. Two ways to Rome from the Indian Ocean. (Redrawn from Seland, E. (2011) The Persian Gulf or the Red Sea? Two axes in ancient Indian Ocean trade, where to go and why. *World Archaeology* 43(3), 398–409, Figure 1.)

kilometres (3504 miles). The Persian Gulf corridor required a trip by ship of 2350 kilometres (1460 miles) from India to the confluence of the Tigris and Euphrates Rivers and then by caravan across the Syrian Desert for 1400 kilometres (870 miles), for a total of 3750 kilometres (2330 miles).

For the Romans, obtaining goods from South East Asia through the Red Sea versus the Persian Gulf corridor had obvious advantages. The total distance of the Red Sea–Nile route was about a third longer than the Persian Gulf–Syrian Desert alternative, but it required a much shorter overland journey. It was cheaper to transport goods by sea than land, and the shorter caravan distance across Egypt to Alexandria likely reduced cost (Wiesehöfer, 2006). The Red Sea route also freed the Romans from the taxes imposed on silk and spices by their bitter Parthian enemies.

In spite of these advantages, the Romans never stopped trading through both corridors. Maintaining two trade links buffered them against shortages caused by weather and political conditions. Due to the monsoon winds, travel across the oceans was possible only once a year along both routes. The Romans essentially had two seasons of delivery, a spring one in Antioch and a late summer one in Alexandria. Goods headed to the Red Sea corridor would have left India between December and January, arrived at the Red Sea in February and got to Alexandria in August. Goods travelling through the Persian Gulf corridor would have left India for the Gulf in November, arrived there between January and February and made their way to Antioch between April and May.

As Seland argues brilliantly:

> the employment of multiple routes ensured that Mediterranean markets would receive Indian Ocean imports at different times of the year and to multiple entrepôts. The reconstruction of passages from the mouth of the Indus to the Mediterranean shows that commodities transported by way of the Persian Gulf and the Syrian Desert were likely to arrive at Antioch or another Mediterranean port by late spring, while goods going by way of Egypt would probably reach Alexandria in the early fall. This would reduce risks connected to weather and political conditions on the Red Sea and Euphrates borders and is also likely to have had a balancing effect.
>
> (Seland, 2011, p. 406)

The First Direct Contact between Rome and China

For centuries, the Romans had only a vague notion of the geography of China. That knowledge came through the contact of sailors with merchants in India and the Malay Peninsula. In the *Periplus of the Erythraean Sea*, its author provides vivid accounts of eastern trade cities in India, speculating that silk arrived there from a great inland city called Sinae by a land route through Bactria to the Ganges. The great Greco-Roman cartographer Ptolemy based his diagrams of the Far East to a large extent on a report of a Greek sailor/merchant named Alexandros who described Roman trade on the north-west Malay Peninsula and a trip along the coast of Vietnam to the trade port of Cattigara, probably Oc Eo.

In Ptolemy's second-century map of the known world, he included the Gulf of Thailand and the South China Sea, marked the location of Cattigara and labelled the surrounding land to the east as 'Sinae' ('land of silk') (McLaughlin, 2010). To the Romans this was the extreme south-eastern corner of the world. The Malay Peninsula was presented as the Golden Chersonese (Golden Peninsula), India was shown as being bordered by the Indus and Ganges Rivers, and Ceylon ('Taprobane') was depicted as an island off the coast of India. Ceylon was drawn with exaggerated size, representing its importance in the world trade of classical antiquity.

There are tantalizing bits of evidence that document early, mostly indirect trade between the Romans and Chinese. Roman coins dating back as early as the first century CE have been found in Xi'an, China and in the ancient settlement at Oc Eo (Ball, 2016). Roman glassware and silverware have also been discovered at other Chinese archaeological sites dating to the Han period. However, the quantity of Roman coins found in China is tiny compared with that found in India, suggesting that trade with India was a much greater focus of the Romans. The Roman coins and glassware found in China most likely got there via intermediate Indian traders.

The first direct contact between mainland China and Rome did not occur until the Roman Empire defeated Parthia in the second century and took control of the Persian Gulf. Before then, the Persian Achaemenid Empire, Kushans and Parthians had done everything in their power to protect their profitable status as middlemen and discourage contact between the two great powers of the East and West. In 97 CE the Chinese Han had even tried to send an ambassador, Kan Ying, to Rome but he only got as far as Mesopotamia before returning home, after being misinformed by the Parthians that the rest of his journey was extremely dangerous and would take another two years (Hill, 2009).

It wasn't until 166 CE that the first Roman envoy made his way to China, sent by the emperor Antoninus Pius or Marcus Aurelius via the Persian Gulf. This diplomatic contact was recorded in the ancient Chinese history of the Han Dynasty, *Hou Hanshu* (*Book of the Later Han*). Other Roman missions found their way to China in 226 and 284 CE, followed by a long absence until a Byzantine embassy was set up in 643 CE.

References

Ardika, I.W. (2018) Early contacts between Bali and India. In: Saran, S. (ed.) *Cultural and Civilisational Links between India and Southeast Asia*. Palgrave Macmillan, Singapore, pp. 19–29.

Ball, W. (2016) *Rome in the East: Transformation of an Empire*, 2nd edn. Routledge, London and New York.

Bellina, B. (2003) Beads, social change and interaction between India and South-east Asia. *Antiquity* 77, 285–297.

Donkin, R.A. (2003) *Between East and West: The Moluccas and the Traffic in Spices up to the Arrival of Europeans*. American Philosophical Society, Philadelphia, Pennsylvania.

Ellen, R.F. (1979) Sago subsistence and the trade in spices: a provisional model of ecological succession and imbalance in Moluccan history. In: Burnham, P.C. and Ellen, R.F. (eds) *Social and Ecological Systems*. Academic Press, London, pp. 43–74.

Flecker, M. (2017) Early Chinese voyaging in the South China Sea: implications on territorial claims. *Journal of Maritime Studies and National Integration* 1, 1–21.

Frederick, W.J. and Worden, R.L. (eds) (1993) *Indonesia: A Country Study, 5th edn*. Area Handbook Series. Library of Congress, Washington, DC.

Gungwu, W. (1958) The Nanhai trade: a study of the early history of Chinese trade in the South China Sea. *Journal of the Malayan Branch of the Royal Asiatic Society* 31, 3–135.

Gupta, S. (2005) The Bay of Bengal interaction sphere (1000 BC–AD 500). *Bulletin of the Indo-Pacific Prehistory Association* 25, 21–30.

Hill, J.E. (2009) *Through the Jade Gate to Rome: A Study of the Silk Routes during the Later Han Dynasty, 1st to 2nd Centuries CE*. BookSurge, Charleston, South Carolina.

Hirst, K.K. (2018) Oc Eo, 2,000-year-old port city in Vietnam. *ThoughtCo*. Available at: https://www.thoughtco.com/oc-eo-funan-culture-site-vietnam-172001 (accessed 28 January 2021).

Kimura, J. (2019) Cross-regional and -chronological perspectives on East Asian seafaring and shipbuilding. In: Schottenhammer, A. (ed.) *Early Global Interconnectivity*

Across the Indian Ocean World, Vol. II. *Exchange of Ideas, Religions, and Technologies*. Palgrave Series in Indian Ocean World Studies. Palgrave Macmillan, Cham, Switzerland, pp. 203–226.

Kingwell-Banham, E., Bohingamuwa, W., Perera, N., Adikari, G., Crowther, A. *et al.* (2018) Spice and rice: pepper, cloves and everyday cereal foods at the ancient port of Mantai, Sri Lanka. *Antiquity* 92, 1552–1570.

Lansing, S., Redd, A.J., Karafet, T.M., Watkins, J., Ardika, I.W. *et al.* (2004) An Indian trader in ancient Bali? *Antiquity* 78, 287–293.

McLaughlin, R. (2010) *Rome and the Distant East: Trade Routes to the Ancient Lands of Arabia, India, and China*. Pen & Sword Books Ltd, Barnsley, UK.

Miksic, J. (1997) Historical background. In: Mathers, W.M. and Flecker, M. (eds) *Archaeological Report: Archaeological Recovery of the Java Sea Wreck*. Pacific Sea Resources, Annapolis, Maryland, pp. 5–33.

O'Reilly, D.J.W. (2006) *Early Civilizations of Southeast Asia*. AltaMira Press, Lanham, Maryland.

Rao, K.P. (2001) Early trade and contacts between South India and Southeast Asia (300 BC–AD 200). *East and West* 51, 385–394.

Schottenhammer, A. (2013) The 'China Seas' in world history: a general outline of the role of Chinese and East Asian maritime space from its origins to *c*. 1800. *Journal of Marine and Island Cultures* 1, 63–86.

Seland, E.H. (2011) The Persian Gulf or the Red Sea? Two axes in ancient Indian Ocean trade, where to go and why. *World Archaeology* 43, 398–409.

Sen, T. (2006) The travel records of Chinese pilgrims Faxian, Xuanzang, and Yijing. *Education about Asia* 11(3), 24–33.

Sen, T. (2019) Buddhism and the maritime crossings. In: Schottenhammer, A. (ed.) *Early Global Interconnectivity across the Indian Ocean World*, Vol. II. *Exchange of Ideas, Religions, and Technologies*. Palgrave Series in Indian Ocean World Studies. Palgrave Macmillan, Cham, Switzerland, pp. 17–50.

Smith, R.L. (2009) *Premodern Trade in World History*. Routledge, London and New York.

Sulistiyono, S.T. and Rochwulaningsih, Y. (2013) Contest for hegemony: the dynamics of inland and maritime cultures relations in the history of Java island, Indonesia. *Journal of Marine and Island Cultures* 2, 115–127.

van Wyhe, J. (ed.) (2015) *The Annotated Malay Archipelago by Alfred Russel Wallace*. NUS Press, Singapore.

Wang, G. (1958) The Nanhai trade: a study of the early history of Chinese trade in the South China Sea. *Journal of the Malayan Branch of the Royal Asiatic Society* 31, 1–135.

Wheatley, P. (1961) *The Golden Khersonese: Studies in the Historical Geography of the Malay Peninsula before AD 1500*. University of Malaya Press, Kuala Lumpur.

Wheeler, M. (1954) *Archaeology from the Earth*. Clarendon Press, Oxford.

Wiesehöfer, J. (2006) *Ancient Persia from 550 BC to 650 AD*. I.B. Tauris Publishers, London.

10
Golden Age of Byzantium

Setting the Stage – Roman Power Shifts

During the third century, Rome was in a constant state of flux with a revolving door of 22 emperors taking the throne. The Roman Empire was plagued with conflict on all its frontiers – with Germans, Goths and Parthians. In an attempt to restore peace and prosperity, the emperor Diocletian divided power into a tetrarchy (rule of four) in 284 CE. Diocletian and Galerius were given rule of the eastern part of the empire, while Maximian and Constantine were given power in the west. This arrangement fell apart in 324, when Diocletian and Maximian retired from office and Constantine took over total power and reunified Rome. To solidify his rule, he moved the Roman capital to Byzantium which he renamed Constantinople. He also made Christianity the official religion of Rome.

Constantine's effort at reunification proved fragile for when he died the eastern and western empires again fractured into two pieces. The Western Roman Empire lasted only another 75 years, collapsing under waves of Germanic tribe movements called the 'Völkerwanderung' or Migration Period. The Visigoths were the first to enter Roman territory in 376, chased by the fearsome Huns. They were first tolerated by the Romans as defenders of the Danube territory, but eventually turned upon Rome and sacked the great city in 410 CE. They then moved on to Spain and established an empire there. The Ostrogoths followed the Visigoths into Italy and conquered it in 474 CE, actually backed by the Byzantines. Also in the fifth century, a fusion of western Germanic tribes called the 'Franks' began their takeover of Gaul and the Anglo-Saxons began their invasion of Roman Britain. Europe was falling into a fragmented 'Dark Ages'.

The Eastern Roman Empire (Byzantine Empire or Byzantium) stayed upright and remained powerful in various forms for almost 1000 years. The Byzantine Empire remained Roman in structure and function, but with an increased religious influence. The Church came to permeate all aspects of

© J.F. Hancock 2021. *Spices, Scents and Silk: Catalysts of World Trade*
(J.F. Hancock)
DOI: 10.1079/9781789249743.0010

Byzantine society including a network of monasteries across rural Asia Minor that essentially controlled peasant life. The emperor was the head of the Church and took an active lead in discussions of Church reform and theological disputes. The state religion remained Christian, although the Greek Orthodox Church of Byzantium diverged significantly from the Roman Catholic Church of the West. It developed many viewpoints alien and repugnant to the western Catholic Church, which ultimately led to their complete separation.

Constantine Builds His City

When Constantine reunited the Roman Empire in 324 CE, he realized he needed a new capital more geographically central than the decaying Rome. He chose the old Greek city of Byzantium on the north-western edge of Asia Minor, located along the Bosporus strait between the Sea of Marmara and the Black Sea. It had a well-protected harbour on the Golden Horn estuary and could be easily defended, as it was surrounded by water on three sides. Most importantly, this ancient city sat right in the middle of all the great Silk and Spice Routes that connected Asia with Europe.

Constantine undertook a massive rebuilding effort to make his city the envy of the world, much like Alexander the Great with Alexandria (Herrin, 2007; Luttwak, 2011; Wasson, 2013). The project took six years. Constantine laid out his city over seven hills (just like Old Rome) and divided it into 14 districts. He crisscrossed it with wide avenues lined with statues of the who's who of history's greatest leaders, including Alexander the Great, Caesar, Augustus, Diocletian, and of course himself, dressed as Apollo. He built two major colonnaded streets that bisected the heart of the city. At the intersection of these two streets he constructed a massive four-tiered arch, the Tetraphylon.

Nearby were two huge public baths called Zeuxippus and Testratoon, and to its north Constantine laid out a large square courtyard and surrounded it with large porticos, leading to a library and two shrines. In the centre of the city, he built many great monuments, including the spectacular domed Hagia Sophia, the Church of Holy Wisdom. Southward he had constructed a magnificent imperial palace, which was opulently decorated with exotic marble and fine mosaics and entered through the monumental Chalke Gate with gigantic bronze doors. Nearby, he also enlarged an ancient hippodrome to host chariot races, replacing an old amphitheatre where gladiator contests had once occurred. Gladiator competitions were now disfavoured by the newly Christian Romans.

To fully fortify the city against attack, Constantine built a wall around the entire city, surrounding an area of eight square miles. To make sure that the city had a reliable water supply even when invaded, Constantine also built the massive Binbirdierek Cistern, along with others, and developed a system of aqueducts and tunnels to bring water into the city during the rainy winter season.

The World Turns

The fall of the Western Roman Empire led to a dramatic shift in the direction of world trade. When the Visigoths breached the walls of Rome in 410 CE and sacked the city, what had been the centre of world trade collapsed after 500 years of domination. No longer were the spices and silks of South East Asia being funnelled to Europe. The focal point of international trade would shift east to Constantinople in the remaining half of the Roman Empire. Constantinople replaced Rome as the spice and silk trading centre of the world.

Constantinople also became the new focus of the grain trade with Egypt and North Africa (Casson, 1991; Paine, 2013). As Rome struggled to survive and its population plummeted, the massive shipments of grain that once went to the Tiber River in Italy were increasingly redirected to the Bosporus strait in north-western Turkey. This was a monumental shift as, at its peak, the free provision of the *cura annonae* had filled the bellies of at least 200,000 people a day in Rome, and another 800,000 expected a regular supply of grain at reasonable prices. It has been estimated that at least 150,000 to 300,000 tons of grain was delivered to Rome each year (Charles and Ryan, 2009).

It was in fact much simpler and cheaper to ship the grain to Constantinople than Rome. The sea lanes to Constantinople were much shorter and didn't require the super-freighters of old. Wheat was now produced in Sicily, Tunisia, Egypt and the plains of Asia Minor. Even from Egypt, a ship could now make two or three trips back and forth a year with smaller cargoes. The only real challenge was successfully navigating the treacherous straits of the Dardanelles. Emperor Justinian solved this problem in the sixth century by building a huge granary on the island of Tenedos near the mouth of the strait that could hold all the cargoes of the vessels (Casson, 1991).

The Exotic Luxuries of Byzantium

The importance of silk in the Eastern Roman Empire was paramount as a powerful symbol of the very wealthy and the Orthodox Church. The Byzantine rulers considered silk a precious commodity and restricted its use to the aristocracy and ecclesiastical ceremony. Richly patterned, colourful silks became a powerful, highly regulated trade commodity, bringing rich customs duties to the royal treasury. Bolts of silken cloth were gifted to foreign ambassadors who visited Constantinople to impress them with the power of the Byzantine Empire.

Purple silk was especially revered in the Byzantine Empire and was restricted to the ecclesiastical royals. Byzantine rulers wore flowing purple robes, adorned with precious stones and gold (Williams, 2012). The reputation of purple came through its rarity, because purple dye originated from only one place in the world: the ancient Phoenician city of Tyre in modern-day Lebanon. To produce the dye, a purple-producing mucus was harvested from a species of sea snail now known as *Bolinus brandaris* and then exposed to sunlight for a precise amount of time. It took as many as 250,000 molluscs to yield just one ounce of usable dye. Tyrian purple was said to carry divine connotations, as it resembled the colour of Christ's clotted blood.

The Byzantines, rich and poor, also required their food to be richly seasoned with pepper as had been the Roman habit for centuries. Chinese ginger and Indonesian nutmeg also became important in their diets, moving from the ancient Romans' medicine cabinets to the Byzantine kitchen. As Robin Trento relates in her blog from the Getty:

> The food of medieval Constantinople was truly a fusion. It combined ancient Roman culinary traditions, local Greek and Anatolian practices, the dictates of the Eastern Orthodox Church, and the influence of cross-cultural exchange. Constantinople (today Istanbul) was and is a vibrant trading center ... The variety of herbs and vegetables available in Constantinople was far greater than that enjoyed by ancient Greeks and Romans – and they were eaten cooked as well as raw, unheard of in Europe. The Byzantines enjoyed meat (when not fasting per the dictates of the Church), primarily pork, lamb, poultry, and certain types of gazelles and donkeys. They were also avid fish-eaters, taking advantage of the rich waters where the Bosphorus joins the Black and Aegean Seas, and they carried on the ancient Roman tradition of seasoning nearly everything with fish sauce, garum, which had long since fallen from favor in Europe.
>
> (Trento, 2014)

They used forks and spoons at their meals, unknown in Europe, which were 'devised to aid Byzantine nobles in protecting their long, ample sleeves from being soiled' (Trento, 2014).

The demand for incense in the Roman world fell dramatically with the introduction of Christianity as the state religion in 323 CE. The burning of frankincense in religious ceremony was banned in the early Christian church and the practice of cremation after death was replaced with simple burials. The burning of incense was no longer necessary to cover the stench of funeral pyres. However, the demand for silk remained high among the elite in the Byzantine world and the desire for spice actually grew.

The Golden Age of the Eastern Roman Empire under Justinian

The golden age of the Byzantine Empire came under Justinian I, who ruled from 527 until his death in 565. During his reign, he established a legal code that would shape the modern concept of the state and he recovered much of the territory of the former Western Roman Empire (Fig. 10.1). He ruled as a complete despot, was treated as a near-god and wielded absolute power.

Under Justinian, Constantinople stood as the trading powerhouse of Europe; its importance far dwarfing all others. It became the major port linking Western Europe with the Far East, and the city itself developed a voracious appetite for all things exotic – spices, silks and perfumes. The backbone of Byzantine trade was the grain feeding the masses, but other trade goods

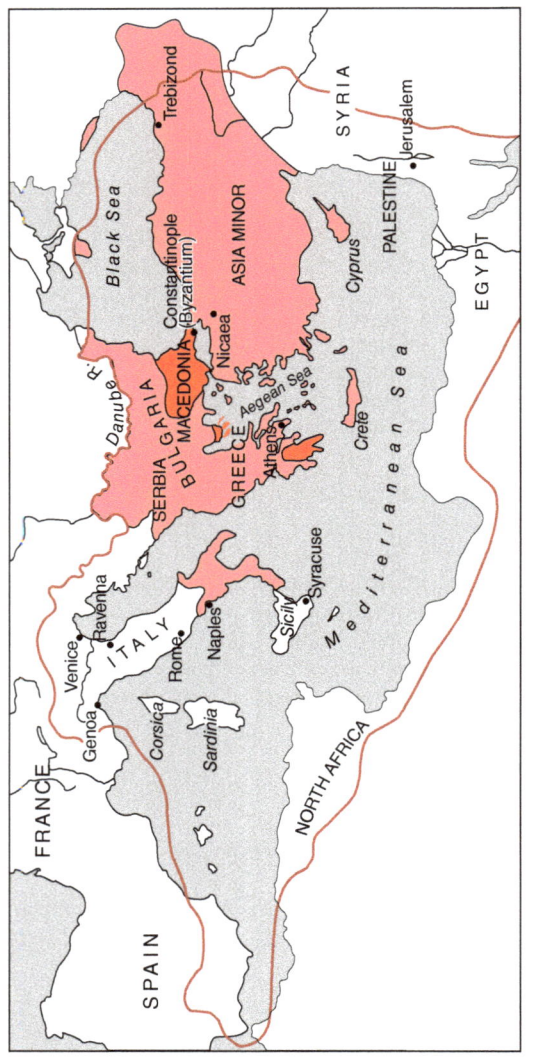

Fig. 10.1. Byzantine Empire under Justinian and in about 1020 and 1360. (Redrawn from Teall, J.L. and MacGillivray Nicol, D. (n.d.) Byzantine Empire. *Encyclopedia Britannica*. Available at: https://www.britannica.com/place/Byzantine-Empire, accessed 18 February 2021.)

of all kinds flowed into the city including slaves, furs, honey and timber from Northern Europe; silk from China; pepper, cloves and gems from South East Asia; ivory and gold from Africa; and wine, olive oil, cork, tin and iron from Western Europe. Byzantine merchant vessels from Constantinople travelled far and wide across the North Sea, Mediterranean and even to the Indian Ocean via the Red Sea. Imperial and private workshops in Constantinople churned out textiles of wool, linen and hemp and metalworks that served local and international markets.

Byzantine Attitudes about Trade

Trade in the Byzantine world was highly regulated by the state. In fact, almost everything in Constantinople was state-controlled (Herrin, 2007; Cartwright, 2018). There was a city 'eparch' whose broad responsibilities included how factories and workshops were run, the wages paid, the way import and export goods were handled and tariffs. The empire was essentially a huge trading organization. Trade goods were generated by hereditary guilds and these goods were transported by state-subsidized merchants called *navicularii*. State-appointed officials known as *kommerkiarioi* were responsible for collecting duties on all commercial transactions and marking goods with a lead seal once they had passed through the system.

Trade was so rigidly controlled by the state that the only lucrative form of investment by private individuals was landed property. Trade and commerce were not highly regarded as a profession in the Byzantine Empire and considered almost undignified by the aristocracy. The Byzantines were happy to purchase trade goods from others but did not want to get their hands dirty moving the goods themselves. As Herrin (2007, p. 148) put it: 'Byzantium inherited from Rome a contempt for trade as an activity not worthy of free men'. Emperor Theophilos (829–842 CE) is said to have burned an entire ship and its cargo when he discovered that his wife Theodora 'was dabbling in commerce and had financial connections with the vessel' (Cartwright, 2018).

State control of trade also included regulation of what could be traded outside the empire. Great care was taken to prevent export of anything that might aid Byzantium's enemies. Prohibited goods included salt, gold, iron for weapons, timber for ships, and a highly inflammable Greek Fire used in warfare. Export of luxurious silks dyed with Tyrian purple was also restricted, being reserved for the imperial court and clergy.

The 'Dollar of the Middle Ages'

Coins were at the centre of the Byzantine economy. The state minted coins to pay its armies and officials, and 'these coins filtered down through all levels of society' (Cartwright, 2017). Coins were used to purchase goods and most importantly to pay taxes. When the economy was strong and wars were being fought, coins were plentiful, issued to pay the armies, their suppliers and the burgeoning bureaucracy. Byzantine coins were spread all over the world in trade, being found as far away as Scandinavia (Adelson, 1960). When times were lean, as in the seventh and eighth centuries, coins became scarcer and barter played a more important role, particularly in the provinces of Asia Minor.

Copper, silver and gold coins were issued with images stamped on them of emperors and Jesus Christ. Justinian II was the first emperor to put Jesus Christ's image on coins in 691 CE with the legend *rex regnantium* ('King of kings'). Coins were one of the ways rulers could disperse their image across their reign and remind everyone of who was in charge and was God's representative on earth. The most important coin in Byzantium for 700 years was the gold *nomisma* (*solidus*), which had to be used to pay annual taxes. When it was first minted by Constantine I in 312 CE, 72 *nomismata* equalled a pound (453 grams) of gold. Each coin carried 4.4 grams of pure gold and was between 21 and 22 millimetres wide. Intermediate in value was the common silver coin called the *miliaresion*, introduced in 720 CE, with 12 of them being worth one *nomisma*. The least valued coin was the copper one called the *follis*, introduced by Anastasius I (r. 491–518 CE) – 24 *folles* were equal to one *miliaresion* and it took 288 to equal one *nomisma* (Cartwright, 2017).

In terms of actual values, Cartwright relates that:

> a labourer would have earned between five and twelve *folles* a day while a middle-ranking official earned some 1000 *nomismata* per year. One *nomisma* could have bought you a pig, three were needed for a donkey, fifteen was the price of a camel, and a slave with a few skills would have set you back 30 gold coins – a purchase to be made with care, clearly. Aristocrats measured their wealth in the thousands of *nomismata* while in the rare cases where a field army's four-year treasure chest was stolen by the enemy, such as the Bulgars in 809 CE or the Arabs in 811 CE, we know that it consisted of 80–90,000 *nomismata* – enough to make a tax collector's eyes water.
>
> (Cartwright, 2017)

Trading with the Enemy

For centuries, the Byzantines' only major antagonist in the East was the Sasanian Empire. The Sasanians had become the major power in the Near East when they defeated the Parthians in 224 CE. They reinstated many aspects of Iranian culture into the fibre of their government, which had been set aside under the Hellenistic rule of the Seleucids and the decentralized rule of the Parthians (Shahbazi, 2012). The Sasanian Empire became a major centre of worldwide thought and knowledge, rivalling the Byzantines. The great Academy of Gundishapur was established, which obtained and translated the great works of Greece, Persia and India. This accumulation of knowledge would prove to be the backbone of the European Renaissance a thousand years hence. The ancient monotheistic religion of Iran, Zoroastrianism, became a fundamental part of the Sasanian Empire's organization and administration, although the government left the trading communities of the other religions largely alone. The period of Sasanian rule is considered by many of today's Iranians as a highlight of their civilization.

The Romans and Sasanians fought long and hard in a 'continuous push and pull' (Cervantes, 2013). The Sasanians held on as tight as they could to the trade routes passing through Mesopotamia and Syria, doing everything in their power to disrupt western trade in the Indian Ocean. This epic struggle would go on until the Muslims routed the Sasanians in the eighth century and Islam became the Eastern Roman Empire's most powerful enemy. Private trade between individuals of the two empires was discouraged and merchandise could only move through designated border towns like Dara and Nisibis, under stringent governmental control. At these locations customs duties were collected and merchants searched to make sure that they were not actually spies. Most of the Iranian goods came from the Silk Route running from China, through Central Asia and northern Persia to Mesopotamia. In addition to silk, spices, precious stones, incense and ivory followed this route.

Aksum and Byzantium's Indian Ocean Connections

From the first to the sixth century CE, the most frequented route from the Roman world to the East was down the Red Sea from the Egyptian ports to the Kingdom of Aksum, along the southern Red Sea (Butzer, 1981). Originating in the highlands of Ethiopia, Aksum became a trading juggernaut that maintained close ties with Rome and ultimately controlled northern Ethiopia, the Sudan and South Arabia. Its port of Adulis on the Eritrean coast of Ethiopia became a gateway city that moved goods from the Abyssinian Plateau and the Sudanese plains into a maritime exchange network connected to India, China, the Black Sea and even Spain. The level of Roman trade through Aksum ebbed and flowed over the years, but for the greater part of seven centuries remained strong. Aksum arose to its greatest power during the third century CE when Rome went through its great period of political instability.

There are good records of what was traded by Aksum. Pliny mentions the goods brought to Adulis by the 'Trogodites and Ethiopians' and a whole chapter of the *Periplus of the Erythraean Sea* is devoted to trade from Adulis (Munro-Hay, 1991). Exports from Aksum included ivory, rhinoceros horns, elephants, hippopotamus hides, gold and slaves. Ivory was an extremely important luxury item in international trade and Aksum competed only with its neighbours in Meroë (Kush) for control of that trade. Many thousands of elephants were reported to have grazed within its borders. Spices and sugarcane flowed into Adulis from the Far East. Aksum also controlled the frankincense trade from about 400 to 600 CE, after it invaded Himyar and southern Arabia.

Aksum came to hold sway over the southern Red Sea trade. As Sidebotham relates:

> At the southern end of the Red Sea many, though certainly not all, cargoes would have been transshipped from the vessels of northern Red Sea origin (Romans, Graeco-Roman-Egyptians and, perhaps, Nabataeans) to those of South Arabs, Indians, Axumites/Ethiopians and others for conveyance to more distant ports in the Persian Gulf and Indian Ocean. It was not all competition. Rome and Axum had an extended period of friendly relations based partly on what became their common religion, Christianity, and on their interests in diverting trade from the Sassanian Persians.
>
> (Sidebotham, 1996, p. 787)

Available evidence suggests that Red Sea trade showed:

> a commerce substantially free of government interference and regulation, whether it be Roman or foreign. There was some regulation of the commerce by the Roman government, but this was chiefly concerned with the collection of taxes and took place only within the empire. As far as can be told, outside the empire traders were perfectly free to go where they liked and pick up whatever goods they preferred, controlled only by very limited local direction. There is no direct Roman control evident in the lands of the Red Sea, let alone further afield, and similarly there is evidence of only limited and local Arab attempts to control the trade in the Red Sea. All things considered, it seems that the merchants of the Red Sea trade purchased their cargoes and brought them to Roman Egypt in what was essentially an unfettered free market.
>
> (Young, 2001, p. 33)

Christians Surrounded by Muslims

Positioned at the centre of southern Red Sea trade, Aksum served as a meeting point of world religions. Its emperors believed their lineage traced back to King Solomon and the Hebrews of the Old Testament. They were sure that the Queen of Sheba came from Aksum and had been impregnated by the Jewish king. Aksum became Christian in the fourth century (Harrower et al., 2019) when a Syrian named Frumentius came to be captured and later hosted by the Aksumite court whose king, King Ezana (r. 320–350), he ultimately converted.

During the early years of Islam several of Muhammad's followers fled to Aksum to avoid persecution and were kindly received there. Aksum fiercely remained Christian, but the caliphate gradually strangled the Aksumite economy by taking over its maritime trade routes. By the tenth century, Aksum had let go of its coastal holdings probably because its capital Adulis was a difficult 12 days' journey from the coast. The profits may have been no longer high enough to make that trip, and Aksum chose to concentrate its trade instead on the interior of Africa.

The Secret of Silk Escapes

The process of sericulture was a guarded secret of the Chinese state for centuries and anyone who revealed the process or smuggled silkworm cocoons outside China did so at the threat of death (Cartwright, 2017). However, the Chinese couldn't keep their secret forever and silk began to be manufactured in Korea around 200 BCE, when waves of Chinese immigrants settled there. Gradually, over the next 1000 years, sericulture spread to Japan, Persia, Byzantium, across the Muslim conquests and to the city states of Italy.

Until the mid-sixth century, most silk found its way to Europe through the Silk Routes across China and the northern steppes of Central Asia. The importance of this trade route changed dramatically around 550 CE when two Nestorian monks showed up at Emperor Justinian's court in Byzantium with an enticing offer. As Snell tells the story:

> They promised the emperor they could acquire silk for him without having to procure it from the Persians, with whom the Byzantines were at war. When pressed, they, at last, shared the secret of how silk was made: it was spun by worms. Moreover, these worms fed primarily on the leaves of the mulberry tree. The worms themselves could not be transported away from India … but their eggs could be. As unlikely as the monks' explanation may have sounded, Justinian was willing to take a chance. He sponsored them on a return trip to India with the objective of bringing back silkworm eggs. This they did by hiding the eggs in the hollow centers of their bamboo canes. The silkworms born from these eggs were the progenitors of all the silkworms used to produce silk in the west for the next 1,300 years.
>
> (Snell, 2019)

The Byzantine church and state created imperial workshops, monopolized production and tried to keep the secret to themselves (Lopez, 1945; Snell, 2019). They built silk factories, known as *gynaecea*, where only women worked as essentially serfs, bound to the factories by law. The Byzantines dominated silk production in Europe throughout most of the Middle Ages. Their only real competition came from the Muslims who learned about silk when they won the eastern Byzantine provinces in the seventh century and started to produce it for themselves. The Byzantines held sway over the European silk industry for the next 700 years until a weaving industry emerged in Italy in the twelfth century as a by-product of the Crusades.

Justinian's Plague

In the middle of Justinian's reign, the world was to suffer its second great plague. Like the Antonine Plague of the second century, Justinian's Plague spread along the great trade routes, emerging first in China and north-east India, travelling to Ethiopia, moving up the Nile to Alexandria and then east to Palestine and across the entire Mediterranean region. Using DNA analysis of human skeletal remains from an early medieval cemetery, it has been determined that the plague was primarily bubonic, caused by *Yersinia pestis* (Harbec et al., 2013). The disease was carried by fleas infesting black rats (*Ratus ratus*) which likely travelled from port to port on grain ships.

Constantinople, at the crossroads of all the trade routes, suffered particularly hard from the plague. The first great outbreak would last about four months, but the disease would persist off and on for the next three centuries. It is estimated that 20 to 40% of the people in Constantinople died of the plague and nearly 25% of the rest of the Byzantine Empire (Horgan, 2014). Total deaths in the empire were in excess of 25 million and could have been as high as 50 million.

The plague greatly weakened the Byzantine Empire in both political and economic ways. The Byzantine army was decimated greatly by the disease and was unable to recruit new soldiers. As the disease spread through the empire, Byzantium's ability to resist its enemies was greatly diminished. Trade throughout the empire became greatly disrupted, as finding enough healthy sailors to man ships became problematic. Grain was in short supply and prices soared. Tax revenues declined, but Justinian steadfastly refused to give up on his ambitious construction projects and continued to fight to hold his kingdom together.

End of the Red Sea Portal

The partnership with Aksum kept Byzantine maritime trade links alive for a couple of centuries, but Sasanian power in the Persian Gulf and Indian Ocean grew to the point in the fifth and sixth centuries that they were almost completely dominating trade with India and were effectively disrupting Roman trade ambitions all across South East Asia (Daryaee, 2003; Jones, 2018). Their strategy was to purchase as much Indian cargo as possible and keep it out of the Byzantine Empire's hands. To do this, the Sasanians established trading colonies far and wide across the Persian Gulf, Malaysia, India and Ceylon. The Eastern Roman Empire, especially during the reigns of Justin (518–527) and Justinian (527–565), clung to the Ethiopians of Aksum to hold on to their link with Indian Ocean trade (Whitehouse and Williamson, 1973).

In 524/5 Justin supported an Ethiopian invasion of the Himyarites in Yemen, which put a ruler in place who was favourably disposed towards Rome. The empire also supported an invasion of Iran, in hopes of challenging the Sasanian stranglehold on Indian Ocean trade. The Sasanians responded by invading Yemen themselves and placing their own ruler on the throne. This gave the Sasanians control of Aden and the entrance to the Red Sea, effectively shutting off the Red Sea portal of the Romans to Indian Ocean trade.

The Byzantine Empire's maritime link with South East Asia through the Red Sea came to its final end when it lost control of Egypt at the end of a devastating series of wars fought with the Sasanian Empire. Persian armies pushed them first out of Mesopotamia, then the Caucasus, much of Anatolia, the Levant and finally Egypt itself in 621. The Romans made several attempts to retake Alexandria, but none had enduring success. Some 1000 years of Greek and Roman rule over Egypt had ended and with it the Red Sea link of Europe with the Asian spice trade.

References

Adelson, H.L. (1960) Early Medieval trade routes. *The American Historical Review* 65, 271–287.
Butzer, K.W. (1981) Rise and fall of Axum, Ethiopia: a geo-archaeological interpretation. *American Antiquity* 46, 471–495.
Cartwright, M. (2017) Byzantine coinage. *Ancient History Encyclopedia*. Available at: https://www.ancient.eu/Byzantine_Coinage/ (accessed 28 January 2021).
Cartwright, M. (2018) Trade in the Byzantine Empire. *Ancient History Encyclopedia*. Available at: https://www.ancient.eu/article/1179/ (accessed 28 January 2021).
Casson, L. (1991) *The Ancient Mariners: Seafarers and Seafighters of the Mediterranean in Ancient Times*, 2nd edn. John Hopkins University Press, Baltimore, Maryland.
Cervantes, A.C. (2013) Sasanian Empire. *Ancient History Encyclopedia*. Available at: https://www.ancient.eu/Sasanian_Empire/ (accessed 28 January 2021).
Charles, M. and Ryan, N. (2013) The Roman Empire and the grain fleets: contracting out public services in antiquity. Available at: https://apebhconference.files.wordpress.com/2009/09/charles_ryan1.pdf (accessed 28 January 2021).
Daryaee, T. (2003) The Persian Gulf trade in late antiquity. *Iran Chamber Society*. Available at: http://www.iranchamber.com/history/articles/persian_gulf_trade_late_antiquity.php (accessed 28 January 2021).
Harbeck, M., Seifert, L., Hänsch, S., Wagner, D.M., Birdsell, D. *et al.* (2013) *Yersinia pestis* DNA from skeletal remains from the 6th century AD reveals insights into Justinianic Plague. *PLoS Pathogens* 9, e1003349.
Harrower, M., Dumitru, I., Perlingieri, C., Nathan, S., Zerue, K. *et al.* (2019). Beta Samati: discovery and excavation of an Aksumite town. *Antiquity* 93, 1534–1552.
Herrin, J. (2007) *Byzantium: The Surprising Life of a Medieval Empire*. Princeton University Press, Princeton, New Jersey and Oxford.
Horgan, J. (2014) Justinian's Plague (541–542 CE). *Ancient History Encyclopedia*. Available at: https://www.ancient.eu/article/782/ (accessed 28 January 2021).
Jones, E.J. (2018) Long distance trade and the Parthian Empire: reclaiming Parthian agency from an Orientalist historiography. Master's thesis, Western Washington University, Bellingham, Washington. Western Washington University Graduate School Collection No. 692. Available at: https://cedar.wwu.edu/wwuet/692 (accessed 28 January 2021).
Lopez, R.S. (1945) Silk industry in the Byzantine Empire. *Speculum* 20, 1–42.
Luttwak, E.N. (2011) *The Grand Strategy of the Byzantine Empire*. Belknap Press of Harvard University, Cambridge, Massachusetts.
Munro-Hay, S. (1991) *Aksum: An African Civilization of Late Antiquity*. Edinburgh University Press, Edinburgh.
Paine, L. (2013) *The Sea & Civilization: A Maritime History of the World*. Vintage Books, New York.

Shahbazi, A.S. (2012) Byzantine–Iranian relations. *Encyclopædia Iranica* IV/6, 588–599. Available at: http://www.iranicaonline.org/articles/byzantine-iranian-relations (accessed 28 January 2021).

Sidebotham, S. (1996) Romans and Arabs in the Red Sea. *Topoi* 6(2), 785–797.

Snell, M. (2019) Silk production and trade in medieval times. *ThoughtCo*. Available at: https://www.thoughtco.com/silk-lustrous-fabric-1788616 (accessed 28 January 2021).

Trento, R. (2014) What did Byzantine food taste like? *Iris Blog*. J. Paul Getty Trust, Los Angeles, California. Available at: http://blogs.getty.edu/iris/what-did-byzantine-food-taste-like/ (accessed 28 January 2021).

Wasson, D.L. (2013) Constantinople. *Ancient History Encyclopedia*. Available at: https://www.ancient.eu/Constantinople/ (accessed 28 January 2021).

Whitehouse, D. and Williamson, A. (1973) Sasanian maritime trade. *Iran* 11, 29–49.

Williams, E. (2012) Trade and commercial activity in the Byzantine and early Islamic Middle East. In: *Heilbrunn Timeline of Art History*. The Metropolitan Museum of Art, New York. Available at: http://www.metmuseum.org/toah/hd/coin/hd_coin.htm (accessed 28 January 2021).

Young, G.K. (2001) *Rome's Eastern Trade: International Commerce and Imperial Policy, 31 BC–AD 305*. Routledge, London.

11 Pan Islamica

Setting the Stage – Rapid Spread of Islam

In the early seventh century, the various warring tribes of the Arabian Peninsula became unified under Muhammad and began to systematically take control of the whole Middle East. Over a period of just a few decades, the Arab tribesmen made rapid conquests of Palestine, Syria, Iraq, Iran and then Egypt (Fig. 11.1).

When Muhammad died, he was succeeded by a series of caliphs chosen from his most loyal companions. It was under the leadership of these early caliphs (632–656) that the conquest of territories outside Arabia began. They marched steadily eastwards into the Sasanian Empire and northwards against the Byzantines. The Sasanians suffered their first crushing defeat at the hands of the Arabs in Iraq at the Battle of al-Qadisiyah in 637 and their capital of Ctesiphon was taken soon afterwards, leading to a rapid disintegration of their empire. The Arabs continued westwards towards Egypt and by 646 Sasanian Alexandria had fallen and north-east Africa was occupied.

In Byzantine territory, the Muslim army won at Yarmuk in 636 near the border of modern-day Syria and Jordan. A newly organized Muslim navy also destroyed the Byzantine fleet at the Battle of the Masts in 655 but were not able to take Constantinople. The Byzantine capital would survive for another 800 years until the Ottoman era.

The first Muslim dynasty called the Umayyad was established by Muawiyah I in 661 CE, just 30 years after Muhammad died. Under its caliphates, the Arabs continued their rapid expansion in all directions reaching eastwards as far as the Indus River in India and westwards across North Africa to Spain and France.

The Byzantines Redirect Their Trade

With their conquests, the Muslims now controlled all the major trade routes from South East Asia and China. All the exotic goods passing through the

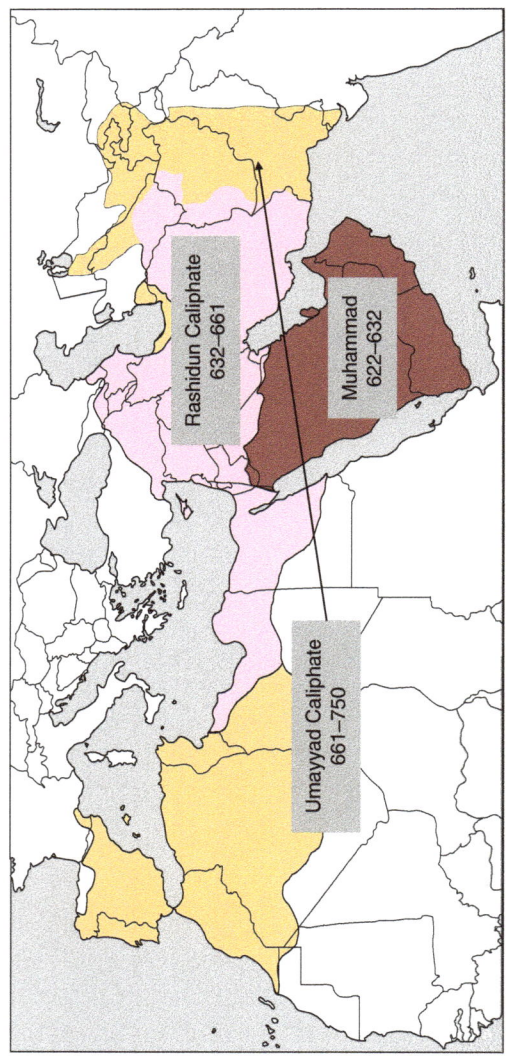

Fig. 11.1. Expansion of Islam (622 to 750 CE). (Redrawn from DieBuche, Public domain, via Wikimedia Commons. File: Map of expansion of Caliphate.svg)

Red Sea, Persian Gulf and across Central Asia were now moving within one Islamic Empire; the Muslim world had become 'an integrated global logistic system available to all parties who recognized the suzerainty of the caliphate' (Bernstein, 2008, p. 75). Not only was the bulk of the Western spice and silk trade in Muslim hands, but they also controlled the massive amounts of grain produced in Egypt and began diverting it from Constantinople to Mecca and Medina.

With the Muslims now controlling the Middle East and North Africa, the remainder of the Byzantine Empire shifted the bulk of its trade from the east to the north. For their grain, the Byzantines began to rely more on their own production in Asia Minor and the ancient breadbasket surrounding the Black Sea. For a while, the Byzantines were also able to obtain grain from Carthage and Sicily, but these sources also fell to the Muslim forces in the eighth and ninth centuries (Teall, 1959). For their spices and silk, the Byzantines focused most of their attention towards the northern steppe routes that wove through Central Asia to the Black Sea. These ancient trade routes became much more critical to Byzantium (and Europe) with the Arabs now controlling both the Red Sea and Persian Gulf trade routes. The northern Silk Routes became the main portal of South East Asian goods to the West.

A New Trading Empire Emerges in the Northern Steppes: the Khazar Khaganate

A powerful empire called the Khazar Khaganate arose around 650 CE and took control of the northern hub of the Silk Route (Gottesman, 2011; Asadov, 2012). For about three centuries ($c.650-965$), the Khazars ruled an area extending from the Volga–Don steppes to the eastern Crimea and the northern Caucasus. This region was outside the old Roman Empire that had extended only as far as the southern Caucasus. The Khazars became the centre of the early medieval spice and silk trade, serving as a crossroads and buffer between the caliphate, the Byzantine Empire and the nomads of Western Europe.

There were three primary trans-Caucus trade routes that travelled out of Khazar (not shown in figure). One went by caravan north of the Caspian Sea to Tana on the Sea of Azov and then by ship across the Black Sea to Constantinople. Another went by caravan to the port of Abaskun on the southern Caspian Sea and then across by ship to Derbent and from there west by caravan to the Black Sea or north by ship to Tana. The other route travelled by caravan south of the Caspian Sea and then north-west to Trebizond on the southern Black Sea coast and then by ship to Constantinople. These trade routes were truly an international affair – the Umayyads controlled Abaskun and Derbent, the Byzantines ruled Trebizond and the Khazars controlled most of everything in between. The Khazar government levied a 10% tariff on all foreign merchants who passed through its territory.

Khazaria was the first feudal state to be established in Eastern Europe and supported Byzantium in its struggles with Persia and then the caliphate. The Khazars aided the Byzantine emperor Heraclius in 641 by sending an army of 40,000 soldiers in his campaign against the Persians in the Byzantine–Sasanian War of 602–628. The eighth-century emperors, Justinian II and Constantine V, took royal Khazar brides. The Khazars also kept Islamic power in check by fighting bloody wars against the caliphate over Transcaucasia and North Caucasus. Religiously, Khazaria became neutral in the early 800s by converting to Judaism and becoming the world's largest Jewish kingdom.

Arab Agricultural Revolution

As the Muslim armies spread from Egypt across northern Africa to Spain, they fostered what has come to be called the 'Arab agricultural revolution' (Watson, 1974; Decker, 2004; Hancock, 2017). In their march towards Spain, they encountered a wide array of crops along the way including sugarcane, coconut, palms, cotton, aubergine, lemon, limes, mangoes, rice, spinach, sorghum, sour orange, watermelon, wheat and yam. They introduced this assemblage of crops to the places they conquered and in so doing altered the agriculture of the whole Mediterranean region. As S.W. Mintz (1985, p. 25) put it: 'The Mediterranean Arab conquerors were synthesizers, innovators transporting the diverse cultural riches of the lands they subjugated back and forth across portions of three continents, combining, intermixing, and inventing new adaptations.'

The Muslims began rotating crops and planting throughout the year, allowing for more intense use of available land. Longer-season crops such as winter wheat might be followed by summer sorghum, or crops with shorter growing seasons like spinach, yam and aubergine might be grown back-to-back in the same year. This multiple cropping depleted the soil of its nutrients more rapidly than single crops, so the Arabs began using manures and other supplements to maintain the soil's fertility. Year-round planting required the development of dependable irrigation systems to provide the crops with sufficient water, particularly during the dry season. To accomplish this, the Muslims restored the old Roman irrigation systems that had fallen into decay and greatly expanded them. They also added new innovations such as the Persian bucket wheel, the water screw and underground water tunnels.

The Centre of the Muslim World Shifts

The pace of Arab expansion stalled in the mid-eighth century, during a period of intense infighting caused by a change in the ethnic composition of the Muslim armies. In the beginning, they were made up solely of Arab tribesmen and the terms 'Muslim' and 'Arab' were inseparable. At the peak of the Muslim expansion, there were simply not enough Arabs to sufficiently man the troops and it became necessary to receive other peoples into Islam to maintain full strength. This inclusion caused great internal strain as the new converts were not fully integrated into Muslim society and felt like second-class citizens. These rising tensions led to a rebellion against the Umayyad Caliphate in 747.

The revolt began in Persia and was headed by descendants of the youngest uncle of Muhammad named al-Abbas, who ultimately established the Abbasid Caliphate. He led an eastern Muslim army that was composed of many Persian converts and infused with Persian culture and pride. The Abbasids claimed to be the true successors of Prophet Muhammad and attacked the moral character of the Umayyads. The hostilities officially began in Merv and within a year the Abbasid forces captured the Umayyad capital of Damascus, thereby taking control of most of the Muslim Empire. All the members of the Umayyad family were murdered, except Abd-al-Rahman, who fled to Spain and founded an independent Umayyad Caliphate there as the Caliphate of Cordoba. The Abbasid dynasty would rule the rest of the Muslim world for another 500 years.

One of the first things the Abbasid Caliphate did upon assuming power was to move the Muslim capital east to Baghdad in Iraq, close to Ctesiphon, putting ancient Mesopotamia once again at the centre of the world's largest empire. The move appeased the Persian support base and reflected a growing reliance on Persian bureaucrats. The capital of the Arab world was now located at the convergence of all the great trade routes, including those winding by sea through the Persian Gulf and by land across Central Asia from China.

The Round City of Baghdad

The original core of the new capital was constructed by Caliph al-Mansur in 762–766 as the official house of the Abbasid court. Mansur brought in architects, designers, carpenters, engineers, blacksmiths and surveyors from all across the Empire to construct his city. Probably 100,000 workers and craftsmen worked on the project at a cost of 18 million gold dinars, worth billions of dollars today. The construction of Baghdad was by far the greatest architectural project of the Islamic world.

The city was constructed as a two-kilometre-wide complete circle surrounded by two massive walls that had four main gates: Bab al-Kufa ('gate of Kufa'), Bab al-Sham, ('gate of al-Sham or Damascus'), Bab al-Khorasan ('gate of Khorasan') and Bab al-Basra ('gate of Basra'). In the centre of the majestic circle were the sprawling caliphal Palace of the Golden Gate and the 'mosque of the palace', the Al-Khulafa Mosque. Early Baghdad was also home to Bayt al-Ḥikmah (House of Wisdom), a major Abbasid public academy and intellectual centre. The city of Baghdad came to rival the structural magnificence of Constantinople and at its peak was just as large with a population of more than a million inhabitants.

Baghdad evolved into the cultural, commercial and intellectual centre for the Islamic world. The Abbasid period became a Golden Age for Islam, with its caliphs sponsoring great artists and scientists who collected and translated the great medical, astronomical and other scientific texts from the classical period in Greece and Rome, ultimately saving them from being lost. As described by Kallie Szczepanski:

> While Europe languished in what was once called its 'Dark Ages,' thinkers in the Muslim world expanded upon the theories of Euclid and Ptolemy. They invented algebra, named stars like Altair and Aldebaran and even used hypodermic needles to remove cataracts from human eyes. This was also the world that produced the stories of the Arabian Nights – the tales of Ali Baba, Sinbad the Sailor, and Aladdin came from the Abbasid era.
>
> (Szczepanski, 2019)

Islam and Medieval Medicine

The spread of Islam across the Middle East and South East Asia led to 'an unprecedented era of ferment in all branches of learning' (Tschanz, 2003). A single language now linked peoples from the borders of China to the south of France; Arabic became the common tongue of the educated in the East like Latin and Greek had been to the West. Medicine became an important focus of Islamic scholars building on the work of the Christian scholars at the Persian university at Jundishapur, the old Sasanian capital of Persia. Here, Islamic physicians learned about the work of Hippocrates, Galen and other Greek physicians, and became familiar with the accumulated medical knowledge of Byzantium, Persia, India and China.

Under the Abbasid caliphs, the Islamic scholars also compiled and translated the entire body of Greek medical works into Arabic so that they were more widely available. The most important of these translators was Hunayn ibn Ishaq al-'Ibadi (809–873), who was reported to have received an equal weight of gold for his manuscripts (Tschanz, 2003). When the first great medical universities began to crop up in Europe in the thirteenth century, their curriculum was based on Arabic medical books translated into Latin. These then spawned hundreds of pharmacology texts in the various European languages (Nam, 2014).

Muslim medical practice was based on Galen's belief that a sick person could be made healthy by consuming the right mixture of foods and spices. When the body's humours were correctly balanced, a person became healthy. Muslim physicians advanced Galen's techniques by developing syrup-based medicines to treat illness and restore proper balance between the humours. Important components of many of these medicines were the spices – pepper, cinnamon, ginger, cloves, nutmeg, saffron and mace. Pepper, cinnamon, ginger and saffron had been known to the ancient Greeks, but cloves, nutmeg and mace were new additions of the Arabs, added through their contacts with Indonesia and China.

The Islamic achievements in medieval medicine went well beyond Galen's treatments. While medieval European medicine was still mired in superstitions and the rigid teachings of the Catholic Church, Islamic doctors were developing new techniques in surgery, founding the first hospitals, began training physicians and produced encyclopaedias of medical knowledge. Islamic doctors were the first to use antiseptics to clean wounds and used opium as an anaesthetic. They understood that some diseases were infectious, including leprosy, smallpox, tuberculosis and sexually transmitted ones, and pioneered methods of quarantine.

Spread of Islam across South East Asia

In the mid-600s, the Muslims began to spread throughout Indonesia largely through trade. This expansion of the Muslim Empire was mostly a peaceful process, and the conversion to Islam did not prove particularly difficult for the local people. To a large part, the acceptance of Islam throughout Inner and South East Asia came from contacts with Muslim merchants. First, a few Muslims would arrive in a port and begin trading. Soon more settlers arrived and brought with them their own religious experts (mullahs), who set up houses of prayer. Inspired by the honesty and integrity of the merchants, a large number of locals would then convert to Islam. Merchants accepting Islam enjoyed increased access to the Muslim trade network, a steadier flow of goods and reduced transaction costs. A critical turning point came when a local ruler also converted to Islam, providing a 'more universal kind of legitimacy that transcended his own group and allowed for the rise of greater polities' (Houben, 2003, p. 153).

As described by Michalopoulos and colleagues:

> Muslim merchants carried the message of Islam wherever they traveled. This was possible because of the Muslim practice of 'direct' trade, one of the most remarkable innovations of Islam. Prior to Muslim conquests, trade was conducted by a network of local merchants who traded exclusively in their homelands. In other words, they played the role of intermediary agents with goods (often spices) being transported from one carrier to another by short journeys, creating a trade-relay. Muslims instead did not rely on Intermediaries and personally travelled the entire length of the journey, crucial for the diffusion of the religion along the trade routes and at the destination. On the receiving end, the new religion appealed to the local merchants because it legitimized their economic base more than most belief systems present at that time. Merchants converting to Islam had clear advantages including (i) cooperation within the Muslim trading network, (ii) valuable contacts to expand their trade, and (iii) rules governing commercial activities naturally favoring Muslims over non-Muslims.
>
> (Michalopoulos *et al.*, 2017, p. 4)

Muslim Expansion Towards China

As the Middle East was becoming unified under the umbrella of Islam, power in China was also being coalesced after a long period of disarray following the Han Dynasty's collapse in 222 CE (Curtin, 1984; Chaudhuri, 1985). During this period, Silk Route travel had slowed to a crawl as short-lived, regional dynasties and barbarians fought for control. This all changed in 589 CE when the Sui Dynasty reunited China and paved the way for the Tang, who led China through a period of great prosperity that lasted from 618 to 907 CE. Under the Tangs, the eastern reaches of the Silk Routes were brought back under unified control and their capital of Chang'an rose to worldwide prominence.

Only a year after the fall of the Umayyads, an Abbasid army was on the doorstep of China fighting against the Imperial Tang in what is now known as Kyrgyzstan in Central Asia. There, on the banks of the Talas River in 751, were set the boundaries between the Muslim and Buddhist/Confucianist worlds. After the Prophet Muhammad had died in 632, the Muslim armies had rapidly spread across Central Asia taking control of Merv from the Sasanians, and then Bukhara, the Ferghana Valley and as far east as Kashgar. The Arabs continued their push into Chinese territory by roundly defeating the Tang forces at Talas, supported by local peoples, but their supply lines were now stretched to the limit, and it was beyond their capacity to continue pushing further through the eastern Hindu Kush mountains and into the deserts of western China. Thus, the boundary between the great empires was set.

With their defeat of the Sasanians and their advances into Central Asia, the Muslims were firmly established as the intermediary of the land-based Silk Routes. Trade moved particularly well across this route in the seventh century, 'albeit through a period of false starts and occasional reversals' (Foltz, 2009, p. 96). The Muslims built a string of embassies along the major trade routes to support that mercantile activity and encouraged their charismatic preachers to go about the Islamization of Central Asia. As in south-west Asia, the Muslim domination of commercial activity encouraged local businessmen to convert to Islam to facilitate their trade activities at home and abroad.

Muslim Maritime Trade with South East Asia

By the seventh century, Arab sailors along with ones from Persia, India and Ceylon were using three maritime trade routes to obtain Chinese and South East Asian goods (Donkin, 2003; Bernstein, 2008). The first journey was to the Isthmus of Kra, half-way down the Malay Peninsula, where Chinese cargo had been deposited either by ship up the Strait of Malacca or transported overland from the other side of the isthmus. The traders could also sail to one of the Indonesian entrepôts such as Palembang, where Chinese ships had deposited goods. The most adventurous continued through the Strait of Malacca to Indochina in the South China Sea and on to China itself.

In the second half of the eighth century, trade flow along the Silk Route was again cut off when the Tibetan tribes in Central Asia revolted and the Tahirids and then Saffarids took control. The Muslims responded by redirecting the bulk of their silk and spice trade with China through the Persian Gulf. This sea trade had always been an important portal to the spices of the Malay Archipelago and India, but when the Han Dynasty collapsed, the Muslims were forced to focus their China trade almost solely on the sea.

India remained the central hinge for global trade in the Islamic period. On the Malabar and Coromandel coasts, colonies of traders from all over came together to trade cottons and silks, spices and perfumes, as well as gold, silver and ivory. Gujarat traders, in particular, went out widely across the Indian Ocean in search of goods, and brought back exotic luxuries from a mix of cultures that made everyone prosperous. The Gupta Empire (320–550 CE) and the Chola Empire (800s to 1300s CE) were major Indian powers with wide influence and prosperity.

Muslim Sea Trade with China

Muslims first travelled to China soon after they had wrestled control of Persia. The rise of the Muslim Abbasid Empire in the Middle East allowed for what Bernstein (2008, p. 75) called 'the Baghdad–Canton express'. Relatively free trade was now possible across Asia to North Africa, something that had been missing since the Roman–Parthian period of the early Christian era.

In the ninth-century Arabic manuscript entitled *Akhbar al-Sin wa'l-Hind* (*Accounts of China and India*), there is an eyewitness account of maritime travel from Baghdad to Canton. The journey started at Basra on the Shatt al-Arab between today's Kuwait and Iran and then proceeded across the Persian Gulf to Oman. From there the sailors headed to the Malabar Coast of India, riding the monsoon winds. They then travelled across the Bay of Bengal to Kedah on the west coast of Malaysia and then south through the Strait of Malacca to Indochina and finally Canton. The *Akhbar* suggests the whole trip took four months, but with weather delays and bureaucratic snafus, the entire trip likely took a year.

During the Tang Dynasty of China, thousands of foreign traders entered China and established settlements. The Chinese became familiar enough with the origin of the Middle Eastern sailors that they referred to them by different names. The Persians were 'Po-ssi' and the Arabs, 'Ta-shih'. Muslims in general were 'Hui'. The distant Eastern Roman Empire was known as 'Fu-lin'. Guangzhou and Yangzhou became the leading commercial centres where significant numbers of Christians, Jews, Muslims, Zoroastrians, Buddhists and others lived and kept houses of worship. There are records that over 100,000 foreign merchants had settled in China by the mid-800s and had become the 'middleman of the Nanhai' (Flecker, 2015).

A turbulent relationship (to say the least) existed between these foreigners and the Chinese (Schottenhammer, 2013). In response to the actions of corrupt port officials, the foreign merchants sacked the storehouses of Yangzhou in 758, burned the city and fled with their plunder. In 760, Chinese rebels overran the city of Yangzhou and killed thousands of merchants and Chinese civilians. In 878, the Arab geographer, Abu Zayd, reported that 'no less than 120,000 Muslims, Jews, Christians and Parsees perished' when the rebel leader Huang-Chao burned and pillaged Guangzhou.

After the massacre at Guangzhou most foreign merchants including the Arabs abandoned China and began restricting their trade to northern Vietnam, the Isthmus of Kra and Srivijaya. In fact, there is evidence that the Muslims played a significant role in convincing the rulers of Srivijaya to establish the entrepôt at Palermo to obtain goods from China without the risk and expense of direct trade with China (Gungwu, 1958).

References

Asadov, F. (2012) Khazaria, Byzantium, and the Arab Caliphate struggle for control over Eurasian trade routes in the 9th–10th centuries. *The Caucasus and Globalization* 6, 140–150.

Bernstein, W.J. (2008) *A Splendid Exchange: How Trade Shaped the World*. Grove Press, New York.

Chaudhuri, K.N. (1985) *Trade and Civilization in the Indian Ocean: An Economic History from the Rise of Islam to 1750*. Cambridge University Press, Cambridge.

Curtin, P.D. (1984) *Cross-Cultural Trade in World History*. Cambridge University Press, Cambridge.

Decker, M. (2009) Plants and progress: rethinking the Islamic agricultural revolution. *Journal of World History* 20, 187–206.

Donkin, R.A. (2003) *Between East and West: The Moluccas and the Traffic in Spices up to the Arrival of Europeans*. American Philosophical Society, Philadelphia, Pennsylvania.

Flecker, M. (2017) Early Chinese voyaging in the South China Sea: implications on territorial claims. *Journal of Maritime Studies and National Integration* 1, 1–21.

Foltz, R. (2010) *Religions of the Silk Road*. Palgrave Macmillan, London.

Gottesman, B. (2011) The empire of the Khazars. *World History Encyclopedia*. Available at: https://www.academia.edu/4810010/The_Empire_of_the_Khazars (accessed 28 January 2021).

Gungwu, W. (1958) The Nanhai trade: a study of the early history of Chinese trade in the South China Sea. *Journal of the Malayan Branch of the Royal Asiatic Society* 31, 3–135.

Hancock, J.F. (2017) *Plantation Crops: Power and Plunder, Evolution and Exploitation*. Routledge, London.

Houben, V.J.H. (2003) Islam: enduring myths and changing realities. *The Annals of the American Academy of Political and Social Science* 588, 149–170.

Michalopoulos, S., Naghavi, A. and Prarolo, G. (2017) *Trade and Geography in the Origins and Spread of Islam*. Working Paper No. 18438. National Bureau of

Economic Research, Cambridge, Massachusetts. Available at: http://www.nber.org/papers/w18438 (accessed 28 January 2021).

Mintz, S.W. (1985) *Sweetness and Power. The Place of Sugar in Modern History*. Penguin Books, New York.

Nam, J.K. (2014) Medieval European medicine and Asian spices. *Korean Journal of Medical History* 23(2), 319–342.

Schottenhammer, A. (2013) The 'China Seas' in world history: a general outline of the role of Chinese and East Asian maritime space from its origins to c. 1800. *Journal of Marine and Island Cultures* 1, 63–86.

Szczepanski, K. (2019) The Abbasid Caliphate. *ThoughtCo*. Available at: https://www.thoughtco.com/what-was-the-abbasid-caliphate-195293 (accessed 28 January 2021).

Teall, J.L. (1959) The grain supply of the Byzantine Empire, 330–1025. *Dumbarton Oaks Papers* 13, 87–139.

Tschanz, D.W. (2003) Arab roots of European medicine. *Heart Views* 4(2), 9. Available at: https://www.heartviews.org/text.asp?2003/4/2/9/64472 (accessed 28 January 2021).

Watson, A.M. (1974) The Arab agricultural revolution and its diffusion, 700–1100. *The Journal of Economic History* 34(1), 8–35.

Spice Trade in the Dark Ages of Europe

12

Setting the Stage – Collapse of the Western Roman Empire

The Western Roman Empire collapsed under waves of Germanic tribe movements during the 'Völkerwanderung' or Migration Period. The Visigoths were the first to enter Roman territory in 376, chased by the fearsome Huns. They were first tolerated by the Romans as defenders of the Danube territory, but eventually turned upon Rome and sacked the great city in 410 CE. They then moved on to Spain and established an empire there. The Ostrogoths followed the Visigoths into Italy and conquered it in 474 CE, backed by the Byzantines. Also in the fifth century, a fusion of western Germanic tribes called the 'Franks' began their takeover of Gaul and the Anglo-Saxons began their invasion of Roman Britain. So, as the great Islamic Empire was amalgamating all of the Middle East and North Africa, Europe was falling into a fragmented 'Dark Ages'.

The Migration Period lasted for centuries and the ensuing chaos resulted in countless deaths through famine, plague and war. Cultural growth came to a standstill as 'barbarians' took control of the former Roman possessions. The city-based way of life enjoyed by the Romans fell into steep decline in many locations. The Catholic Church became the sole unifying cultural force that remained after the fall of the Western Roman Empire. A large part of Europe was briefly unified under Charlemagne and the Carolingians in the 800s, but 'their empire crumbled under the weight of succession disputes and repeated invasions by Scandinavian Vikings and eastern European Magyars. From 850 onwards, Europe was again ripped apart by political fragmentation, warfare and unrest' (Asbridge, 2010, p. 6).

Spice Use in Europe During the Dark Ages

With the collapse of Rome, spice flow into Europe certainly must have diminished but it by no means stopped. There are a number of clear references in the historic literature on the use of exotic spices in Europe after Rome's fall (Chevallier, 2014). In a fifth-century 'Excerpta' added to the cookbook attributed to the Roman gourmet Apicius, there is a list of seasonings that a good cook should keep on hand, including saffron, pepper, ginger and clove. Clove is a particularly interesting inclusion as it came late to Roman cuisine and was not originally mentioned in *Apicius*. A Byzantine physician and ambassador named Anthimus describes in his *De observatione Ciborum* (*On the Observance of Foods*) the spices used in sixth-century Gaul and makes many references to pepper, along with clove and ginger. A monk named Marculfe in what food blogger Chevallier calls 'something very like a medieval secretary's handbook' describes what local hosts were obliged to provide for travelling officials in Gaul in the seventh century and mentions cumin, pepper, costus, clove, spikenard, cinnamon and mastic. In a charter given to the great monastery at Corbie in Picardy by the Frankish kings Chlothar III (655–673) and Childeric II (673–675), among the supplies the cellarer was to receive were 30 pounds of pepper, 150 pounds of cumin, 2 pounds of clove, 1 pound of cinnamon, 2 pounds of nard and 30 pounds of costus. These records show that spices still played an important role in the daily lives of at least the monks and nobility in early medieval Europe.

Level of Western Trade in the Early Medieval Age

For many decades, there has been a contentious controversy on the economic state of Western Europe after the fall of the Roman Empire (Pirenne, 1922; Dopsch, 1937; Mango, 2009). As summarized by Lelis:

> On the one hand, some have maintained and continue to argue that the transition from late Roman to post-Roman involved little drastic change in the overall infrastructure and prosperity of the western provinces from the fifth to the eighth centuries. On the other hand, many have held that the end of centralized Roman administration in the West was accompanied by a slide into infrastructural decadence and economic squalor, which conditions were hardly remedied before the eleventh century.
>
> (Lelis, 2013, p. 17)

These proponents feel that for most people in Europe, life did not extend beyond the tiny villages where they were born and lived until they died.

Overall, the evidence suggests that effect of the Germanic tribes' migrations on Western trade varied widely (McCormick, 2001; Wickham, 2005; Davis and McCormick, 2008). Although most of the wealthy Romans in Britain and north-eastern Gaul did lose their estates, most Roman elites living in Spain, Italy and southern Gaul figured out how to come to terms with the migrants and learned to coexist with them. In general, the Germanic 'barbarians' were not opposed to Roman civilization and Roman landowners were often allowed to keep their properties. After all, the Germanic peoples had lived in close proximity to the Romans for decades and many had already been absorbed as soldiers and colonists. The Germanic peoples had their own monarchical form of government and the fall of Rome led to a change in management, not a total collapse in government. Roman lifestyles were largely restored soon after the Ostrogoths, Lombards and Franks took control of large swaths of Italy and Western Europe.

Germanic tribes proved capable of maintaining the parts of Roman-style government that were relevant to them and merchants still operated in many parts of Western Europe where urban traditions had not been extinguished. Aristocrats in the barbarian empires still desired the luxuries that had been enjoyed by Rome and 'as soon as the first waves of these invaders had settled down in the new successor states, the Byzantine merchants revived western trade' (Adelson, 1960, p. 285). Likewise, the Byzantine world still craved the products of Northern Europe such as fur, honey, wax, timber and particularly slaves, which Europeans were happy to trade for coins, manufactured goods and spices.

Mediterranean Trade in the Early Medieval Period

Trade in the western Mediterranean continued, even as the Western Roman Empire staggered and slowly died under the relentless pressure of the Barbarians. In fact, there was more shipping going on in the western than the eastern Mediterranean between 320 and the mid-fifth century, as evidenced by the number of shipwrecks found in the two regions (Kingsley, 2009). The fourth-century archaeological deposits in both Spain and Italy are dominated by North African amphorae, the large pottery containers used to transport foodstuffs like olive oil and wine.

The western shipwrecks from the fourth century contain only goods from North Africa, but in fifth-century shipwrecks eastern amphorae begin to reappear, signalling a recovery in wider Mediterranean trade (Decker, 2009). Eastern products were once again major parts of the economies of Spanish centres by the sixth century. In fact, trade across the Mediterranean was almost one directional: eastern cargoes were much more common in the west than the reverse.

As Michael Decker describes:

> Throughout the late 5th and the 6th centuries, the Iberian Peninsula was largely under the control of the Visigoths. Society there continued to be heavily impacted by the residue of the Romanized elite, and this group, more so perhaps than the Goths, had links and contacts with the eastern Mediterranean through family and friends, and were probably more intellectually drawn to the centre of imperial power at Constantinople. Communications with the eastern Mediterranean were apparently relatively facile: whatever the threat posed by the Vandal fleet or other piratical elements, travel from one end of the Mediterranean world to the other was common.
>
> (Decker, 2009, p. 240)

Western trade remained significant throughout the sixth century, as Justinian had retaken most of the Roman Empire in North Africa and had even grabbed a slice of Spain. The Byzantines had regained the area extending from Ceuta at the Strait of Gibraltar to the Red Sea port of Aila.

Western trade across the Mediterranean remained strong until the early decades of the seventh century, when:

> long-distance exchange very clearly contracted. The incessant flow of produce that had survived the political disruptions of the Goths, Vandals and Slavs stopped abruptly ... maritime trade in the western Mediterranean seems to have been the last expression of classical antiquity to die out, outliving by some 150 years the transformation of elite residences into fishermen's huts in the late 5th century.
>
> (Kingsley, 2009, p. 33)

The abrupt downturn in eastern Mediterranean trade in the early 600s was associated with the Byzantines' struggles in the East, first with the Persians and then Muslims. The Mediterranean became a hostile zone controlled by Arab pirates who dominated the sea for the next 300 years. Long-distance trade from the East to the West became restricted to the land-based Silk Routes passing through the European Steppe. Mediterranean Sea trade with Western Europe did not re-emerge until the tenth century, when the Italian cities of Genoa, Pisa, Amalfi and Venice began establishing trading colonies across the Mediterranean and aggressively routing out pirate bases.

Early Medieval Trade in Europe

There is much evidence that trade continued in the early medieval period across the rivers of Western Europe (Adelson, 1960). Western Europe was strongly linked with the Mediterranean through the Rhône–Saône corridor, which weaved its way from the south-eastern corner of France through its heart to Paris. Goods travelled along the Seine through northern France, the Aquitaine in south-western France and along the Rhine from Switzerland north through Germany and the Netherlands. Vigorous trade was conducted all across north-western continental Europe, the North Sea and the Baltic.

Marseille on the southern coast of France, with its superior deep-water port, had long been an important entry point of goods from the East, and remained so even after its capture by the Visigoths in the fifth century. It was a key European emporium, linking the Mediterranean world with Western Europe. Trade goods from all over the world arrived in Marseille, from the Near East and even as far as Indonesia. Marseille had overseas connections to the ports of the Levant from Alexandria to Constantinople and Antioch. Most of this shipping was directed by Syrian and Jewish merchants in Carthage and other western Mediterranean ports.

Active trade also occurred via alpine passes from northern Italy to the Rhine. There is a trail of Byzantine coins (lightweight, gold *solidi*) found across Western Europe in archaeological sites that bears testimony to the breadth of trade occurring in the sixth and seventh centuries. A concentration of them follows a route from northern Italy over the Alps and down the Rhine to Frisia, a historic region of the Netherlands fronting the North Sea. Other silver coins minted by the Ostrogoths 'came to be adopted by the cities of the southern Gallic region, and from there spread throughout the Frankish realm and the rest of the European successor states' (Adelson, 1960, p. 283).

Lelis describes trade in early medieval Europe and the Mediterranean at four distinct levels:

> The top level comprised the Byzantine prestige production and distribution system ... gold–garnet decorations and special helmets produced and distributed locally within northwestern Europe ... The second level consisted of what might be termed 'ordinary exotic goods': non-imperial silks, spices, aromatics, specialty wines, and the like. These continued to travel across the Mediterranean and to arrive in western ports such as Marseille throughout the sixth and seventh centuries, with distribution northwards. The third level was the continued production and distribution of bulk goods such as pottery and olive oil within the Mediterranean, at least in coastal areas, through the seventh century. This level was active as well in Francia and the Rhine basin, where commercial pottery production continued throughout, even if on reduced scales, along with the production of wine ... Finally, there is the level of local production and exchange, the scope and quality of which varied extremely from area to area.
>
> (Lelis, 2013, p. 339)

The Radhanites, Medieval Tycoons

A remarkable group of Jewish merchants called the Radhanites kept the Mediterranean and Levantine trade routes operating throughout the Dark Ages in Europe (Gil, 1974; Gottesman, 2011). Jewish-owned ships were seen as neutral and were allowed to travel anywhere. The earliest record of the Radhanite merchants is found in the ninth-century account of Ibn Khordadbeh in Arabic, *The Book of Roads and Kingdoms*. Khordadbeh was the Director of Posts and Intelligence in the province of Jibal in north-western Iran and served as a postmaster, sheriff and spy for the Abbasid Caliphate. According to this account, the Radhanite merchants operated along four trade routes that stretched from central France to the Arabian Peninsula, North Africa, Egypt, Byzantium, Palestine, Syria, the northern parts of India, central Europe and China. No one merchant travelled all the way from France to China, but the goods were handed off to a string of Radhanite traders along the routes.

In Ibn Khordadbeh's account:

> These merchants speak Arabic, Persian, Roman, the Frank, Spanish, and Slav languages. They journey from West to East, from East to West, partly on land, partly by sea. They transport from the West eunuchs, female slaves, boys, brocade, castor, marten and other furs, and swords. They take ship from Firanja (France), on the Western Sea, and make for Farama (Pelusium). There they load their goods on camel-back and go by land to al-Kolzum (Suez), a distance of

twenty-five farsakhs. They embark in the East Sea and sail from al-Kolzum to al-Jar and al-Jeddah, then they go to Sind, India, and China.

On their return from China they carry back musk, aloes, camphor, cinnamon, and other products of the Eastern countries to al-Kolzum and bring them back to Farama, where they again embark on the Western Sea. Some make sail for Constantinople to sell their goods to the Romans; others go to the palace of the King of the Franks to place their goods. Sometimes these Jew merchants, when embarking from the land of the Franks, on the Western Sea, make for Antioch (at the head of the Orontes River); thence by land to al-Jabia (al-Hanaya on the bank of the Euphrates), where they arrive after three days' march. There they embark on the Euphrates and reach Baghdad, whence they sail down the Tigris, to al-Obolla. From al-Obolla they sail for Oman, Sindh, Hind, and China.

These different journeys can also be made by land. The merchants that start from Spain or France go to Sus al-Aksa (in Morocco) and then to Tangier, whence they walk to Kairouan and the capital of Egypt. Thence they go to ar-Ramla, visit Damascus, al-Kufa, Baghdad, and al-Basra, cross Ahvaz, Fars, Kerman, Sind, Hind, and arrive in China.

Sometimes, also, they take the route behind Rome and, passing through the country of the Slavs, arrive at Khamlidj, the capital of the Khazars. They embark on the Jorjan Sea, arrive at Balkh, betake themselves from there across the Oxus, and continue their journey toward Yurt, Toghuzghuz, and from there to China.

(Adler, 1987, pp. 2–3)

In essence, the Radhanites operated across the four major trade routes that had historically linked Europe with South East Asia and China: (i) the southern Silk Route passing from China through Central Asia to Syria; (ii) the Northern Silk Route leaving China and veering north through the Caucasus and across the Black Sea; and the two Spice Routes from India, (iii) one travelling through the Red Sea and (iv) the other through the Persian Gulf. The Radhanites were able to leapfrog their trade goods across the bitterly rivalled territory of the Islamic caliphates, orthodox Christian Byzantines and Rus', Jewish Khazars and Roman Catholic Europeans.

The origin of the Radhanites is a bit of a mystery. Some historians suggest that the name comes from the Persian phrase *Rah Daan* ('he who knows the way'), while others believe it traces to the Rhône river valley in France (called *Rhodanus* in Latin). The leading authority on the Radhanites, Moshe Gil (1974), asserts they came from the district of Radhan in southern Mesopotamia, which he considers their central base of operations.

The Radhanite age ended in the tenth century when the political stability of China and Central Asia became too tenuous to maintain international trade linkages. The Silk Routes became dangerous and unstable when the Tang Dynasty of China fell in 908. This situation was exacerbated by the Turkic invasions into Anatolia and the disintegration of the Abbasid Caliphate in the Middle East into a collection of feuding Muslim principalities. Trade across Central Asia was also disrupted by the attacks of the Kievan Rus' on the Central Asian kingdom of Khazar Khaganate. As the Radhanite trade networks unravelled, the spice trade in Europe came to a standstill and prices rose precipitously. This would remain the case until the great mercantile Italian city states of Venice, Genoa, Pisa and Amalfi arose and became the next great international trading powers.

Rise of the Gotlanders

In the 700s, a powerful mercantile people began to spread throughout the Baltic region along the Russian rivers from the Island of Gotland located off the east coast of Sweden (Gannholm, 2013; Haywood, 2015). This area was largely uncivilized and also well outside the old Roman Empire. These traders were as much warriors as businessmen, and advanced into new areas via fortified outposts. Once the local people were pacified, new settlers were recruited to create towns and trading cities. This process was repeated over and over again as the Gotlanders moved further east until their sphere of influence touched the Byzantine and Islamic worlds.

The Gotlanders came to be called Varangians, a name probably derived from the Old Norse word *var*, which means 'union by promise' in English. The Gotlandic merchants kept themselves together by making mutual oaths for defence and profit sharing. The word 'Varangian' later became the synonym for Gotlanders who were employed as mercenaries to the rulers of the Byzantine Empire. It should be noted that the Varangians are separate from the peoples often referred to as 'Vikings'. Vikings were warriors from Denmark, western Sweden and Norway. The Elbe River was the border between Vikings and Varangians: east of the Elbe were the Varangians and west, the Vikings (Gannholm, 2013).

The Varangians became wealthy trading slaves, furs (beaver and black fox) and swords, and Arab dirhams became their favoured medium of exchange. In fact, Gotland has been pinpointed as the origin of these Eastern European traders by the extraordinary amount of Islamic silver coins that have been uncovered in Gotlandic soil (Noonan, 1998; Gannholm, 2013). This silver also found its way to the Iberian Peninsula and was widely traded with the Franks of Western Europe. The silver used to mint these coins came from mines discovered in the 600s within the Muslim-controlled provinces in Central Asia.

The Gotlandic merchants who settled in the East Slavic area of northern Russia were called *al-Rus* by the Arabs. *Rhos* (Rus) came from the old Norse word *ruotsi*, which meant 'expedition of rowing boats'. The Rus' are thought by many historians to have been organized in a polity named the Rus' Khaganate, although the existence of this entity is largely hypothetical until the tenth century. 'Khaganate' is a term used by Turkic and Mongolian peoples to represent an empire. Over time, the Rus' assimilated with the native Slavs along the Russian rivers and lost their distinct Gotland identity. Kiev emerged in the tenth century as the centre of the Rus' polity, and their leader Prince Vladimir's official adoption of Christianity in 988 tied them to Constantinople and the rest of eastern orthodox Christianity.

Rus' Trade with the Muslims and Byzantines through Khazaria

From the second half of the 700s, Rus' traders began moving south down the waterways controlled by the Khazars and made regular commercial trips to the Khazar capital of Atil in pursuit of the Arabic silver coins and silk, spices, wine, jewellery, glass and books from the Byzantine Empire. In return, the Rus' traded captured Slavs from the Eurasian Steppe and offered fur, honey, wax and timber. Silk movement can be traced from Constantinople, or Rayy in Iran, to Kiev and Novgorod, then into the Baltic and Scandinavia, and finally into England (Shepard, 1995; Fleming, 2007). Two major trade routes evolved: one down the Volga and across the Caspian Sea to the Muslim-held lands as far as Baghdad; and the other across the Black Sea to the Byzantine Empire (Fig. 12.1). As the mercantile fleets of the Rus' passed through the Khazar Kingdom they were tithed.

The city of Baghdad must have been an amazing sight to the Rus'. Between 903 and 913, the Arab writer Ibn Rustah wrote an eyewitness account that recorded the Rus' had 'no villages, no cultivated fields' and that 'their only occupation is trading in sable and squirrel and other kinds of skins, which they sell to those who will buy them' (Gabriel, 1999). Baghdad was now the crown

Fig. 12.1. Rus' trade routes. (Redrawn from File: Varangian routes.png. Wikimedia Commons. https://commons.wikimedia.org/w/index.php?title=File:Varangian_routes.png&oldid=545242759.)

jewel of the Islamic Empire, a lavishly embellished city with expansive green parks and gardens, marble palaces, promenades and finely built mosques.

Rus' Attacks on the Islamic and Byzantine Worlds

While most interactions between the Rus' and Baghdad, Khazaria and the other Muslim lands were peaceful, there were some notable exceptions (Gabriel, 1999; Watson, 2001). The Rus' were so fearless that they periodically raided Byzantine and Muslim strongholds. Records of these confrontations have been left to us by several medieval Arab chroniclers including Ibn al-Athīr, who wrote a comprehensive 11-volume history of the world in c.1231, and *The Russian Primary Chronicle*, a Kievan Rus' historical work attributed to the monk Nestor that gives a detailed account of the early history of the Eastern Slavs.

In the first major confrontation in 860, a fleet of about 200 Rus' vessels sailed down the Bosporus and attacked the suburbs of Constantinople. The attack took the Greeks completely by surprise in what Saint Photios the Great of Constantinople called 'a thunderbolt from heaven'. The city was largely undefended at the time, as the Emperor Michael III was away fighting the Abbasid Caliphate and his navy was confronting Arab pirates in the Mediterranean Sea. After landing, the Rus' went on a rampage, setting homes and churches on fire, as well as drowning and stabbing the residents. They subsequently retreated, for some unknown reason, after pillaging the suburbs.

The Rus' launched another attack on Constantinople in 941, this time with disastrous consequences for them. They sent a massive fleet of 1000 ships to the city but were defeated by a small fleet of 15 old Byzantine warships fitted with Greek Fire projectors that spewed burning chemicals at the invaders. A large number of Rus' ships and soldiers were set ablaze, and any soldiers who jumped overboard to escape the flames drowned, weighed down by their armour.

The Rus' also launched a number of raids across the Caspian Sea into Muslim lands. In the largest in 913, a fleet of 500 Rus' ships attacked the Gorgan region, in the territory of what is now Iran, where they plundered goods and took women and children as slaves. An army of enraged Khazars and some Christians eventually attacked and defeated them, leaving few survivors.

During another campaign in 943, a large Rus' armada attacked the prosperous trading city of Barda on the south shore of the Caspian Sea. The local people fought them mightily by throwing stones and hurling abuse, but the Rus' rounded them up and slaughtered 5000. As described by the Arab chronicler Ibn al-Athīr':

> After this lasted for a long time, they ordered the people of the town to depart and [they said that] they would not attack the townsmen for an interval of three days, and an individual was free to leave with whatever possessions he could carry. Most of the townsmen remained [in Barda'a] after the appointed time, and the Rūs then killed many people, and they took some ten thousand souls captive. They gathered those who remained in the Friday Mosque, and they said to the remaining townsmen: 'You can either ransom yourselves or we will kill you.' A Christian

came forth and settled on twenty dirhams for each man. But the Rūs did not keep to their bargain, except for the sensible ones, after they realized that they would not receive anything for some townsmen. They massacred all of those [for whom they could receive no ransom], and only a few fled from the massacre. The Rūs then took the valuables of the people and enslaving the remaining prisoners, and took the women and enjoyed them.

(Watson, 2001, p. 434)

The Rus' used Barda for several months as a base for plundering the adjacent areas, but eventually were forced to leave when they were greatly weakened by an outbreak of dysentery from tainted fruit and had become isolated for defence in the citadel of Barda. They left the fortress at night, carrying on their backs what they could of their plundered treasure, gems and fineries.

In recognition of their fearlessness, the Rus' were recruited half a century later by Byzantine Emperor Basil II to defend Constantinople (Cartwright, 2017). About 6000 Rus' mercenaries were formed into an elite 'Varangian Guard' to protect Constantinople and serve as the emperor's personal bodyguards. These mercenaries fought in every major Byzantine campaign from then on until Constantinople was captured by Crusaders in 1204.

References

Adelson, H.L. (1960) Early Medieval trade routes. *The American Historical Review* 65, 271–287.

Adler, E. (1987) *Jewish Travellers in the Middle Ages*. Dover Publications, New York.

Asbridge, T. (2010) *The Crusades: The Authoritative History of the War for the Holy Land*. HarperCollins, New York.

Cartwright, M. (2017) Varangian guard. *Ancient History Encyclopedia*. Available at: https://www.ancient.eu/Varangian_Guard/ (accessed 28 January 2021).

Chevallier, J. (2014) Spices in France in the Dark Ages. *Les Leftovers: sort of a food history blog*. Available at: https://leslefts.blogspot.com/search?q=medieval+spices (accessed 28 January 2021).

Davis, J.R. and McCormick, M. (2008) *The Long Morning of Medieval Europe: New Directions in Early Medieval Studies*. Ashgate Publishing, London.

Decker, M. (2009) Export wine trade to West and East. In: Mungo, M.M. (ed.) *Byzantine Trade, 4th–12th Centuries: The Archaeology of Local, Regional and International Exchange. Papers of the Thirty-Eighth Spring Symposium of Byzantine Studies, St. John's College, University of Oxford, March 2004*. Ashgate Publishing, London, pp. 239–252.

Dopsch, A. (1937) *The Economic and Social Foundations of European Civilization*. Harcourt Brace, New York.

Fleming, R. (2007) Acquiring, flaunting and destroying silk in late Anglo-Saxon England. *Early Medieval Europe* 15(2), 127–158.

Gabriel, J. (1999) Among the Norse tribes: the remarkable account of Ibn Fadlan. *Aramco World: Arab and Islamic Cultures and Connections* 50(6). Available at: https://archive.aramcoworld.com/issue/199906/among.the.norse.tribes-the.remarkable.account.of.ibn.fadlan.htm (accessed 12 April 2021).

Gannholm, T. (2013) *Gotland: The Pearl of the Baltic Sea: Center of Commerce and Culture in the Baltic Sea Region for over 2000 years*. Stavgard Förlag AB, Stånga, Sweden.

Gil, M. (1974) The Rādhānite merchants and the land of Rādhān. *Journal of the Economic and Social History of the Orient* 17, 299–328.

Gottesman, B. (2011) The Radhanite merchants. *World History Encyclopedia*. Available at: https://www.academia.edu/4810020/The_Radhanite_Merchants (accessed 28 January 2021).

Haywood, J. (2015) *Northman: The Viking Saga* AD *793–1241*. Thomas Dunne Books, New York.

Kingsley, S. (2009) Mapping trade by shipwrecks. In: Mungo, M.M. (ed.) *Byzantine Trade, 4th–12th Centuries: The Archaeology of Local, Regional and International Exchange. Papers of the Thirty-Eighth Spring Symposium of Byzantine Studies, St. John's College, University of Oxford, March 2004*. Ashgate Publishing, London, pp. 31–37.

Lelis, A.A. (2013) We are not the periphery: barbarian economies and Northern Europe in the exchange patterns of Western Eurasia, 1800 BC–AD 900. PhD thesis, University of Minnesota, St Paul, Minnesota.

McCormick, M. (2001) *Origins of the European Economy: Communications and Commerce* AD *300–900*. Cambridge University Press, Cambridge.

Mango, M.M. (ed.) (2009) *Byzantine Trade, 4th–12th Centuries: The Archaeology of Local, Regional and International Exchange. Papers of the Thirty-Eighth Spring Symposium of Byzantine Studies, St. John's College, University of Oxford, March 2004*. Ashgate, London.

Noonan, T.S. (1998) *The Islamic World, Russia and the Vikings, 750–900: The Numismatic Evidence*. Ashgate Publishing, London.

Pirenne, H. (1922) Mahomet et Charlemagne. *Revue belge de Philologie et d'Histoire* ˉ, 77–86.

Shepard, J. (1995) Constantinople – gateway to the north: the Russians. In: Mango, C. and Dagron, G. (eds) *Constantinople and its Hinterland: Papers from the Twenty-Seventh Spring Symposium of Byzantine Studies, Oxford, April 2003*. Routledge, Aldershot, UK, pp. 243–260.

Watson, W.E. (2001) Ibn al-Athīr's accounts of the Rūs: a commentary and translation. *Canadian/American Slavic Studies* 35, 423–438.

Wickham, C. (2005) *Framing the Early Middle Ages: Europe and the Mediterranean, 400–800*. Oxford University Press, Oxford.

The Eastern Roman Empire and The Rise of Venice

13

Setting the Stage – Eastern Roman Empire Struggles to Control Italy

The last emperor of the Western Roman Empire, Romulus Augustulus, was deposed by a Hun mercenary general named Odoacer in 476, who ruled as King of Italy for 13 years until he was overthrown by the Ostrogoth Theodoric the Great. Even though Western Romans were no longer ruling the Italian territory, the Byzantine emperors still considered the peninsula to be part of the Roman Empire. When Odoacer died his kingdom devolved into chaos and was followed by a series of weak and self-serving monarchs. In 535, the Byzantine emperor Justinian I decided to take action and sent his general Belisarius against the Ostrogoths. After landing in Sicily, Belisarius marched steadily up through Italy, taking Naples and Rome by 536 and Ravenna in 540. In desperation, the Ostrogothic nobility offered Belisarius rule of Italy for himself, but he refused and declared all of the Ostrogoth Empire and its treasury in Justinian I's name. Justinian made Italy a province with its capital in Ravenna and placed it under the authority of an *exarch*, the Greek word for 'governor'.

Justinian held control of the bulk of Italy for the next 25 years, but at his death in 565 the peninsula was being threatened by a Germanic people, the Longobards ('long beards') or Lombards. By late 569 the Lombards had conquered most of northern Italy and soon added much of central and southern Italy. They were mostly unopposed by the Byzantines who were now drained of resources after the Gothic Wars and the plague. The Lombards held on to most of the Italian Peninsula for the next 200 years, with the Byzantines hanging on only to Ravenna and the coastal wetlands along the northern Adriatic Sea containing Venice.

Byzantine Empire after Justinian

When Justinian died in 565 CE the Byzantine Empire was by far the largest and most powerful state in Europe. However, the plague and the massive debts incurred to recover the Western Roman territory had left the empire's treasury in dire financial straits (Herrin, 2007; Luttwak, 2011). To keep the empire afloat, his successors had no choice but to heavily tax its citizens as revenues diminished. The imperial army became stretched extremely thin and was unable to maintain most of the territory reconquered during Justinian's rule. It was forced to endure attacks not only from the Lombards, but also the Persian Empire, the Slavs and then the Muslims. By the end of the seventh century Byzantium had lost most of Italy, Syria, all of the Holy Land, Egypt and finally North Africa.

During the eighth and early ninth centuries, the empire also went through a period of considerable religious conflict. The Byzantine emperor Leo III launched a movement known as 'Iconoclasm' that rejected the holiness of religious images and banned their worship. This movement led to a break in the already strained relations between the Eastern and Western Christian churches. Iconoclasm waxed and waned throughout the eighth and ninth centuries and didn't end until 843, when a Church council under Emperor Michael III restored icon veneration.

Islam in the Ninth Century

As the ninth century dawned, the Islamic world had completed the bulk of its expansion. Baghdad was flourishing now as the cultural and economic capital of the world. Arab merchants travelled widely across the Indian Ocean and were capturing most of the trade with India, East Africa, Indonesia and China. Their merchant colonies had sprung up along the Malabar Coast, Ceylon, Java, Sumatra, the East African coast and China. Islam had grown rich supplying itself and the rest of the world with exotic trade goods.

Various Muslim groups also controlled large swaths of the eastern Mediterranean. Crete fell into the hands of Muslims under Abu Hafs in the 820s, who established a piratical emirate there that lasted until the year 965. Muslims had their first settlement in Sicily by 827 and they ruled all of the island from 902 until 1061. Muslims made repeated conquests of Sardinia from the late eighth century to the tenth, and gained complete control in 902, which lasted until the thirteenth century. Large parts of Cyprus were also conquered by Muslims in the mid-600s but had come under joint rule in 688 when Justinian II and the caliph Abd al-Malik agreed to share proceeds from the island. This arrangement mostly held until the late eleventh century.

Rule of Europe after the Fall of Rome

In the years following the fall of Rome, the rule of the European continent devolved into a complex array of fiefdoms with localized power (Holmes, 2001; Asbridge, 2010). The only unifying force was the Catholic Church and the path to secular power came through alliances with it. The Church was highly structured, with its power base stretching down from the Pope to cardinals, archbishops, bishops, priests, monks and nuns. The bishops were at the base of the power chain, controlling a 'diocese' of many churches in a kingdom. Their power extended to kings, who commonly took orders from the bishop and consulted with him on all affairs of the state. The fear of God was very real among both peons and nobility in medieval society.

The first great dynasty to emerge and rule large swaths of Europe was the Carolingians based in Francia (Belgium and France). They became the pivotal power supporting the Pope in Rome against the Lombards, as the Byzantines had become too weak to help. To protect Pope Stephen II from the Lombard onslaught in the mid-700s, the Frankish king Pepin the Short (751–768) invaded northern Italy and established the Papal States. Stephen rewarded Pepin by anointing him as King of the Franks and bestowed upon him the title of *patricius Romanorum* ('Patrician of the Romans'). This action signalled the Pope now understood that his future support would have to come from the Frankish West rather than the Byzantine East.

The first post-Roman leader to take control of the whole European continent was the Frankish king Charlemagne. By 800, he controlled much of modern France, Germany, Italy, Belgium and the Netherlands, laying claim to the title 'Emperor of the West'. Pope Leo III proclaimed him the 'Emperor of the Romans' to solidify his relationship with the Catholic Church. Charlemagne's realm became known as the Holy Roman Empire.

Charlemagne's dynasty held a large part of Europe together for a while, but it too fell apart in the 850s due to successional disputes and invasions from Scandinavian Vikings and European Magyars. Europe came to be fragmented once more into what Asbridge describes as:

> many smaller polities, ruled over by warrior-lords, most of whom were bound by only loose ties of association and loyalty to a crown monarch ... these men bore titles such as *dux* and *comes* (dukes and counts) that harkened back to Roman and Carolingian times, and were drawn from the ranks of a nascent military aristocracy – the increasingly dominant class of well-equipped, semi-professional fighting men who came to be known as knights.
>
> (Asbridge, 2010, p. 7)

Resurgence of Byzantine Empire in the Tenth and Eleventh Centuries

Even though it was greatly reduced in size from its peak under Justinian, the Byzantine Empire was able to hold fast to its position at the centre of world trade throughout the early Middle Ages. Its capital of Constantinople was by far the largest city in Christendom and was the crossroads of all the trade routes radiating across the world. It supported a diverse culture of merchants speaking a cornucopia of worldwide languages. Into the city flowed spices and sandalwood from India and South East Asia, silk from China, grain and wool from Central Europe, ivory and gold from Africa, wine and cork from Western Europe, and fur and honey from Northern Europe.

The Byzantines were trading far and wide across what Shepard (2006) called their three overlapping spheres: (i) the 'Byzantine Commonwealth' representing Constantinople and the surrounding lands under direct Byzantine control; (ii) the 'Christian–Islamic Orient' representing territories in North Africa and the Levant that had once been under Greek and Roman control, but were now largely Muslim-held; and (iii) 'Latin Christendom' representing Europe and what was once part of the Western Roman Empire.

The Byzantine Empire entered a golden age in the late tenth and early eleventh centuries, under the dynasty founded by Basil, Michael III's successor. The Byzantine Empire was now much smaller than it was at its peak in the seventh century, but it had much greater wealth through trade and higher international acclaim. Its capital Constantinople stood at the centre of all the world's great trade routes. The imperial government actively patronized art and began major restorations of churches, palaces and other cultural institutions. Greek became the official language, and the study of ancient Greek history and literature was widely promoted. Monks played a central role in everyday life as administers of orphanages, schools and hospitals, and served as missionaries among the Slavic peoples of the central and eastern Balkans and Russia.

Emperor Basil II (976–1025) led a resurgence in Byzantine power by securing strong physical paths to all three of Shepard's overlapping circles (Small, 2012). His Bulgarian campaigns re-established the ancient Roman road, the *Via Egnatia*, an important trade route which wove across the Balkans from Constantinople to the coast of Albania at Durazzo with clear sailing from there to Italy and the West. He captured Crete in 961 and Cyprus in 965, which cleared the Aegean Sea routes of most pirate activity and opened up the Islamic eastern Mediterranean to Byzantine trade. His empire was also well connected to the Crimea and the Eurasian Steppe through the Don and the Dnieper Rivers. As Small puts it:

> With its geographic position astride important communication routes between the west, the Eurasian steppe and the Near East, the Byzantine Empire and Constantinople in particular, was physically able to communicate with all three spheres of Shepard's overlapping circles.
>
> (Small, 2012)

Byzantine Trade in Silk and Spices with the Muslims

In the ninth century, Trebizond, which sat on the Black Sea at the crossroads between the Byzantine, Armenian and Muslim states, served as a major maritime outlet for the oriental spices. Trebizond was frequented by Muslim merchants who were particularly interested in obtaining Byzantine silks, since trade along the Silk Routes had largely dried up at this time with the fall of the Tang Dynasty.

Trebizond lost its central role in the tenth century when spices were increasingly diverted from the Persian Gulf, now plagued by political instability, to the Red Sea and Alexandria. In response, the Byzantines developed an active trading relationship with the Fatimids who had taken control of most of Egypt and the Levant. The Fatimids had benefited greatly from the shift in the spice trade from the Persian Gulf to the Red Sea resulting from the political instability of the Abbasid Caliphate. Merchants from Byzantium, and to a lesser extent from Venice and other emerging Italian city states, traded actively in Alexandria for spices and in return provided timber, slaves, pharmacological plants, silk textiles, furniture and even cheese (Small, 2012). Cyprus became a key centre of trade between Byzantium and the Fatimids, and there was likely a colony of Egyptian merchants in Constantinople (Jacoby, 2000). Trade between the Muslims and Christians continued mostly unabated throughout the tenth century except when Emperor John Tzimiskes placed an embargo on timber exports to the Fatimids in 971 when he was campaigning in Syria.

Constantinople in the tenth century also maintained significant trade links with the Muslim world through various Levantine ports. Byzantine access to Middle East trade was greatly facilitated by the conquests into Syria of emperors Nikephoros and John Tzimiskes and the recapture of Aleppo, Damascus, Beirut, Acre, Sidon, Caesarea and Tiberias.

The Rise of Venice

Venice first emerged in the fifth century as a haven for Italian refugees fleeing the successive waves of Hun and then Germanic invasions into northern Italy. The refugees banded together in the islands of the Venice Lagoon for mutual defence. When Ravenna and most of northern Italy fell to the Lombards in 751, Venice was able to survive, and over the next 700 years rose steadily from being a subject of the Byzantine Empire to becoming its equal partner and ultimately playing a role in its overthrow (Norwich, 1982; Crowley, 2012).

The story of Venice as a mighty trading power began in 565 when Justinian awarded it protection and trading privileges throughout his empire. In return, Venice kept its ports open to Byzantine imperial vessels and stood at the ready to protect the interests of Byzantium when needed. For centuries afterward, Venice served as a critical Byzantine outpost in a hostile northern Italy controlled mostly by the Lombards.

The Venetians become the primary merchant middlemen between the East and the West. Venice's nearness to the Frankish empires and favourable relationship with Byzantium put it at the very centre of world trade. Byzantine merchants would bring spices, silk and other luxury goods from the East to the Venetian markets and then the Venetians would distribute them to the

European markets in Italy, France and Germany. The Byzantines were paid with salt and fish obtained locally by the Venetians and slaves, timber and metal that the Venetians had bartered from Europe.

Over the centuries, Venetian sea power grew steadily while that of the Byzantine Empire gradually diminished. In fact, the Byzantines came to depend on the Venetian fleet to protect them from their foes. The Byzantines were more than happy to let the Venetians:

> patrol the Adriatic, transport Byzantine troops to and from Italy and Sicily and provide naval support to Imperial expeditions in the same region as early as 827 when the Muslims were besieging Syracuse. And it was the Empire's strategic position, obliging her to fight in two distant operational theatres of war – Asia Minor and the Balkans, along with its limited resources in money and manpower that prompted the use of diplomacy, bribery, fraud and other means to avoid war! In other words, the Byzantines were more willing to have others to fight their wars than send naval detachments in a region far away from their main operational theatres closer to the capital. And as long as they provided them with rewards, the Venetians were more than willing to play that role.
>
> (Theotokis, 2018)

In the ninth and tenth centuries the Venetians played a critical role in warding off the Slavs and Muslims when they threatened Constantinople. This support greatly elevated the standing of Venice in Byzantine eyes. Venice was no longer considered a subservient power but instead a valuable ally. In 879, Basil I sent a diplomatic delegation to Venice to strengthen their bond of cooperation rather than demanding the Venetians come to him (Nicol, 1992). The ambassadors took with them valuable presents for the doge and conferred the high imperial title of *Protospatharios* upon him. In 1005, Emperor Basil II encouraged the marriage of the doge's eldest son Pietro to Maria Argyropoulina, a member of the royal household:

> The emperors needed their Venetian partners to act as a military and economical presence for them in a region which was far removed from the strength of the Empire's heartland in the east. With Venetian assistance, Constantinople was assured a continued existence in Italy and that part of the Mediterranean well into the twelfth century.
>
> (Echebarria, 2013, p. 27)

When a huge Norman force blockaded Constantinople in 1082 and was poised to attack, the Venetians again came to the Byzantines' rescue, forcing the Normans to retreat in a fierce naval battle. A grateful Emperor Alexius made a decree or *chrysobull* ('golden bull') giving the Venetians a commercial monopoly throughout the Byzantine Empire.

The Venetians were allotted a number of shops, factories and houses in Constantinople 'with free access to and egress from the district' and those living in the city were awarded 'annual financial grants and grand titles'. The Venetian merchants were given the right 'to trade in all manner of merchandise in all parts of his empire free of any charge, tax, or duty payable to his treasury' (Nicol. 1992, p. 61).

The golden bull gave the Venetians free rein across a burgeoning eastern trade network. The chrysobull of 1082 contains a list of ports 'encountered by ships on their way to Constantinople, in the east from Laodikeia along the seaboard of northern Syria and Asia Minor, and in the west from Dyrrachion along the coast of the Peloponnesus' (Jacoby, 2008, p. 684). As Crowley eloquently describes the impact:

> [the] Golden Bull of 1082 was the golden key that opened up the treasure-house of eastern trade for Venice. Its merchants flocked to Constantinople. Others started to permeate the small ports and harbors of the eastern seaboard. By the second half of the twelfth century, Venetian merchants were visible everywhere in the eastern Mediterranean. Their colony in Constantinople grew to around twelve thousand and, decade by decade, the trade of Byzantium imperceptibly passed into their hands. They not only funneled goods back to an avid market in continental Europe, they acted as intermediaries, restlessly shuttling back and forth across the ports of the Levant, buying and selling.
>
> Their ships triangulated the eastern seas, shipping olive oil from Greece to Constantinople, buying linen in Alexandria and selling it to the Crusader states via Acre; touching Crete and Cyprus, Smyrna and Salonika. At the mouth of the Nile, in the ancient city of Alexandria, they bought spices in exchange for slaves, endeavoring at the same time to perform a nimble balancing act between the Byzantines and the Crusaders on one hand and their enemy, the Fatimid dynasty in Egypt, on the other. With each passing decade, Venice was sinking its tentacles deeper into the trading posts of the East; its wealth saw the rise of a new class of rich merchants. Many of the great families of Venetian history began their ascent to prominence during the boom years of the twelfth century. The period heralded the start of commercial dominance. With this wealth came arrogance – and resentment.
>
> (Crowley, 2012, p. 17)

Rise of Other Italian Mercantile States

There were many other maritime republics that arose in Italy during the Middle Ages other than Venice. These included Genoa, Pisa, Ragusa (now Dubrovnik), Gaeta, Ancona and Noli. The most powerful were Genoa, Pisa and Amalfi, who carried on extensive trade across the Mediterranean and built strong navies for protection and conquest. Venice came to dominate Adriatic trade, while Pisa and Genoa focused their trade more heavily on Western Europe.

Amalfi, on the western coast of Italy, was the first of the maritime republics to make serious inroads into the Mediterranean trade monopoly held by the Arabs, beginning about 850. They founded mercantile bases widely across Italy and the Middle East in the tenth century. They were particularly important as grain merchants, but also traded heavily in salt, slaves, timber, spice and silk. Their fleets played a key role in the defence of Pope Leo IV against the Muslims in 848 when they tried to establish a foothold on the Italian mainland. Amalfi remained strong for centuries until a series of natural disasters weakened it in the twelfth century and it was overrun and sacked by Pisans in 1137.

Pisa and Genoa, both on the Tyrrhenian Sea, came into power in the eleventh century. Both of these mercantile cities developed strong trade linkages between Constantinople, Muslim-held ports in the Middle East and the western Mediterranean. Pisa also became powerful in the Black Sea. Like Venice, Pisa and Genoa were granted privileges in Byzantium, but they were more limited. These mercantile states traded salt, slaves and timber to Egypt and Syrian ports for gold dinars, which they then used to buy Byzantine silks to resell in the West. They also moved grain to Europe. Pisa and Genoa initially teamed up to block Arab expansion into the Mediterranean, but ultimately became bitter enemies after the First Crusade. They fought mightily between themselves and Venice for more than a century for control of the lucrative Mediterranean trade routes.

Western Maritime Trade in the High Middle Ages

In the eleventh and twelfth centuries, the Mediterranean and the Black Sea:

> constituted two separate commercial regions. Each of these regions handled different goods and had its own navigation conditions, trade patterns and shipping networks. At their junction, Constantinople served as destination or point of departure for trade and shipping in one or the other region, and as transit and transshipment station for commodities sailing between them.
>
> (Jacoby, 2008, p. 686)

Byzantine traders largely dominated the Black Sea and Sea of Azov basin, while the Venetians and the other Italian mercantile states emphasized the Mediterranean.

The Black Sea area was particularly important to Constantinople as a source of grain, fish and salt, and to a much lesser extent spices and silks (Karpov, 2005). Its importance as a source of spices and silk had been greatly diminished over the last century due to the unrest in Central Asia leading to the breakup of the Silk Routes and the shift in the spice trade from the Persian Gulf to the Red Sea resulting from the political instability of the Abbasid Caliphate.

There has been considerable debate on whether the Byzantines did not allow Italian trade in the Black Sea region or the Italians were simply not interested in that market due to their deep involvement in Mediterranean trade. The Byzantines did control trade at various times in some commodities such as timber, gold, grain and silk, but in general Italian merchants were not denied access to Black Sea ports. It is more likely that the Italian merchants had little presence in the region because:

> the volume and variety of Black Sea goods they wished to acquire did not warrant the expense and risks involved in frequent sailings beyond Constantinople. Rather, the Italian merchants appear to have been mostly content to purchase these goods from intermediaries in Constantinople, although they were more expensive than if bought around the Black Sea. As a result, they displayed only marginal

interest in trade and transportation in that region, and at best operated there on a limited scale before 1204.

(Jacoby, 2008, p. 698)

A Byzantine Cry for Help

As the eleventh century progressed, the Muslim forces surrounding the Eastern Roman Empire were gaining strength and squeezing the Byzantine Empire. A new Turkic people, the Seljuk Turks, had invaded and captured Baghdad from the Abbasids in 1055 and then went on to take the Holy Lands in Syria and Palestine from the Byzantines. They then invaded Byzantine Asia Minor and by 1081 had captured almost all that region. A huge slice of the old Eastern Roman Empire was now under Muslim rather than Christian control.

Fearing for the survival of the Byzantine Empire, Emperor Alexius Comnenus appealed to the West for help and in 1095 Pope Urban II agreed. He called for the First Crusade at the Council of Clermont to 'liberate the Church of God'. He saw the campaign as a way for the growing population of knights in Europe to direct their energies towards a spiritually meritorious act. To encourage their participation, he promised them forgiveness for all the sins they had confessed and save them from eternal damnation. After Urban II's rallying speech, an army came together numbering around 60,000 men in total and some 6000 knights (Cartwright, 2018a).

This First Crusade was remarkably successful and over a period of seven years much of the Levant was conquered and four 'Crusader States' were established: (i) the Kingdom of Jerusalem (including a narrow coastal strip from Jaffa to Beirut); (ii) the County of Edessa (modern Urfa in south-east Turkey); (iii) the County of Tripoli (modern-day Lebanon); and (iv) the Principality of Antioch (between Edessa and Tripoli along the coast). This Christian foothold in the mostly Muslim Middle East survived until 1291, although its size and structure ebbed and flowed over time under continual harassment from Muslim armies. There would be eight official Crusades throughout the twelfth and thirteenth centuries CE, all of them meeting with more failure than success. The Crusader States were perpetually hamstrung by dynastic rivalries, too few troops and only sporadic, often ineffective, support from the West in the form of Crusades.

Influence of Crusader States on European Trade

Having a Christian foothold in the largely Muslim Middle East had a very significant impact on Western trade. As Thomas Asbridge put it:

> For both Islam and the west, perhaps the most striking transformation wrought by the crusades related to trade. Levantine Muslims already maintained some

commercial contacts with Europe before the First Crusade through Italian seaborne merchants, but the volume and importance of this interaction was revolutionized in the course of the twelfth and thirteenth centuries, largely as a result of the Latin settlement of the eastern Mediterranean. The crusades and the presence of the crusader states reconfigured Mediterranean trade routes ... and played a role in solidifying the power of the Italian mercantile cities of Venice, Pisa and Genoa.

(Asbridge, 2010, pp. 665–666)

In spite of the constant warfare that occurred during the long period of the Crusades, the Italian merchant cities maintained active trade with many ports in the Levant (Cartwright, 2018b). The larger cities became active mercantile centres with traders in residence from Arabia, Iraq, Byzantium, North Africa and Italy. Specialized markets arose where locals and foreigners could purchase a wide array of goods from silks and spices to basic foodstuffs, leather goods, cloth, furs and other manufactured goods. Much higher quantities of exotic goods began entering Europe than ever before including the spices (especially pepper and cinnamon), sugar, dates, lemons, cotton cloth and Persian carpets. The Italian states of Venice, Genoa and Pisa grew very wealthy through their trade and they gained a lot of additional support transporting Crusader armies and pilgrims to the Holy Land.

To get to and from the centres of trade along the eastern Mediterranean, Christian, Jewish and Muslim traders moved remarkably freely across hostile lands. The great Muslim chronicler Ibn Jubayr who travelled in the Middle East during the twelfth century wrote:

> One of the astonishing things that is talked of is that though the fires of discord burn between the two parties, Muslim and Christian, two armies of them may meet and dispose themselves in battle array, and yet Muslim and Christian travelers will come and go between them without interference. In this connection we saw at this time, that is the month of Jumada al-Ula [in the Islamic calendar], the departure of Saladin with all the Muslims troops to lay siege to the fortress of Kerak, one of the greatest of the Christian strongholds lying astride the Hejaz road [the pilgrimage route to Makkah] and hindering the overland passage of the Muslims. Between it and Jerusalem lies a day's journey or a little more. It occupies the choicest part of the land of Palestine, and has a very wide dominion with continuous settlements, it being said that the number of villages reaches four hundred. This Sultan invested it, and put it to sore straits, and long the siege lasted, but still the caravans passed successively from Egypt to Damascus, going through the lands of the Franks without impediment from them. In the same way the Muslims continuously journeyed from Damascus to Acre (through Frankish territory), and likewise not one of the Christian merchants was stopped or hindered (in Muslim territories). The Christians impose a tax on the Muslims in their land which gives them full security; and likewise the Christian merchants pay a tax upon their goods in Muslim lands. Agreement exists between them, and there is equal treatment in all cases. The soldiers engage themselves in their war, while the people are at peace and the world goes to him who conquers.

(Broadhurst, 1952, pp. 300–301)

The Byzantine Empire Struggles with the Rising Power of Venice

Venice and Byzantium were strong partners in an alliance that had held solidly for centuries. Venice's economy relied on the trade passing through Constantinople, the gateway for the products produced in India, South East Asia and China. The chrysobull of 1082 gave the Venetians huge 'economic and geographic power against all their rivals and eventually, power against the Byzantines themselves' (Echebarria, 2013, p. 42). The Byzantine merchants became mostly speculators, allowing the Venetians and other Italian mercantile states to do the actual trading. The Byzantines used the Venetian navy as a sort of a buffer zone in the Adriatic, providing transportation for imperial expeditions and patrolling the Adriatic against enemies. The Byzantine navy had become too weak by the second half of the eleventh century to do these jobs itself.

The Venetians became very wealthy and powerful from these endeavours, and they grew increasingly arrogant. As time went on the Venetians' crude, mercantile ways began to increasingly grate on the Byzantines. The Venetians also played a dangerous game of trading with both the Byzantines and their bitter Muslim enemies in Egypt and the Middle East. They insisted on maintaining their own foreign policy, always weighing the pros and cons of which side to take and being careful to not alienate either party.

One of the Venetians' most important assets was their commercial quarter in Constantinople, which was awarded to them by Alexius' chrysobull in 1082. The number of Venetian merchantmen in the city came to approach 10,000. As more and more Venetians packed into Constantinople, the native residents became increasingly appalled at what they perceived to be the uncouth behaviour of the Venetians and their arrogance. The Byzantine historian Choniates described the Venetians as 'beggars, cunning in thought ... and surrounding themselves with wealth, they pursue insolence and impudence' (Theotokis, 2018).

The animosity between the Byzantines and the Venetians grew until it finally erupted during the reign of John II (1118–1143) when he failed to renew his father's privileged trade agreement with the Venetians (Madden, 2017). John was tired of the Venetians, who acted as the senior partner in their long-term relationship, and he decided to prove that he had the upper hand. He was forced to back off from this decision when the Venetians used their powerful Crusader fleet to launch punitive naval raids in the Ionian and Aegean Seas between 1122 and 1125. John had to relent when it became clear to him that the Byzantine Empire no longer had the power to control the seas surrounding Constantinople. Not being able to face another adversary and needing all the allies he could find, the emperor ratified a new treaty in 1126. The balance of power had now swung mightily towards the Venetians and the Byzantines had lost control of their long-term ally.

The power struggle between Constantinople and Venice continued to simmer until it boiled over again during the reign of Manuel I (1143–1180).

In an attempt to reduce the power of Venice, Manuel entered into agreements with its rivals – Pisa, Genoa and Amalfi – and gave them quarters next to the Venetians in the northern part of Constantinople. The Italians proved to be unruly, were impervious to imperial rule and began to squabble among themselves. In early 1171, the Venetians attacked and almost completely destroyed the Genoese quarter. In response, Manuel I decided to rid the city of the Venetians once and for all, ordering all 20,000 Venetians living on Byzantine territory to be arrested and their property confiscated. An incensed Venice sent a fleet of 120 ships against Byzantium, only to fail and limp home after being decimated by the plague.

The Venetians remained at war with the Byzantine Empire for almost a decade, but instead of direct confrontations, sponsored Serb uprisings, besieged Ancona (Byzantium's last stronghold in Italy) and joined in a treaty with the Norman Kingdom of Sicily (Echebarria, 2013). Relations between the empire and the Venetians warmed briefly after the death of Manuel I in 1180, when his Latin widow, Maria of Antioch, became the regent to her infant son Alexios II Komnenos. However, this pro-Italian attitude proved ephemeral when she was overthrown in April 1182 by Andronikos I Komnenos.

Victorious Andronikos I entered the city in a wave of popular support and:

almost immediately, the celebrations spilled over into violence towards the hated Latins, and after entering the city's Latin Quarter a mob began attacking the inhabitants ... The ensuing massacre was indiscriminate: neither women nor children were spared, and Latin patients lying in hospital beds were murdered. Houses, churches, and charities were looted. Latin clergymen received special attention, and Cardinal John, the papal legate, was beheaded and his head was dragged through the streets at the tail of a dog.

(Nicol, 1992, p. 107)

A semblance of order was eventually restored to the Latin quarter; however it wasn't until 1186 that the Venetians and the Byzantine Empire arrived at a reconciliatory agreement from the events of 1171. This agreement let off some of the pressure between the empire and Venice, but the undercurrent of hostility would persist and fester, leading to the brutal sack of Constantinople in 1204 during the Fourth Crusade.

The Fourth Crusade and the Sacking of Constantinople

Between 1174 and 1187, the Ayyubid sultan Saladin conquered most of the Crusader States and returned them to Muslim control. The Holy City Jerusalem fell in 1187 to the great consternation of the Christian world. A third Crusade was launched in 1189 to recover it, and even though most of the Crusader States were reclaimed, Jerusalem itself remained in Muslim hands.

During the Third Crusade, tensions flared mightily between the Byzantine and Holy Roman Empires when the Byzantines refused to transport the troops of Holy Roman Emperor Frederick I across the Dardanelles, fearing he was conspiring with breakaway republics in the Balkans. Frederick came very close to invading Constantinople until the Byzantine emperor relented.

In 1202, Pope Innocent III called for a fourth Crusade to recapture the Holy City from the Muslims. The grand plan was first to invade and conquer Egypt and then go after Jerusalem (Madden, 2017). The Venetian navy was contracted to move an expected Crusader force of 35,000 men and 10,000 horses.

Only about 10,000 Crusaders wound up assembling in Venice and the leadership of the Crusade had no way to cover the Venetians' enormous costs. To obtain partial payment, the Venetians convinced the Crusaders to sack Zara, a Catholic city on the Adriatic that was currently under Hungarian control. The city was subsequently conquered and brought under Venetian control. A furious Pope Innocent excommunicated the Crusader army for this action, but they continued on.

While wintering in Zara, the now rogue Crusader leadership was convinced by the Byzantine prince Alexios Angelos to divert the Crusade to Constantinople and restore his family to rule. Alexios offered the Crusaders 200,000 silver marks (51 tons of silver) and promised to fully resupply the Crusader army once himself and his father were returned to power. The Crusaders agreed and a portion of their army subsequently sailed to Constantinople and was able to easily overwhelm the weak Byzantine forces. Alexios desperately tried to come up with the promised funds to repay the Crusaders, but after stripping his enemies of all their wealth and robbing the Church of many of its gold and silver relicts, he fell far short. In response, the Crusader army sacked the city to recover its investment.

The sacking of the great city of Constantinople was truly appalling. As Madden in his book on Istanbul describes it:

> For three days the victorious Westerners feasted on the bloated corpse of New Rome. The crusaders and the Italian refugees were merged into a hideous mob driven by greed, lust and hate. Oaths sworn to the Gospels to leave women and ecclesiastical buildings unmolested were forgotten in the frenzied anarchy. Many fanned out to the richest houses and palaces where they found their wealthy owners waiting. They stripped the homes of all their wealth, forced the owners to reveal their hidden treasures and threw them all out taking the dwellings for themselves. With nowhere to go these ragged lords streamed out of the city, leaving it to its western conquerors.
>
> In the holy sanctuaries of Constantinople, the Europeans stripped the altars of all precious furnishings, smashed icons for the sake of their silver or gems, and defiled the consecrated Eucharist and Precious Blood. Patens were used as bread dishes, and chalices as drinking cups. Radiant Hagia Sophia was stripped of everything of value. The priceless main altar was hacked to pieces and divided among the looters. So much wealth was found in the great church that mules were brought in to carry it all away. Unable to keep their footing on the slick marble floors, some of the beasts tumbled to the ground and were split open by the sharp objects they carried, defiling the church with their excrement and blood.
>
> (Madden, 2017, pp. 205–206)

When the rape of Constantinople was over, all that was left of the once proud Byzantine Empire was three rump states centred around Nicaea, Trebizond and Epirus. The Crusaders founded several other states on the rest of Byzantine territory, but Jerusalem remained under Muslim control. After almost 1000 years of European dominance, the Byzantine Empire had evaporated. The ancient Roman Empire was now truly finished, and Venice with its huge trading empire was left as the most powerful Christian state.

References

Asbridge, T. (2010) *The Crusades: The Authoritative History of the War for the Holy Land*. HarperCollins, New York.

Broadhurst, R. (1952) *From Muhammad ibn Ahmad Ibn Jubayr, travels of Ibn Jubayr: being the chronicles of a Medieval Spanish Moor concerning his journey to the Egypt of Saladin, the Holy Cities of Arabia, Baghdad the City of the Caliphs, the Latin Kingdom of Jerusalem, and the Norman Kingdom of Sicily*. Cape, London.

Cartwright, M. (2018a) Crusader states. *Ancient History Encyclopedia*. Available at: https://www.ancient.eu/Crusader_States/ (accessed 28 January 2021).

Cartwright, M. (2018b) The crusades: consequences & effects. *Ancient History Encyclopedia*. Available at: https://www.ancient.eu/article/1273/the-crusades-consequences--effects/ (accessed 28 January 2021).

Crowley, R. (2012) *City of Fortune: How Venice Ruled the Seas*. Random House, New York.

Echebarria, D. (2013) *Byzantine sorrow and Venetian joy: the failure of Byzantine diplomacy and the expansion of trade in the Mediterranean, 700–1200*. MA thesis, University of Nevada, Reno, Nevada.

Herrin, J. (2007) *Byzantium: The Surprising Life of a Medieval Empire*. Princeton University Press, Princeton, New Jersey and Oxford.

Holmes, G. (ed.) (2001) *The Oxford History of Medieval Europe*. Oxford University Press, Oxford.

Jacoby, D. (2000) Byzantine trade with Egypt from the mid-tenth century to the Fourth Crusade. *Thesaurismata* 30, 30–31.

Jacoby, D. (2008) Byzantium, the Italian maritime powers, and the Black Sea before 1204. *Byzantinische Zeitschrift* 100(2), 677–699.

Karpov, S. (2005) The Black Sea region, before and after the Fourth Crusade. In: Laiou, A. (ed.) *Urbs Capta: The Fourth Crusade and Its Consequences*. Realités Byzantines, Vol. 10. Éditions Lethielleux, Paris, pp. 285–294.

Luttwak, E.N. (2011) *The Grand Strategy of the Byzantine Empire*. Belknap Press of Harvard University, Cambridge, Massachusetts.

Madden, T.F. (2017) *Istanbul: City of Majesty at the Crossroads of the World*. Penguin Books, New York.

Nicol, D.M. (1992) *Byzantium and Venice: A Study in Diplomatic and Cultural Relations*. Cambridge University Press, Cambridge.

Norwich, J.J. (1982) *A History of Venice*. Knopf, New York.

Shepard, J. (2006) Byzantium's overlapping circles. In: Jeffreys, E. (ed.) *Proceedings of the 21st International Congress of Byzantine Studies, London, 21–26 August 2006*. Vol. 1, *Plenary Papers*. Ashgate, London, pp. 15–55.

Small, A.M. (2012) Was Byzantium a major commercial force in the tenth-century eastern Mediterranean? Available at: https://www.academia.edu/4899914/Was_Byzantium_a_major_commercial_force_in_the_tenth_century_eastern_Mediterranean (accessed 28 January 2021).

Theotokis, G. (2018) Byzantium and Venice: the rise and fall of a medieval alliance. *Medievalists.net.* Available at: https://www.medievalists.net/2018/12/byzantium-venice-medieval-alliance/ (accessed 24 February 2021).

Medieval Shifts in the Balance of Power

Setting the Stage – The Late Medieval European Economy

Agricultural productivity in Europe increased dramatically in the High Middle Ages and the populations of cities and towns grew significantly in response. From 1000 to 1300, the population of Europe at least doubled; Paris came to have over 200,000 occupants, Milan, Genoa and Venice about 100,000, and London, Cologne, Novgorod, Ghent and Prague more than 40,000 (Chandler and Fox, 1974).

One of the consequences of European population growth was, as Jean Favier tells it:

> a new need to mark the social differences between the classes. This could be achieved by eating, dressing or building in a different way from the local peasants. Although they always had a place in life, religious worship, and adornments of the rich and powerful, from the eleventh century onward luxury goods such as ivory, silk and spices were increasingly in demand as an outward demonstration of the owner's strength, wealth and success. As people again began to undertake long journeys – pilgrimages and crusades being the most notable examples – there came into fashion a certain exoticism that was to be an enduring characteristic of the ruling classes.
>
> These new needs revived trade in the Mediterranean and were to make Italy's fortune. This time it was no longer a matter of allowing a few colonies of Syrians to arrive, as they had done in the sixth century, bring the products of Asia to the West. Now it was Europe, and particularly Italy that took the initiative and traveled through the Black Sea to Antioch and Alexandria in Egypt to meet the caravans bearing silks from China and spices from India. Since luxury was an important element in court life, and the court a symbol and means of asserting political power, the rivalry of the great families was closely linked to that between the supplying cities. Everything conspired to push up demand. From the eleventh to twelfth century, there was a rapid movement away from casual peddling to an established trade and permanent routes.
>
> (Favier, 1998, pp. 19–20)

Spices in Medieval Cuisine

Spice popularity in both cuisine and medicine reached its historical peak during the Middle Ages in Europe. Spices were not only thought to be healthy, but also were widely used to enrich the natural qualities of food (Freedman, 2008). Food in medieval households was highly processed and richly spiced. Uncooked food was rarely eaten, even vegetables and fruit. The spices were used to season all types of food including meat, fish, soups, sweet dishes and even wine. It was also common in medieval banquets to pass around a 'spice platter' from which guests could choose extra seasonings like pepper for their already richly accented meals.

The noted expert on medieval gastronomy, Paul Freedman (2007, p. 50), tells us that 'spices were omnipresent in medieval gastronomy' and 'something on the order of 75% of medieval recipes involves spices'. In the updated edition of *Pleyn Delit: Medieval Cookery for Modern Cooks*, Sharon Butler, Constance Hieatt and Brenda Hosington provide 131 medieval recipes of which 92 include spices (Butler *et al.*, 1996). In a cookbook produced for the King of Naples in 1500, of 200 recipes, cinnamon was used in 125 and ginger in 76 (Freedman, 2007). 'This passion for spices is all the more notable given the modern European abandonment of almost all spices other than pepper, except for a certain modest presence in desserts' (Freedman, 2007, p. 50).

Historical records are filled with references to the copious use of spices among the wealthy in medieval Europe (Jenkins, 1976; Toussaint-Samat, 1994; Shaffer, 2013). When the King of Scotland visited the King of England in 1194, he received among other gifts a daily allotment of 4 pounds of cinnamon and 2 pounds of pepper (surely more than he could consume in a day). Lamprey, adored by English aristocracy, was slathered in a peppery sauce. The story goes that King Henry I of England was killed in 1135 by consuming a huge meal of pepper-smothered lamprey, although food poisoning was probably the culprit. A sauce served at the Feast of St Edward in 1264 was prepared from 15 pounds of cinnamon, 12½ pounds of cumin and 20 pounds of pepper. When the Duke of Bavaria-Landshut got married in 1476, the banquets honouring the event required 205 pounds of cinnamon, 286 pounds of ginger and 85 pounds of nutmeg. At a single banquet for 40 in medieval England the food was spiced with 1 pound of columbine powder, ½ pound of sugar, 1 ounce of saffron, ¼ pound of cloves, ⅛ pound of nutmeg and ⅛ pound of pepper.

Scholars have long debated why spices became so popular in medieval cuisine. It is commonly asserted that they were used to preserve meat or mask the flavour of decomposing produce, but their effects would be far less than the common practices of salting, smoking or pickling. Some have asserted that the copious use of spice in food preparation was fuelled by Galen's medical theories promoting their effects on health. However, there is no reason to believe that medieval diners stuck to healthy diets any more than people today. Most likely, the exotic origins of spices and their costliness made them a status symbol which fuelled their copious use.

Spices in Medieval Medicine

Asian spices played a pivotal role in the medicine of the Middle Ages. Their curative and healthy properties were acclaimed by physicians and the masses, both poor and wealthy alike (Freedman, 2008). Native plants were also incorporated into the European medical texts, but patients continued to prefer Asian spices if they could afford them. The drugs outlined in herbals and medical texts of the Middle Ages were almost all composed of Asian spices (Riddle, 1965). The widespread belief was that the spices must be more powerful medicines if they came from a distant 'paradise'.

When the first great medical universities began to crop up in Europe in the thirteenth century, their curriculum was based on Arabic medical books translated into Latin. These then spawned hundreds of pharmacology texts in the various European languages (Kuk, 2014). Most of these books catalogued the warm, cold, dry and wet properties of individual medicines and how they could be used to balance the humours. In all of them, Asian spices were considered to be the most powerful remedies.

Through the proliferation of texts based on Galen's philosophies, the Asian spices came to play a central role in the medicine of the Middle Ages. Among the medieval pharmacology texts, probably the most famous one was *Circa Instans* written in the twelfth century by Matthaeus Platearius, an esteemed doctor of the Salerno Medical School in southern Italy. Based on the work of Galen, this book was translated into French, English and German and became the elementary medical text used by physicians and apothecaries across all of Europe. It detailed the properties of specific medicines, their composition and their effects according to Pliny.

In the French version of *Circa Instans* called *Livre des simples médecines*, the properties and effects of the Asian spices were described as follows.

> Pepper – 'It is hot to the beginning of the fourth degree and dry to the middle of the fourth. Item, the wine in which black pepper has cooked with figs removes the thick sticky phlegm from the chest and respiratory organs, and it is very effective for asthma caused by cold. Item, powered pepper gnaws away the dead flesh of sores.'

> Ginger – 'It is hot to the third degree and moist to the first. Item, the wine which has been cooked with figs and raisins de Carême is good for cough of cold cause and for cold of the chest. Item, the wine in which ginger has cooked with cumin is good for stomach pain provoked by flatulence, and this procures good digestion. Item, for tenesmus, put powdered ginger on a cloth onto the anus. The powder put into food is good for heart weakness and syncope.'

> Cinnamon – 'It is hot in the third degree and dry in the second. It fortifies the brain by its good smell, it forms scars and knits together the wounds. Item, for weakness of the stomach and liver and to help digestion weakened by cold, put into the patient's food and sauces powdered cinnamon and caraway. Item, to restore appetite lost through humors of the stomach, make a sauce of cinnamon powder, sage, parsley, and vinegar and use it. Item, for recently cracked lips and other sores, put powdered cinnamon on the sore. Item, for heart ailments and

syncope, give powdered cinnamon with powdered leaves of the clove tree. Item, the large kind of cinnamon is used in strong medicine.'

Cloves – 'They are hot and dry in the third degree. Item, by their good smell they fortify, and by their quality they spread and disperse humors. Item, to aid digestion, give the wine in which these cloves have been cooked with fennel seed. Item, for suffocation of the womb, make a preparation of powdered cloves and strongly perfumed wine and apply to the fundament or make a pessary on it. Item, for diarrhea caused by a very strong medicine, boil 9 or 10 cloves in a glass phial with rose water and mastic, and give it to the patient slightly warmed. Item, for heart pains and syncope, give powdered cloves with borage juice.'

Nutmeg – 'It is hot and dry to the second degree. It fortifies by its good smell and qualities. Item, for cold of the stomach and bad digestion, and to improve the color of the face, give in the morning 1 dram of nutmeg, this is very good. Item, the wine in which nutmeg has cooked with anise and cumin, cures pain of the stomach and intestines caused by flatulence. Item, the wine in which nutmeg has cooked with mastic should be given to convalescents so that they recover their good spirits. To fortify the brain and respiratory organs, let the patient breathe in the fragrance of this nut, it is good.'

Mace – 'It is hot and dry to the second degree. Item, to aid bad digestion and cold of the stomach, the wine in which mace has cooked is good. Item, to purge the brain of excess of humors, chew some mace and keep it in the mouth for a long time so that its scent may rise to the brain. For weakness of the stomach and liver from cold cause, for the dropsy we call leucophlegmacy, for gripes, asthma, and other diseases of the chest caused by thick phlegm, cook mace in fennel juice. Finally, when it is cooked, add a little wine, strain, and give to the patient, it is a sovereign remedy.'

(Kuk, 2014, p. 327)

Silk in Medieval Europe

By the Middle Ages, silk had come to play a central role in European aristocratic lives. As told by Wagner:

Everything and everyone it touched immediately became eminent. Draped over altars or fashioned into curtains, silk separated spaces within a church or a palace. Beginning in the sixth and seventh centuries in Western Europe, saints' relics were wrapped in silk, stored and displayed in elaborate metalwork and jeweled reliquaries … For hundreds of years … silk's qualities of luxury, versatility and scarcity perpetuated its status as a prized material. The luminescence and softness of the fabric always impressed those who saw or felt it … As a commodity, silk was considered at times to be more valuable than gold.

(Wagner, 2016)

Fleming also relates:

Anglo-Saxon kings not only wore banded costumes, but they sometimes sported clothes made entirely from silk. By this time kings across Europe had adopted all-silk clothing. Indeed, numerous textual witnesses describe the use of all-silk clothes at the Ottonian court and there is interesting material evidence as well. The tunic, for example, that the German emperor Henry II (d. 1024) is said to have worn for his coronation is still extant. The body of the tunic is white, patterned-silk damask, and its trim is fashioned from broad bands of red silk, embroidered with metallic-gold and purple-silk thread. Two magnificent eleventh-century robes associated with Henry II are also still at hand, both heavily embroidered silks. The king of Hungary's coronation mantle, too, survives.

(Fleming, 2007, p. 137)

The World System in the Thirteenth Century

The 'kingpin' in the thirteenth-century world trade system was the land bridge between the eastern Mediterranean ports, the southern outlets to the Indian Ocean and the Central Asian routes to China. As Abu-Lughod describes it:

> the central 'hinge' on which the world system balanced was the link between Italy and the eastern Mediterranean. There were two paths westward: one from the Arab world which controlled the sea passages to the east, the other from the shrunken Byzantine state which controlled access to significant land routes across Asia ... From Italy, especially Venice but also Genoa, the linkage also went to (peripheral) northern Europe which, in the early thirteenth century, was fully dependent upon Italian traders for its trade with the Orient. To the east, the central 'hinge' on which the world system balanced was the Indian Ocean where Arab traders controlled access to India and then to the Hinduized Straits of Malacca which, in turn, acted as an intermediary for China ... Finally, there was at least one but perhaps two subsystems that organized overland traffic between Turkey/Iraq/Iran and northern China, passing through the territories of the various Tartar/Mongol empires, finally reaching Constantinople.
>
> (Abu-Lughod, 1987, p. 11)

Throughout most of the medieval period, the Italian city states of Venice, Genoa, and to a lesser extent Pisa, jockeyed for control of the European and Asian markets (Fig. 14.1). Venice dominated trade in the eastern Mediterranean from the tenth to the twelfth century as an affiliate of the Byzantine Empire. In the 13th century, its power grew even greater when it abandoned its long-time partner during the Fourth Crusade, supported the occupation of Constantinople and took control of most of the Adriatic ports of the Byzantines. To this bounty was added Corfu in 1207 and Crete in 1209. This gave the Venetians a trade network that extended from the Golden Horn of Constantinople to the major ports in Syria and Egypt, and with it access to all the spices flowing into the Middle East from South East Asia and India (Norwich, 1982; Crowley, 2013). Venetian merchants were not the only Italians to 'muscle in on the Mediterranean spice trade. The Genoese and even the Pisans gave them a good run for their money. Still in the end, the fishermen from the boggy lagoon prevailed' (Krondl, 2007, p. 43).

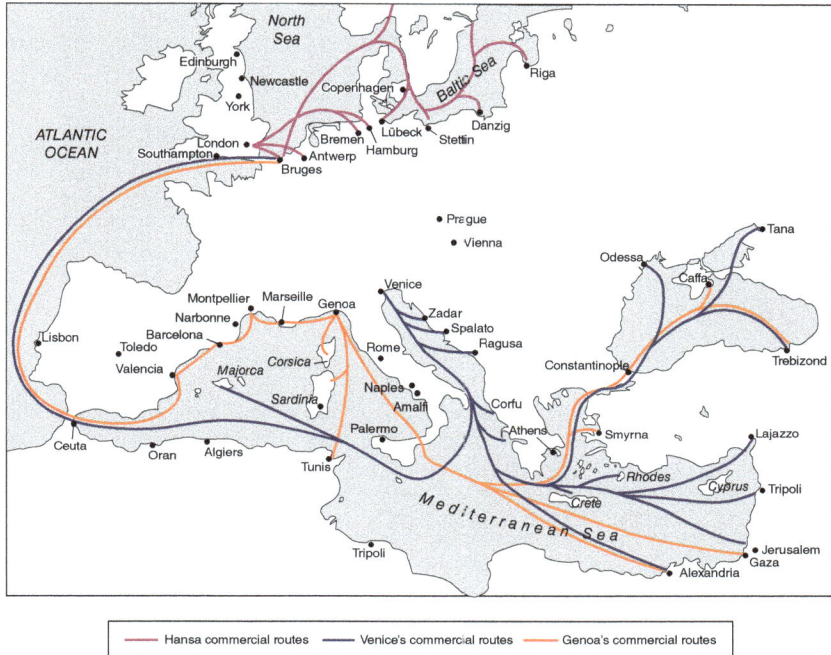

Fig. 14.1. European trade routes in the 13th and 14th centuries. (Redrawn from file made available under the Creative Commons CC0 1.0 Universal Public Domain Dedication. HansWinther. File: Trade routes during Venetian-Genoese wars (1256-1381).jpg)

The Venetian Trading Empire

The location of Venice was perfect to control trade with the spice markets of the Black Sea, Levant and Egypt (Fig. 14.1). As Krondl describes:

> Venice is positioned at the very northwest corner of the Adriatic, the largest gulf in the Mediterranean and just across the Alps from the German-speaking lands. From Venice it is more or less a direct shot down the eastern Adriatic coast, skimming mainland Greece, past the island of Crete, and then straight down to Egypt. This voyage is easily the most direct path between the spice emporia of the orient and the silver mines of the heart of Europe.
>
> (Krondl, 2007, p. 44)

Controlling the Adriatic route became of primary importance to the rulers of Venice from almost its inception. They started by establishing trading communities all along the Dalmatian coast, then making them protectorates and ultimately taking control of them as colonies. The Venetians readily fought with anyone in their way, particularly with the other Italian city states. Their wars with the Pisans over the eastern Mediterranean were particularly notable. Of overriding importance was their spice trade, whether it came from the Black Sea, the Levant or Egypt. Armed galleys were designed and built for this expressed purpose. The Venetians traded in a wide array of 'spices' including black pepper, long pepper, ginger, nutmeg, mace, cloves, cardamom and cinnamon, but two commodities reigned supreme – black pepper and ginger – which made up 50–65% of their bulk purchases (Krondl, 2007). Of these two, pepper was by far the king with 5 pounds of pepper being imported for every 2 pounds of ginger.

The Catalonian Trade Networks

Another trading power emerged in the eastern Mediterranean in the twelfth century: the Catalonian trade network. It encompassed Catalonia, Mallorca and Valencia in north-eastern Spain and was centred in Barcelona. Catalan commerce was able to find a niche in a western Mediterranean dominated by the Genoese, Pisans and Venetians:

> with dedication and effort; [and] they were no doubt able to take advantage of the gaps left in the markets at certain times by the wars among their competitors ... starting in the 13th century they also had the steadfast support of the monarchy both diplomatically and in terms of trade regulations, with protective measures and provisions that created a favorable institutional and legal framework.
>
> (Ferrer, 2012, p. 30)

The savvy Catalonian merchants carefully analysed potential markets and encouraged the production of those agricultural and artisanal products that were most likely to be accepted in foreign markets. They started with wheat, oil, honey, wine, Muslim slaves, weapons (very high-quality swords and knives) and cordovan leathers. Over time they added to their offerings saffron, dried fruit, coral, woollen cloth, ceramics, crafted hides, glue, tallow and glass objects. Their trade networks extended to France, Italy, north-west Africa, the Levant and along the Atlantic coast including Andalusia, Portugal, England and Flanders.

The Hanseatic League

In the twelfth century a great trading league called the Hanseatic League (Hanseatic Union, Hansa or Hanse) also emerged in Europe that linked all the major cities surrounding the Baltic and North seas (Fig. 14.1). It began as a federation of German cities centred in Lübeck and grew to encompass all of Northern Europe. The Hanseatic League was 'an unusual entity' (Liggio, 2007, p. 134), consisting of a series of cities embedded in the middle of the Holy Roman Empire that had purchased charters of self-government from the emperor.

The cities were run by senates composed of powerful merchants and each had its own armed naval fleet. The Hanseatic League did not have its own officials, finances or fleet, but was built on powerful alliances between cities to protect their common interests and monopolize trade. All 'members swore to abide by the Lübeck Law which stipulated that each would protect and defend another in the league, placing their personal armies at each other's disposal' (Mark, 2019). Members traded in copper, fish, flax, furs, grain, honey, iron, resin, salt and textiles.

The league grew steadily in power throughout the thirteenth century and at its peak:

> the Hansa was comprised of almost 200 maritime and interior cities (along rivers). It extended from Bruges and Ghent in Flanders and London in the west to the Republic of Novgorod in western Russia and Tallinn on the Gulf of Finland in the east; from Bergen in the north to middle Germany in the south. [Hansa activities even] extended to Venice ... where German merchants lived and warehoused their goods.
>
> (Liggio, 2007, p. 134)

Internal European Trade and the Champagne Fairs

> Trade and commerce in the medieval world developed to such an extent that even relatively small communities had access to weekly markets and, perhaps a day's travel away, larger but less frequent fairs, where the full range of consumer goods of the period was set out to tempt the shopper and small retailer. Markets and fairs were organised by large estate owners, town councils, and some churches and monasteries, who, granted a license to do so by their sovereign, hoped to gain revenue from stall holder fees and boost the local economy as shoppers used peripheral services. International trade had been present since Roman times but improvements in transportation and banking, as well as the economic development of northern Europe, caused a boom from the 9th century CE.
>
> (Cartwright, 2019)

As the High Middle Ages progressed, many overland trade routes came to link the north-western European cities with the Italian mercantile states. On the North Sea coast, a dense network of trading towns emerged in Flanders; while Italian cities such as Venice, Genoa, Milan and Florence grew wealthy moving goods north-west. Exotic goods were transported laboriously up the Po and Rhône valleys into central and northern France, where they were brought together with those coming south-west from Flanders and the North Sea. The maritime trade routes of the North Seas came to be fully joined with the Mediterranean ones.

International trade fairs became important in the twelfth and thirteenth centuries at the confluence of these trade routes in France, England, Flanders and Germany. The most famous were held in the towns of Champagne, in north-east France. Here Northern European merchants brought furs, woollen cloth, tin, hemp and honey to trade for wine from France and the silks, sugar and spices of the Far East. Goods moving from Italy were carried there by caravans of pack mules, crossing the Alps through the Mont Cenis Pass, along the ancient route called the *Via Francigena*. It took a month to make the trip from Genoa to the French fair cities.

There was a series of fairs staggered across the year in Champagne that lasted six weeks each: Lagny-sur-Marne beginning in January, Bar-sur-Aube held during Lent, Troyes in June and October, and Saint Ayoul in May and September (Cartwright, 2019). All the fairs had a similar schedule. They began with an eight-day *entrée* when merchants set up, followed by a series of days allotted for a cloth fair, a leather fair and then the sale of spices and other things sold by weight. Accounts were settled over the last four days of the fairs. The Counts of Champagne covered the costs of the fairs and policed them. The host towns constructed huge warehouses to hold the trade goods.

The Champagne Fairs remained the focal point of European trade until the end of the thirteenth century, when a direct sea connection was established from the Mediterranean to the North Sea along the Atlantic (Abu-Lughod, 1987). The Flemish city Bruges became the major world market city, located at the crossroads of the northern Hanseatic League and southern trade routes. The first fleet from Genoa arrived in Bruges in 1277 and the first Venetian one in 1314.

The sea route was a much cheaper alternative to the land-based trip over the Alps and it led to a rapid decline in the towns which had hosted the fairs. Merchants from far and wide were welcomed in Bruges, including those dealing in wool from Castile, England and Scotland, grain from Normandy, wine from Gascony and spices from Mediterranean ports. A new connection between the East and West had been forged.

Genghis Khan Reopens the Silk Route

Over the centuries, the amount of movement along the Silk Routes rose and fell depending on the politics of the kingdoms through which it passed. The coming of Genghis Khan and the Golden Horde blew new life into trade along the ancient Silk Routes. When the Tang Dynasty collapsed and lost control of Central Asia, the region had devolved into a disarray of fractured politics. Many thriving oasis towns along the Silk Route, supporting monasteries and grottoes, began to disappear as the glacier-fed streams that had supported them since the end of the Ice Age began running dry or changed course. The spread of Islam from the Middle East also played a key role in the disappearance of the Buddhist communities along the Silk Route, as the followers of Allah defaced and destroyed the Buddhist relics or left them to crumble. Only those previously buried in the sands survived for later discovery by European archaeologists – 'foreign devils' to China.

This all changed dramatically when Genghis Khan and the Mongols established their vast empire in the thirteenth century. He seized northern China between 1211 and 1215, invaded what is now Uzbekistan and Pakistan between 1219 and 1227, then he and his successor son Ogedei moved deep into Eastern Europe, conquering the Rus' and the major city of Kiev by 1240. Only the death of Ogedei outside the city of Vienna stalled the Mongols' advance into Europe.

The Mongols also took control of a goodly portion of the Middle East in the thirteenth century, including the areas of present-day Iran, Iraq, the Caucasus, and parts of Syria, Turkey and Palestine as far as Gaza. The Mongols sacked Baghdad in 1258. Their Middle Eastern conquests were not stopped until 1260, when the Muslim Mamluks finally beat them in the Battle of Ain Jalut in the southern part of Galilee. When it was all said and done, the Mongols controlled the largest empire in world history, spanning from the Pacific Ocean to the Black Sea (Fig. 14.2). As Jack Weatherford tells it:

> In twenty-five years the Mongol army subjugated more lands and peoples than the Romans had conquered in four hundred years. Genghis Khan, together with his sons and grandsons, conquered the most densely populated civilizations of the thirteenth century. Whether measured by the total number of people defeated, the sum of the countries annexed, or by the total area occupied, Genghis Khan conquered twice as much as any other man in history. The hooves of the Mongol warriors' horses splashed in the waters of every river and lake from the Pacific Ocean to the Mediterranean Sea.
>
> (Weatherford, 2004, p. xviii)

No longer was there a series of regional civilizations whose borders impeded trade across the Silk Routes – its entire span was now under the roof of one empire. The Mongols were strong advocates of open trade, allowing products from China and South East Asia to freely make their way to Eastern Europe and the Middle East, and vice versa. The Mongol conquests consisted of savage military action, the subjugation of cities and even the extermination of entire populations, but after the fight they were content to leave their subjects alone to establish their own local economies, apart from taking a modest cut of the proceeds. Governors were appointed who did not overly intervene with the day-to-day life of their subjects and allowed for religious and cultural freedom. To encourage international trade, foreign merchants were often given tax exemptions, loans and guaranteed safe passage along the Silk Route. The Mongols also actively encouraged trade by establishing garrisons along the Silk Route which were patrolled by Mongol soldiers and served as way stations and rest stops for travellers (Beckwith, 2009).

The geopolitical landscape of the world was altered dramatically by the Mongol conquests. From about 1240 to 1360, exotic goods such as silk, porcelain and spices could move easily along the Silk Routes all the way to Europe. Trade activity along the Silk Routes was not only revived by the Mongols but expanded to unprecedented heights. As Kinoshita describes the new world system:

> The Mongol conquests reconfigured Latin Europe's knowledge of Asia with breathtaking speed. When the future saint Francis of Assisi died in 1226 and

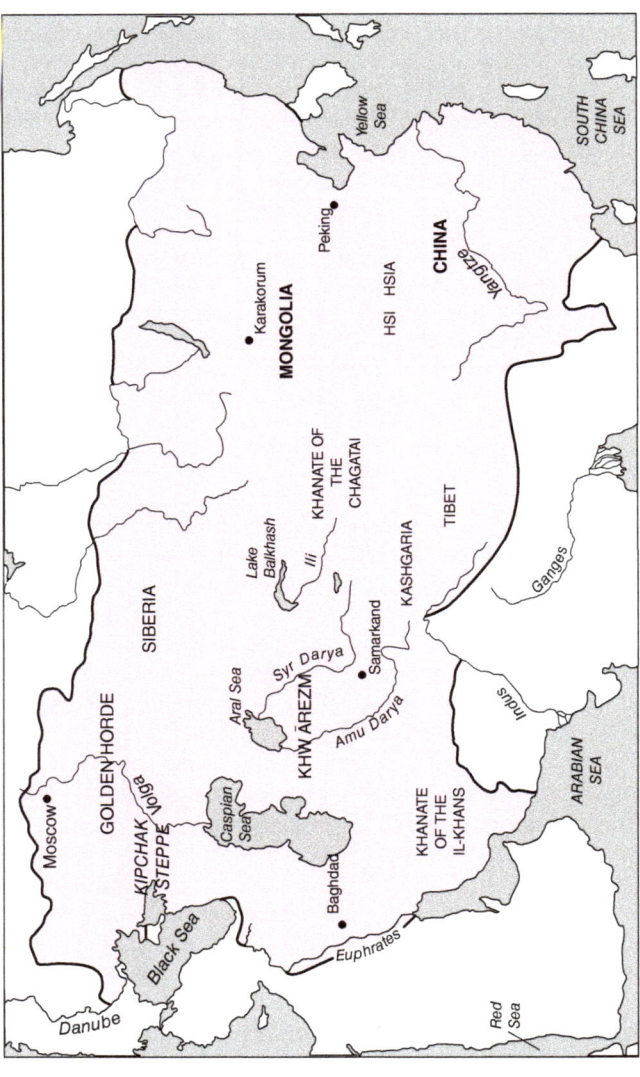

Fig. 14.2. The Mongol Empire at its furthest reaches. (Redrawn from https://www.britannica.com/place/Mongol-empire)

Chinggis Khan [Genghis Khan] in 1227, the two inhabited separate worlds. Within less than two decades, however, these worlds came together: shortly after the Mongols withdrew from central and eastern Europe, Pope Innocent VI dispatched Friar John of Plano Carpini, one of Saint Francis companions and followers on a diplomatic and reconnaissance mission that reached the Mongol court just in time to attend the acclamation (1246) of the new great khan, Chinggis' grandson Güyük.

(Kinoshita, 2016, pp. xx–xxi)

Numerous contacts soon followed between Latin Europe and the Mongol Empire. French King Louis IX sent an ambassador to the Great Khan in the early 1250s. Marco Polo's father Niccoló and his uncle Maffeo embarked on their first journey east in 1260, and Marco began his own epic trip in 1271, described in his famous *Description of the World*. Travel into Central Asia and China became common, and merchants moved freely in both directions. In the early 1400s, a Florentine merchant, Francesco Pegolotti, left us a long account about the ease of travel along the Silk Route. After giving advice on the best route through China and the places along the way, he tells us that 'the journey overland from Tana on the Crimean Peninsula should take about nine months'. And he boasts that the route 'was quite safe by both by day and by night' (Freedman, 2008, p. 170).

End of the Crusader States and Muslim Trade

In April 1291, the Mamluk sultan al-Ashraf Khalil marched on Acre from Cairo determined to 'snuff out the infidel presence within the lands of Islam' (Crowley, 2013, p. 148). The city's population of about 40,000 people came from all over Europe and included merchants from Venice and Pisa, and it was protected by the fabled Templars and Hospitallers. Khalil brought a huge army outside the walls of the city along with a number of giant catapults he had dragged from Cairo.

Under non-stop bombardment the Europeans fought valiantly for five weeks, but to no avail. As Roger Crowley tells the story:

day after day the continuous bombardment relentlessly degraded the defensive ring. Attempts at a negotiated truce were rebuffed. The sultan was implacable. He remembered a massacre of Muslim merchants in the city the previous year and pressed on. On Friday May 18, he ordered a final assault on the stricken town. To the sound of arrows whipping through the air, the crash of rocks, the beating of drums, and the blaring of trumpets, the Mamluk army forced its way into the city and put it to the sword. The final hours of Acre were pitiful and squalid. The Templars and Hospitallers went down almost to a man. Women and children, young and old, rich and poor crowded the quays as the Muslims advanced over the bodies of the indiscriminately slain. At the waterfront, civilization collapsed. Venetian merchants clutching their gold begged for passage, but there were not enough vessels to take them off ... Those unable to pay were left pitifully at water's edge, waiting to be killed or enslaved. When Acre fell, the sultan systematically reduced it to ruins. The remaining Christian strongholds, Tyre, Sidon, Beirut, and Haifa were all stormed or surrendered in quick succession. The Muslims scorched the whole coast against the possibility of a Christian return. They razed the cities to the ground. After two centuries, the crusading footholds in the Holy Land had been swept away.

(Crowley, 2013, p. 149)

Shock waves resounded through Europe after the fall and papal recriminations followed. Venice and Genoa had long been buying silk, spices, flax and cotton in the Islamic world. The wood for the catapults and the soldier slaves that had played a critical role in the destruction of Acre could very well have come from the Italian traders.

Pope Boniface VIII decreed in 1302 a trading ban with the Mamluks in Egypt and Palestine. This meant that the Venetians and Genoese would have to totally bypass the Muslim middleman to obtain the exotic goods from India and South East Asia that Europe so demanded. Now that the Mongols had reopened the Silk Routes, this alternative route proved to be the Black Sea through Trebizond, the Byzantine rump state on the south-eastern corner of the Black Sea, and Tana on the Sea of Azov (Karpov, 2011).

The Alternative Black Sea Route

In the twelfth and thirteenth centuries, almost all the silk and spice obtained by the Italian merchants came from eastern Mediterranean ports and there was little movement of these commodities through the Black Sea. Black Sea trade consisted of mostly grain and slaves and was dominated by the Byzantines. The Black Sea was open to the Venetians after the Fourth Crusade, but they made little effort to penetrate there, being content to solidify their new opportunities in the eastern Mediterranean. Their imprint in the Black Sea region consisted of merchant offices in Soldia (modern Sudak) on the Crimean Peninsula and a few colonies across the northern Black Sea shores.

Genoa and Pisa had chosen not to be major players in the Fourth Crusade, and as a result had not reaped the economic benefits of the Byzantine Empire's fall (Favier, 1998; Madden, 2017). However, Genoa was able to cut deeply into the Venetian trading empire in 1261 when it helped the Byzantine rump state Nicaea recapture Constantinople and re-establish much of the former empire. A grateful Emperor Michael VIII Palaeologus granted the Genoese free trade rights, the use of the most important ports in the Aegean and control of the straits leading to the Black Sea. The control of the Black Sea became of paramount importance when the Mamluk sultan of Egypt conquered the Crusader States and the Pope put sanctions on trade through Alexandria. This shifted most of the spice trade to Trebizond and the other Black Sea ports where Genoa now reigned supreme. They would hold on to this powerful position for almost the next 100 years.

Genoa moved energetically into the Black Sea region and set up its first major colony at Caffa (modern Feodosia), on the Crimean Peninsula, in 1266. It quickly followed Caffa with a series of settlements all along the coasts of the Black Sea and the Sea of Azov, then established a trading station in Trebizond. Caffa soon became a major international centre of commerce:

> connecting Western Europe, Italy, Central Europe, Latin Romania, the Byzantine Empire, the Trebizond Empire, the Muslim Near East, and the entire Eastern Mediterranean with Eastern Europe, Caucasus, steppes of Rumania and the Golden Horde, Middle and Eastern Asia by its traffic routes.
> (Khvalkov, 2015, p. 15)

The city became '"a multicultural colonial urban centre," uniting Latino-Christian, Byzantine-Greek, Slavic and Russian, nomadic Turkic and Tatar, Caucasian, Armenian, Jewish, and Eastern Mediterranean cultures' (Khvalkov, 2015, p. 17).

At its peak, Caffa held a population of nearly 20,000, protected by two formidable concentric walls (Wheelis, 2002). It served as the key transit point for trade in spices and silk coming from the East, as well as grain, fish, caviar, timber, salt, flax, hemp, leather and meat. Caffa also became particularly important as a hub of the worldwide slave trade, exporting slaves from its Black Sea colonies to the metropolis, the rest of Italy, other regions of the western Mediterranean, Constantinople, Asia Minor, the Near East, North Africa and Mamluk Egypt.

Venice Moves into the Black Sea

Venice had been officially barred from the Black Sea in the second half of the thirteenth century but became increasingly aware of its lucrative trade potential. This desire grew particularly hot with the papal decrees to stop trading with the Middle Eastern Muslims, who had been the major supplier of spices and silks to the Venetians.

The Venetians decided to force their way into the Black Sea and Genoa fought mightily to hold on to its favoured position. Numerous pitched battles spilled out all across the eastern Mediterranean and Black Sea. The confrontations 'involved random, chaotic, and opportunistic acts of piracy across all zones of commercial competition' (Crowley, 2013, p. 155). Ships were burned, cargoes pirated, and settlements pillaged. Huge, armed mercantile fleets slugged it out, often without a clear winner.

It wasn't until 1333 that Venice was able to establish its first real toehold in the Black Sea region, when it sent an ambassador to the Muslim overlord Muhammad Uzbeg Khan who allowed the Venetians to establish a colony at Tana on the north-east corner of the Sea of Azov. The Genoese were also present at Tana, but their large galleys had difficulty in the shallow Sea of Azov. The shallow-drafted, lagoon-adapted galleys of the Venetians were able to negotiate the thin Azov waters much easier than those of the Genoese (Crowley, 2013).

Tana, along with the Empire of Trebizond, became Venice's major gateway to the East, connecting the Venetians to the trans-Asiatic 'Mongol' routes carrying the precious oriental commodities: spices, silk, pearls, raw cotton, gold, furs and jewellery. Venetian merchants became wealthy from that trade, much as the Khazars and Byzantines had at Tana and Trebizond centuries earlier. These points of contact with the ancient Silk Routes remained important to the Venetians throughout the fourteenth century.

Periodic Confrontations with the Mongols

While Venetian and Genoese trade generally boomed across the Black Sea in the fourteenth century, it was periodically disrupted by conflicts with the local Muslims. Relations with the Golden Horde were always uneasy at best, and dependent on the whims of the local leadership (Wheelis, 2002). In 1307, Toqta Khan of the Golden Horde besieged Caffa in anger over the Genoese trade in Turkic slaves. The Genoese held out for a year but finally were forced to burn and abandon the city. They returned in 1312 when Toqta died and his successor, Özbeg, welcomed them back.

In 1343, the Mongols under Özbeg's successor Janibeg attacked Tana in retaliation for the Italians brawling with locals. The Italian merchants were forced to flee to Caffa with its better fortifications. Janibeg then besieged Caffa, where the Italians were able to hold out for almost two years. The war ended when Janibeg's forces began to die mysteriously covered with black boils that oozed blood and pus. In desperation, Janibeg tried to force the Italians to give up by catapulting the dead bodies over the walls of Caffa in the first recorded use of biological warfare. The Italians held tight, leaving Janibeg no choice but to give up in 1347, stop the siege and allow the Italians to return to Tana.

What had killed Janibeg's soldiers was the bubonic plague spread by the bacillus *Yersina pestis*. It was the same disease that had decimated Constantinople and the rest of the world during Justinian's reign in the 540s. When the 'Great Pestilence' arrived in the Black Sea region, it had for almost a decade been carving a deadly path across the great trade routes of China, India, Persia, Syria and Egypt (Cartwright, 2018; Snell, 2019).

Black Death Ravages the World

It didn't take long before the population of the whole Black Sea region was decimated by the disease, and from there the disease spread to Europe in 1347. Over the next five years, almost a third of Europe's population – more than 20 million people – would die. The disease likely jumped port to port, carried by sick and dying sailors who had been bitten by pestilence-carrying fleas. It spread from Caffa to the Sicilian port of Messina and from there to Marseille in France and the port of Tunis in North Africa. It then moved to Rome and Florence, and by the middle of 1348, the Black Death had struck Paris, Bordeaux, Lyon and London.

It really wasn't until the late 1300s that Europe recovered from the devastation of the plague. Its effects were exacerbated by an unusually cool period of weather that affected agricultural productivity, resulting in famines in 1358 and 1359. Less severe outbreaks of the plague also surged through Europe in 1362/63, and again in 1369, 1374 and 1390. The end of the fourteenth century found the continent with far fewer people and a social structure that had essentially broken down. Finding enough workers to produce sufficient food was a problem and the cost of manufactured goods skyrocketed. The whole system of serfdom broke down as those who could work were now in a position to ask for wages. There was a communal loss in faith and the authority of the Church was questioned. As the Middle Ages came to a close, the shock waves rippling from the Black Death led to dramatic social change in Europe and a Renaissance. When it was all said and done, Europe burst into an age of exploration and colonization, initiated by Portugal's discovery of the route to India and South Asia.

References

Abu-Lughod, J. (1987) The shape of the world system in the thirteenth century. *Studies in Comparative International Development* 22(4), 3–25.

Beckwith, C. (2009) *Empires of the Silk Road*. Princeton University Press, Princeton, New Jersey.

Butler, S., Hieatt, C. and Hosington, B. (1996) *Pleyn Delit: Medieval Cookery for Modern Cooks*, 2nd edn. University of Toronto Press, Toronto, Canada.

Cartwright, M. (2018) Black Death. *Ancient History Encyclopedia*. Available at: https://www.ancient.eu/Black_Death/ (accessed 28 January 2021).

Cartwright, M. (2019) Trade in medieval Europe. *Ancient History Encyclopedia*. Available at: https://www.ancient.eu/article/1301/ (accessed 28 January 2021).

Chandler, T. and Fox, G. (1974) *3000 Years of Urban Growth*. Academic Press, London and New York.

Crowley, R. (2013) *City of Fortune: How Venice Ruled the Seas*. Random House, New York.

Favier, J. (1998) *Gold and Spices: The Rise of Commerce in the Middle Ages, tr.* C. Higgitt. Holmes and Meijer, New York and London.

Ferrer, M.T. (2012) Catalan commerce in the late Middle Ages. *Catalan Historical Review* 5, 29–65.

Fleming, R. (2007) Acquiring, flaunting and destroying silk in late Anglo-Saxon England. *Early Medieval Europe* 15(2), 127–158.

Freedman, P. (2007) Some basic aspects of Medieval cuisine. *Annales Universitatis Apulensis, Series Historica* 11(I), 44–60.

Freedman, P. (2008) *Out of the East: Spices and the Medieval Imagination*. Yale University Press, New Haven, Connecticut and London.

Jenkins, M. (1976) Medicine and spices, with special reference to medieval monastic accounts. *Journal of Garden History* 4, 47–49.

Karpov, S.P. (2011) Main changes in the Black Sea trade and navigation, 12th–15th centuries. In: *Proceedings of the 22nd International Congress of Byzantine Studies, Sofia, 22–27 August 2011*. Bulgarian Historical Heritage Foundation, Sofia, pp. 417–429.

Khvalkov, E. (2015) The colonies of Genoa in the Black Sea region: evolution and transformation. PhD thesis, European University Institute, Florence, Italy.

Kinoshita, S. (2016) *Marco Polo: The Description of the World*. Hackett Publishing Company, Inc., Indianapolis, Indiana and Cambridge, Massachusetts.

Krondl, M. (2007) *The Taste of Conquest; The Rise and Fall of the Three Great Cities of Spice*. Ballantine Books, New York.

Kuk, N.J. (2014) Medieval European medicine and Asian spices. *Korean Journal of Medical History* 23, 319–342.

Liggio, L.P. (2007) The Hanseatic league and freedom of trade. *Journal of Private Enterprise* 23, 134–141.

Madden, T.F. (2017) *Istanbul: City of Majesty at the Crossroads of the World*. Penguin Books, New York.

Mark, J.J. (2019) Hanseatic League. *Ancient History Encyclopedia*. Available at: https://www.ancient.eu/Hanseatic_League/ (accessed 28 January 2021).

Norwich, J.J. (1982) *A History of Venice*. Knopf, New York.

Riddle, J.M. (1965) The introduction and use of eastern drugs in the early Middle Ages. *Sudhoffs Archiv für Geschichte der Medizin und der Naturwissenschaften* 2, 185–198.

Shaffer, M. (2013) *Pepper: A History of the World's Most Influential Spice*. Thomas Dunne Books, St Martin's Press, New York.

Snell, M. (2018) The arrival and spread of the Black Plague in Europe. *ThoughtCo*. Available at: https://www.thoughtco.com/spread-of-the-black-death-through-europe-4123214 (accessed 28 January 2021).

Toussant-Samat, M. (1994) *History of Food*, tr. A. Bell. Blackwell Publishers Ltd, Oxford.

Wagner, S. (2016) Impact of silk in the Middle Ages. *Textile Society of America*. Available at: https://textilesocietyofamerica.org/6326/the-impact-of-silk-in-the-middle-ages (accessed 28 January 2021).

Weatherford, J. (2004) *Genghis Khan and the Making of the Modern World*. Broadway Books, Penguin Random House LLC, New York.

Wheelis, M. (2002) Biological warfare at the 1346 siege of Caffa. *Emerging Infectious Diseases* 8, 971–976.

Monsoon Islam 15

Setting the Stage – Ottoman Takeover of the Middle East

In the late 1200s, the Seljuk Turk Empire collapsed and control of Anatolia shattered into a patchwork of Turkish principalities called the Anatolian Beyliks. One of the Beyliks was led by Osman I who established the Ottoman Empire and began to take control of Byzantine territory in Anatolia and the other states belonging to the former Seljuk Empire. Osman's son, Orhan, captured the city of Bursa in 1326 from the Byzantines, effectively ending their control over north-western Anatolia. The Ottomans then moved into the Balkans, taking Thessaloniki from the Venetians in 1387 and Kosovo from the Serbs in in 1389. The last large-scale Crusade of the Middle Ages was sent to stop the advance of the Ottoman Turks in 1396, only to be routed at the Battle of Nicopolis.

The Ottoman advance was slowed in the early 1400s by some military defeats and a civil war, but in 1413 Mehmed I emerged as the sultan and restored the empire's power. He led a 'Period of Great Expansion', during which the Ottomans took control of at least ten different European and Middle Eastern states. Another weaker Crusader army was defeated in 1444 at Varna near the Black Sea coast.

In 1453, Mehmed II (1451–1481) finally captured Constantinople after his predecessors had failed in 1394 and 1422. He did it with a force of perhaps 100,000 soldiers and a number of fearsome cannons, the largest 'measuring 9 metres long with a gaping mouth one metre across. Already tested, it could fire a ball weighing 500 kilos over 1.5 km' (Cartwright, 2018). The city fell after many weeks of resistance when a small gate in the Land Walls was accidently left open and the invaders were able to squeeze into the city and fight their way to the main gate where the rest of their comrades were allowed to flood in.

As the fifteenth century ended, the Ottomans had a stranglehold on the spice trade passing through the Persian Gulf and Black Sea. In response, the European merchants began to rely more and more heavily on the Mamluk ports in Egypt and Syria, where they could still obtain spices and other oriental goods through the Red Sea trade routes. In 1517, the Red Sea routes also came under Ottoman control when they defeated the Mamluks, and Egypt and Syria became provinces within their empire. This now gave the Ottomans complete control of all the routes carrying spice and silk in the world system.

The Sea that Fed the Ottoman Spice Routes

By the fifteenth century, there was a vibrant international trade network in the Indian Ocean that medieval Europeans had no idea existed. Its expanse and wealth were well beyond their imagination. As Crowley describes it:

> For thousands of years the Indian Ocean had been the crossroads of the world's trade, shifting goods across a vast space from Canton to Cairo, Burma to Baghdad through a complex interlocking of trading systems, maritime styles, cultures and religions, and a series of hubs: Malacca on the Malay Peninsula, larger than Venice, for goods from China and the further Spice Islands, Calicut on the west coast of India for pepper, Ormuz, gateway to the Persian Gulf and Baghdad, Aden at the entrance to the Red Sea and the routes to Cairo, the nerve center of the Muslim world.
>
> (Crowley, 2012, pp. 64–65)

Three powerful Muslim empires ringed the Indian Ocean (Dale, 2012). The Ottomans in the West occupied the territory once held by the Byzantine Empire and controlled the Red Sea trade route linking South East Asia with Venice. In the centre was the Safavid Dynasty, who controlled the Persian Gulf route. They followed the Shia doctrine in opposition to the Ottomans' Sunni orthodoxy. In the East was the Mughal Empire, covering most of India but still contending with powerful Hindu governments including the kingdom of Calicut and the Vijayanagara Empire in southern India. Ceylon was ruled by Buddhists.

The huge Indian Ocean trading network was also under mostly Muslim control. Muslim traders had spread far and wide from Arabia, settling in mercantile communities across Africa, India, Ceylon, Indonesia and South East Asia. The spread of Islam throughout South East Asia has already been described (Chapter 11, this book). As Boxer relates:

> These various Muslim trading communities grew and flourished; and their richest and most influential traders were sooner or later granted the right to build mosques in the ports where they lived. They then sent for Mullahs, or religious teachers, who in their turn helped to attract other Muslims. In this way the followers of the Prophet spread their creed and their trade to the Swahili coast of East Africa to the spice islands of the Indian Ocean.
>
> (Boxer, 1977, p. 45)

As the Muslim communities grew strong, they became trading empires led by powerful sultans. These included Malacca on the Malaya Peninsula, the islands of Ternate and Tidore in the Moluccas, and a series of rich city states that stretched along the south-eastern coast of Africa.

Ibn Majid and Indian Ocean Navigation

For centuries, Arab sailors ruled Indian Ocean trade. They became expert navigators, well versed on the winds, currents and monsoons of the Indian Ocean. The greatest of all the seafaring Arabs was Ahmad ibn Majid, born in today's Yemen in 1421 (Lunde, 2005a). Known as the 'Lion of the Sea', he rose from humble beginnings to become a legendary master mariner. Nautical historians consider him to be by far the dominant source of knowledge on sixteenth-century Indian Ocean travel and trade.

Ibn Majid left a rich legacy of seagoing literature. His most important work was *Kitab al-Fawa'id fi Usul 'Ilm al-Bahr wa'l-Qawa'id* (*Book of Useful Information on the Principles and Rules of Navigation*), written in 1490, in which he described the history of navigation, key port locations in the Arab trading empire and the intricacies of sailing in the high seas versus along the coast. He also discussed the effects of hurricanes, typhoons and monsoons on long-distance trading. For centuries, it was the primary source of information utilized by Arab sea captains on the ports of the Asian and African continents, the prevailing winds, tidal patterns and celestial navigation.

Ahmad ibn Majid was revered as a great scholar, learning Greek, Tamil, Farsi and many of the East African dialects. Part of his legacy was as a romanticist, and much of his writing was in verse. He was also an inventor, making improvements to several navigational instruments including the astrolabe, used to measure latitude by determining the altitude above the horizon of a celestial body. Ibn Majid died in 1500, just a few years after the Portuguese had exploded into the Indian Ocean. Their navigational success came to depend almost immediately on their finding navigators well versed in the teachings of Ahmad ibn Majid.

Tomé Pires

Much is known about early sixteenth-century trade in the Indian Ocean and South East Asia because of an in-depth report written by Tomé Pires for Manuel I of Portugal between 1512 and 1515. Remarkably, it was left unpublished and lost in the Bibliothèque de la Chambre des députés in Paris until 1944, when it was discovered by historian Armando Cortesão and translated. In this masterpiece entitled *Suma Oriental*, Pires describes his journey from Egypt to the Malay Archipelago and provides a detailed historical and ethnographic account of the Indian Ocean emporium trade.

Pires arrived in Cochin (Kochi), India in 1511 after stopovers in Aden and Ormuz. Little is known of his early life, although he might have been a second-generation apothecary to the royal family, and he was likely commissioned by Manuel I to write the report. His first job in India was as 'factor of drugs' in Cochin but was soon sent to Malacca as a registrar of the Portuguese entrepôt.

From there he visited Java, Ceylon, the Moluccas and ultimately was sent to China where he died in 1524 in a failed diplomatic mission.

Pires left us with vivid descriptions of the major points of trade in the Indian Ocean. It is the first Portuguese description of the Malay Archipelago and the trade network stretching east all the way to Japan. It is a remarkable compilation of the trade activities including historical, geographical, ethnographic, economic and commercial information. Through its pages one can follow the entire complex trading network that crossed the Indian Ocean.

The central role of trade in the Indian Ocean was abundantly clear to Pires As he describes in his preface:

> And in this Suma I shall speak not only of the division of the parts, provinces, kingdoms and regions and their boundaries, but also of the dealings and trade that they have with one another, which trading in merchandise is so necessary that without it the world would not go on. It is this that ennobles cities that brings war and peace.
>
> (Cortesão, 1944, p. 4)

India at the Centre

India was at the centre of Indian Ocean trade for centuries (Fig. 15.1). Among the most important mercantile cities were Hindu-controlled Calicut (today Kozhikode), Cannanore (Kannur), Cochin, Quilon (Kollam) and Muslim Goa along the south-western Malabar Coast, and Muslim-controlled Cambay (Khambhat) of Gujarat in the north-western corner of the Kathiawar Peninsula. By the end of the fifteenth century Gujarati sailors were rivalling the Arabs as dominant traders across the Indian Ocean.

Calicut was by far the most important entrepôt of India and was the world's number one source of pepper. For centuries it was a primary destination of all Indian Ocean traders from Aden, Ormuz, Malacca and China. It also came to be renowned for what the European traders called 'calico' cloth, from which it derived its English name. Calicut was ruled by a powerful Hindu hereditary leader called the *zamorin*, who cooperated closely with Muslim traders to facilitate trade. The other great Indian spice cities of the Malabar Coast (Cannanore, Cochin, Quilon) were unhappy feudal vassalages of the Zamorin of Calicut and paid tribute to him. The Zamorin would prove to be a formidable opponent in Portugal's attempt to block Muslim commerce in the Indian Ocean.

For centuries, the territory of Goa was ruled by the Muslim Shah of Bijapur. The early Portuguese chronicler Duarte Barbosa wrote:

> The city was inhabited by Moors, respectable man and foreigners and rich merchants; there were many great Gentile Merchants and other gentlemen, cultivators, men at arms. It was a place of great Trade. It has a good port which flock

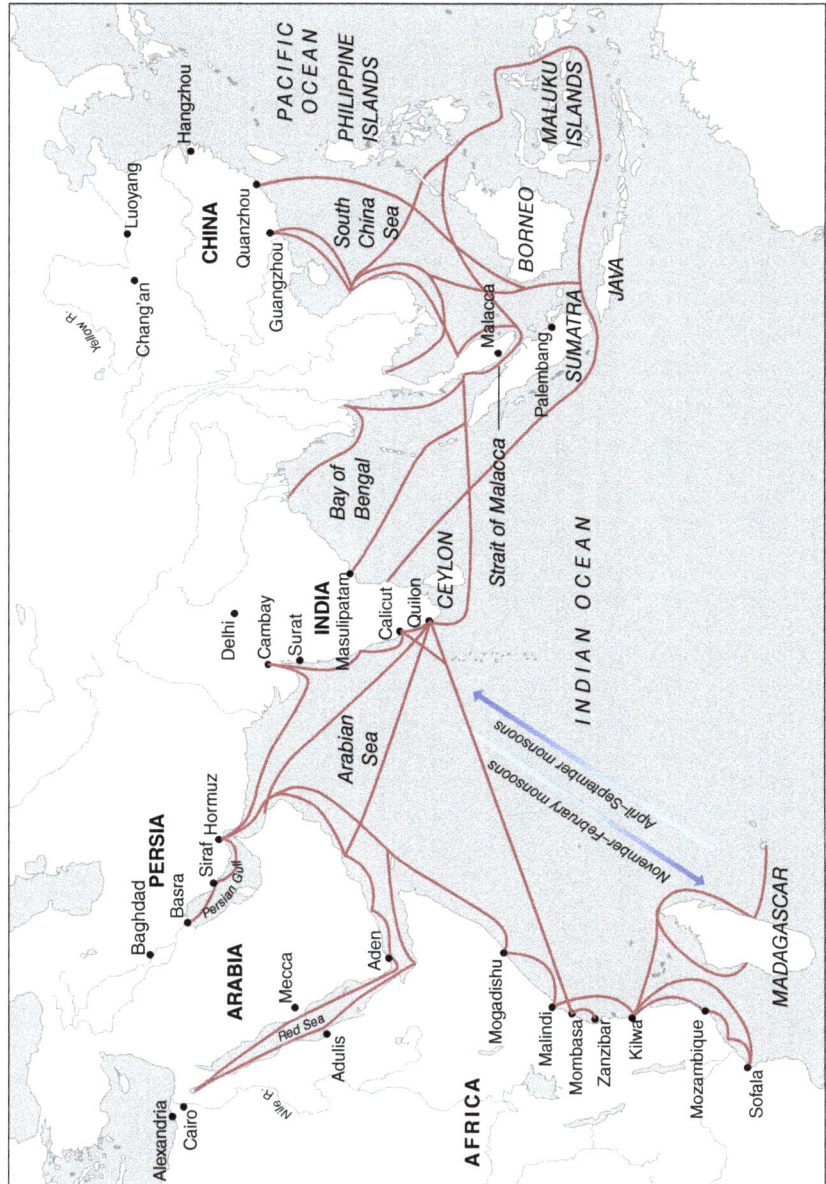

Fig. 15.1. Monsoon Islam before arrival of Portuguese in the Indian Ocean. (Redrawn from Elizabeth_Finn-Indian Ocean Trade. https://i.pinimg.com/originals/33/36/1e/33361e8cf2118b740b8f675aade4f09d.jpg)

many ships from Mekkah, Aden, Hormuz, Cambay Malabar Country. The town was very large with good edifices handsome streets surrounded by walls and towers.
(Fernandes, 1987, p. 284)

Cambay in Gujarat was the home to what came to be the world's most widely travelled sailors. As Tomé Pires observed:

> There is no doubt that these people have the cream of the trade. They are men who understand merchandise; they are so properly steeped in the sound and harmony of it, that the Gujaratees say that any offence connected with merchandise is pardonable. There are Gujaratees settled everywhere. They work some for some and others for others. They are diligent, quick men in trade. They do their accounts with figures like ours and with our very writing. They are men who do not give away anything that belongs to them, nor do they want anything that belongs to anyone else; wherefore they have been esteemed in Cambay up to the present.

> Cambay chiefly stretches out two arms, with her right arm she reaches out towards Aden and with the other towards Malacca, as the most important places to sail to ... They sail many ships to all parts, to Aden, Ormuz, the kingdom of the Deccan, Goa, Bhatkal, all over Malabar, Ceylon, Bengal, Pegu, Siam, Pedir, Pase (Paefe) and Malacca, where they take quantities of merchandise, bringing other kinds back, thus making Cambay rich and important.
> (Cortesão, 1944, p. 42)

Aden: Ottoman Portal to Indian Ocean Trade

The Mamluks' and then the Ottomans' portal to Indian Ocean trade was through Aden, at the opening of the Red Sea. Aden's history as a key Red Sea link with the Indian Ocean went all the way back to antiquity and the Egyptians, Greeks and Romans. Now a well-fortified Arab Muslim enclave, all trade goods from the Indian Ocean found their way there for shipment to Egypt.

In the words of Tomé Pires:

> Aden lies at the foot of a mountain, almost flat on the plain, a little town, but very strong, both in walls, towers and ramparts, as well as in all the paraphernalia of gun towers, loopholes, much ordnance and many warriors – for there are always many people of the country paid to fight, apart from the fact that at any alarm a large number [of the people] from inland rush to help. Inside the city there is a beautiful fortress.

> This town has a great trade with the people of Cairo as well as with those of all India, and the people of India trade with it. There are many important merchants in the city with great riches, and many from other countries live there also. This city is a meeting place for merchants. It is one of the four great trading cities in the world, and it has dealings inside the straits with Jidda, to which it trades most of the spices and drugs in exchange for the said [merchandise]. It trades cloth to Dahlak [islands near todays Eritrea] and receives seed pearls in exchange; it trades coarse cloths and various trifling things to Zeila and Berbera [todays Somalia] in exchange for gold, horses, slaves and ivory; it trades with Socotra, sending cloth,

straw of Mecca, aloes and dragon's-blood; it trades with Ormuz, whence it brings horses; and out of the goods from Cairo it trades gold, foodstuffs, wheat, and rice if there is any, spices, seed pearls, musk, silk and any other drugs.

It trades with Cambay, taking there the merchandise from Cairo and opium, and returning large quantities of cloth, with which it trades in Arabia and the Islands, and seeds, glass beads from Cambay, many carnelians of all colors, and chiefly spices and drugs from Malacca, cloves, nutmeg, mace, sandalwood, cubeb, seed pearls and things of that sort. It takes a great quantity of madder and raisins to Cambay, also to Ormuz [Ormus]; it trades with the kingdom of Goa, and takes there all sorts of merchandise and horses both from [Aden] itself and from Cairo, and receives in return rice, iron, sugar and quantities of gold; it trades with Malabar in India, where the main market was Calicut, whence it took pepper and ginger; and it traded merchandise from Malacca with Bengal in return for many kinds of white cloths, and it traded the merchandise from Malacca also with Pegu [Myanmar] in exchange for benzoin, musk and precious stones, rice also from Bengal, rice from Siam, and merchandise from China which comes through Ayuthia [Thailand].

(Cortesão, 1944, p. 17)

Ormuz: Safavid Portal to Indian Ocean Trade

The Safavid portal to Indian Ocean trade was at Ormuz between the Persian Gulf and the Gulf of Oman. Like Aden was to the Egyptians, it had long served as a vital link between the Persian world and the Indian Ocean. The region's importance was even noted in the *Periplus of the Erythraean Sea*:

At the upper end of these Calaei islands is a range of mountains called Calon, and there follows not far beyond, the mouth of the Persian Gulf, where there is much diving for the pearl-mussel. To the left of the straits are great mountains called Asabon and to the right there rises in full view another round and high mountain called Semiramis; between them the passage across the strait is about six hundred stadia; beyond which that very great and broad sea, the Persian Gulf, reaches far into the interior. At the upper end of this gulf there is a market-town designated by law called Apologus, situated near Charaex Spasini and the River Euphrates.

(Schoff, 1912, p. 36)

As Tomé Pires tells it:

The city of Ormuz is in an island which is almost joined on to Persia, about a league away. This kingdom, besides being rich and noble, is the key to Persia ... It has walls, houses with terraces, towers, and ramparts in it ... Ormuz trades with Aden and Cambay and with the kingdom of Ormuz. Deccan and Goa and with the ports of the kingdom of Narsinga [India] and in Malabar. The chief things the Ormuz merchants take are Arabian and Persian horses, seed pearls, saltpeter, sulphur, silk, alum ... copper, vitriol [sulfuric acid], quantities of salt, white silk, many silver coins and musk, sometimes amber, and a great deal of dried fruit, wheat, barley and foodstuffs of that kind. They bring back pepper, cloves, cinnamon, ginger and all sorts of other spices and drugs, which are greatly in demand in the land of Persia and Arabia ... They also bring back as much rice as they can, white cloth, and iron.

(Cortesão, 1944, p. 19)

Swahili Coast of Africa

A series of rich Muslim-controlled city states stretched along the south-eastern coast of Africa from Sofala (in today's Mozambique) in the south to Mogadishu (in modern Somalia) in the north. In between were Mombasa, Gedi, Pate, Lamu, Malindi, Zanzibar and Kilwa. A strong merchant class had migrated to these cities from all over the Islamic world, in particular Persia, and mixed with the locals to produce the unique Swahili language and culture. The term *Swahili* comes from the Arabic word *sahil* ('coast') and translates to 'people of the coast'. At the peak of their culture from 1100 to 1500, they traded far and wide deep into Africa's interior through the trans-Saharan trade network and across the Indian Ocean to Arabia, Persia, India, South East Asia and China.

The social structure of the Swahili city states was a complex of native African and mixed Arab–African blood. As Cartwright quotes the historian H. Neville Chittick:

> The inhabitants of the towns could be considered as falling into three groups. The ruling class was usually of mixed Arab and African ancestry ... such also were probably the landowners, merchants, most of the religious functionaries and the artisans. Inferior to them in status were the pure-blooded Africans, probably mostly captured in raids on the mainland and in a state of slavery, who cultivated the fields and no doubt carried out other menial tasks. Distinct from both these classes were the transient or recently settled Arabs, and perhaps Persians, still incompletely assimilated into the society.
>
> (Cartwright, 2019a)

The most powerful of the African states were Mombasa and Kilwa, followed by Malindi. They traded ivory from the south, gold and slaves from the western interior and frankincense and myrrh from northern Africa. Kilwa and Mogadishu also produced their own textiles for sale and extracted copper from nearby mines. All of the states produced pottery and iron objects for both local use and trade. The international merchants traded with them mostly cotton, silk and porcelain.

Strait of Malacca and the City of Malacca

As the sixteenth century dawned, the city of Malacca (Melaka) along the Strait of Malacca on the Malay Peninsula had also become a centre of world trade. It was the great international clearing house for pepper, nutmeg and cloves, where East met West.

The history of the region as a trade centre began in about 300 BCE, when small Hindu kingdoms began springing up on Java and Sumatra under the influence of traders from India. These were joined by Buddhist ones in the early centuries CE. In the seventh century, the powerful Buddhist kingdom of Srivijaya arose on Sumatra and took control of most of the Malay Archipelago until it was conquered in 1290 by the Hindu Majapahit Empire from Java. The Majapahit Empire would become so powerful that it refused to pay tribute to China and was able to defeat a force sent by the emperor Kublai Khan to teach them a lesson. While these powerful empires were stretching their muscles, Islamic traders began moving into the region and introduced their faith to Indonesians. As previously described, Islam slowly spread throughout Java and Sumatra.

In about 1402, a Muslim sultanate was established at a fishing village named Malacca by Parameswara, also known as Iskandar Shah. It was located at the narrowest point of the Malacca Strait and was accessible in all seasons. Malacca became the major clearing house for all of the spices produced across Indonesia and ultimately displaced the power of the Majapahit Empire (Hall, 2006). It was the commonest point of contact between East and West, and linked all the major Indian Ocean trading communities.

Malacca's strategic position caught the attention of Ming China, who made it a protectorate and supported it against attacks from Siam, Java and Majapahit. This support helped Malacca become the major trade centre between China, India, the Middle East and Africa. It became the main trade connection between the Indian Ocean and the South China Sea, and almost all East–West trade passed through this narrow strait, creating rich trade kingdoms on its shores. As Tomé Pires tells it:

> No trading port as large as Malacca is known. Nor anywhere they deal in such fine and such highly-praised merchandise. Goods from all over the west are sold here. It is at the end of the monsoons, where you can find what you want, and sometimes more than you are looking for.
>
> (Cortesão, 1944, p. 228)

Sailors from all across the Indian Ocean and South China Sea converged in Malacca to trade pepper, cloves, nutmeg and mace, and it became a major urban centre filled with many residential communities of internationals, among them Indian, Chinese and Javanese. Among the most prevalent were the Gujarat from Cambay. As Tomé Pires further relates:

> There were a thousand Gujarat merchants in Malacca, besides four or five thousand Gujarat seamen, who came and went. Malacca cannot live without Cambay, nor Cambay without Malacca, if they are to be very rich and very prosperous. All the clothes and things from Gujarat have trading value in Malacca and in the kingdoms which trade with Malacca; for the products of Malacca are esteemed not only in this [part of the] world, but in others, where no doubt they are wanted ... If Cambay were cut off from trading with Malacca, it could not live, for it would have no outlet for its merchandise.
>
> (Cortesão, 1944, p. 45)

Sumatra and Java

When the Portuguese arrived at Malacca in the early 1500s, there was a series of warring kingdoms scattered across Sumatra and Java that also had major ports of call for spices. They were not as important as Malacca but were still quite significant.

In Java, the most dominant ones were the Hindu Sunda Kingdom in the west which was by the 1500s in decline and the Demak Sultanate on the north central coast which was ascending. Tomé Pires related 'should de Albuquerque [the latter Portuguese invader] make peace with the Lord of Demak, all of Java will almost be forced to make peace with him ... The Lord of Demak stood for all of Java' (Cortesão, 1944, p. 195). The Sultan of Demak also had control the important ports of Jambi and Palembang in eastern Sumatra.

Demak grew rich importing spices and exporting rice to Malacca and the Molucca Islands. It was strategically located at the end of a channel (now filled) that separated Java and Muria island. From the fifteenth century until the eighteenth, the channel was an important route for ships travelling along the northern Javanese coast to the Moluccas. The channel was linked to the Serang River, which gave access to the rice-producing interior of Java.

On Sumatra was the Muslim harbour kingdom of Samudera Pasai, which in the thirteenth to sixteenth century was located at the northern tip of the island. Northern Sumatra was rich in gold and forest produce and was cultivating pepper by the fifteenth century. It was a common stopover of the Chinese and local merchants of the archipelago who wanted to trade with ships from the Indian Ocean.

The Kingdom of Aceh replaced Samudera Pasai in the sixteenth century as the major regional power. Tomé Pires referred to Aceh as the 'very rich kingdom of Baros' that was also known to people from many nations as 'Panchur' or 'Pansur' (Cortesão, 1944, p. 161). The port city of Baros in west central Sumatra, along with Pariaman, were particularly important in the trade of pepper, gold and other products of the interior. At its peak, Aceh became a formidable enemy of Portuguese-controlled Malacca on the Malayan Peninsula.

Ceylon

An important stopover for merchants on the way to and from Malacca was Buddhist Ceylon (today's Sri Lanka), where the world's finest cinnamon could be obtained along with gems, pearls, ivory, elephants, turtle shells and cloth (Sudharmawathei, 2017). Ships from across the world came to Ceylon for its native products and goods brought from other countries for re-export. The islanders also sent their own ships to foreign ports. The most important items imported were horses from India and Persia, and from China came gold, silver and copper coins, silk and ceramics.

Ceylon held a key strategic position in the Indian Ocean between East and West, being located next to India and along the sea routes that connected the Mediterranean and Middle Eastern worlds with East Asia. There were numerous bays and anchorages dotted along the coast of Ceylon, which provided calm harbours and facilities for ships. At the end of the fifteenth century, the most important port was Colombo, filled with Muslims who had settled down in this country to carry on its trade activities. Three bitter rival kingdoms ruled Ceylon, all under the protection of China through the tribute system.

Tomé Pires describes:

> The beautiful island of Ceylon ... is large; it must be three hundred leagues in circumference, much longer than it is wide. It is very populous; it has many towns and large houses of prayer, with copper pillars, and with roofs covered with lead and copper ... It has all kinds of precious stones, except diamonds, emeralds, turquoises ... It has a great abundance of elephants and ivory; it has cinnamon ... Ceylon trades elephants, cinnamon, ivory and areca with the whole of the Choromandel and Bengal, [and] Pulicat, taking rice, white sandalwood, seed-pearls, cloth and other merchandise in return. Rice, silver, copper, a little quicksilver, rosewater, white sandal-dise of wood and Cambay cloths.
> (Cortesão, 1944, p. 86)

Moluccas

At the far eastern terminus of the Indian Ocean trade network in the East Indian Archipelago were the Moluccas or Spice Islands from whence came cloves, nutmeg and mace. Though far from the main trade routes supplying China, India, Persia, Arabia and Africa, these tiny islands were the only place on earth where these commodities could be obtained.

The earliest mention of the Banda Islands is found in Chinese records dating as far back as 200 BCE. Banda was never settled by Muslim traders and its trade was controlled by a small group of what the Indonesians called *orang kaya* or 'rich men'. Before the arrival of the Europeans the Bandanese had an active and independent role in trade. They carried their cloves in outrigger canoes to Malaysia and the larger islands of Indonesia for trade with Chinese and Indian mariners.

Tomé Pires describes:

> The islands of Banda are six; five produce mace and one has fire [a volcano] ... These [islands] have villages; they have no king; they are ruled by cabilas and by the elders. Those along the sea-coast are Moorish merchants. It is thirty years since they began to be Moors in the Banda islands ... In all there must be between two thousand five hundred and three thousand persons in these islands. The mace is a fruit like peaches or apricots, and when it is ripe it opens and the outer pulp falls, and that in the inside turns red, and this is the mace on the nutmeg, and they gather them and put them to dry. This fruit is ripe all through the year; it is gathered every month. About five hundred bahars of mace must be produced

every year in the islands as a whole, or even six hundred; and there must be six or seven thousand bahars of nutmeg, and that every year, sometimes more, some-times less ... Another island is called Neira. This is a port where the Javanese anchor; it is called Port Neira. It produces mace. And the other three islands, to wit, Pulo Ai (Aee) and Pulo Run (Rud) and Pulo Bomcagy, are three small islands which produce mace. They have no ports in which you can anchor. They bring their mace to the island of Banda. All are in sight of and near to one another.

(Cortesão, 1944, p. 206)

Muslim traders arrived in Ternate and Tidore in the early 1500s and by the end of the century rival sultanates had emerged on the two islands that fought over supremacy in the nutmeg trade with the Chinese and Indonesians. They became bitter rivals who wasted much of the great wealth they amassed from the clove trade by fighting each other. When the European traders arrived on the islands in the sixteenth century, they were able to play Ternate off against Tidore, to get an edge in the clove trade.

Tomé Pires writes:

The Molucca islands which produce cloves are five, to wit, the chief one is called Ternate and another Tidore ... The chief island of all the five is the island of Ternate. The king is a Mohammedan [Muslim]. He is called Sultan Bern Acora-_a?-. They say he is a good man. His island produces at least a hundred and fifty bahars of cloves every year. Two or three ships can anchor in the port of this island; this is a good village. This king has some foreign merchants in his country. They say that the island must contain up to two thousand men, and up to two hundred will be Mohammedans. This king is powerful among his neighbours. His country is abundant in foodstuffs from the land, although many foodstuffs come to the Molucca kings from other islands ... The island must be six leagues round. There is a mountain in the middle of this island, which yields a great deal of sulphur, which burns in great quantities. This king has half the island of Motir (Motei) for his own ... The people of Ternate are knights among those of the Moluccas. They are men who drink wines of their kind. Ternate has good water. It is a healthy country with good air.

The country produces cloves ... Cloves have six crops a year; others say that there are cloves all the year round, but that at six periods in the year there are more. After flowering it turns green and then it turns red; then they gather it, some by hand and some beaten down with a pole, and red as it is they spread it out on mats to dry, and it turns black. They are small trees. Cloves grow like myrtle berries, a great many heads grow together. All this fruit is in the hands of the natives, and it all comes through their hands to the sea-coast.

Although this island of Ternate is the most distant of all from Amboina, and the next in order ought to be the nearest to Amboina, which is Bachian (Pachdo), yet as Ternate is the best, it has been described first, and also because the king is a vassal of the King our lord; and now I will go towards Amboina, describing the islands.

Tidore is an island which is about ten leagues round. The king of this island is a Mohammedan, an enemy of the king of Ternate and his father-in-law. This king has about two thousand men in his country, about two hundred of whom are Mohammedans and the others are heathens. The king is called Raja Almangor.

He has many wives and children. His country produces about one thousand four hundred bahars of cloves a year. There is no port in this island where ships can anchor. He is as powerful a king as the king of Ternate. He is always at war. These two are the most important in the Moluccas, and they compute that this king must have eighty paraos in his country ... His country produces many foodstuffs: rice, meat, fish. They say he is a man of good judgement.

(Cortesão, 1944, pp. 214–215)

Far East Asian Connections

Gujarati sailors did not travel to the Spice Islands and relied on Indonesian merchant sailors to deliver these spices to Malacca, where the Gujaratis then traded for them and dispersed them across the rest of the Indian Ocean. In the fifteenth century, trade between Malacca, northern Borneo and the Philippines was mostly conducted by Brunei merchants from the north coast of Borneo, and the link to the Moluccas was operated primarily by sailors from the southern Philippines.

The Moluccas were reached by two routes: Java and China. As Ptak describes them:

> Ships following the first route would sail from the Malayan peninsula or western Indonesia to the Java and Flores Seas, calling at the ports of northern Java, southern Borneo, Sulawesi and the islands to the east of Madura and Bali, and would then enter the Banda Sea to proceed to the Banda Islands or Ambon ... Other ships would start their voyage in China, travel to Luzon, whence they would go south, to Mindoro or the Calamianes Islands, and enter the Sulu Sea from the north. From the Sulu Sea, if the ship wanted to continue its voyage and sail on to the North Moluccas, it had to pass through the Sulu archipelago, probably keeping close to Basilan, and then, having entered the Celebes Sea, it had to follow the southern Mindanao coast, proceed to Tinaca Point, and thereafter turn south until it reached the Sangihe group.

(Ptak, 1992, p. 27)

Early Chinese Ming and the Indian Ocean

In the early Ming era of the fourteenth century, Chinese junks were not permitted by the emperors to sail directly to South East Asia's ports. Instead, luxury products from South East Asia were delivered to the Chinese elite through a tributary trade system. Foreign regions would send envoys to China to present samples of the goods available in their ports and ask the Ming court for access to the Chinese markets.

As Cartwright explains it:

> The traditional presentation of tribute to Chinese emperors by other, smaller states in Southeast Asia was given to prevent invasion or achieve a theoretical

promise of protection in the case of invasion by a third party or because diplomatic missions giving that tribute were permitted to conduct trade while in China. The tribute, usually far less valuable than the goods which the emperors gave out, had always been a badge of approval to the Chinese, indicative that their emperor was indeed the Son of Heaven and the most powerful ruler on earth. It also confirmed the Chinese vanity that their own culture was superior to all others.

(Cartwright, 2019b)

The tributary missions were generally represented by overseas Chinese, who had moved to South East Asian ports before the Ming era. These Chinese were expected to maintain their Chinese identity and remain loyal to the Ming court. Overseas Chinese communities were located all across South East Asia including Manila in the Philippines, Java's north coast ports, Brunei in Borneo, Ayudhya in Siam, Campa in Vietnam, Malacca and Sumatra's east coast ports (Hall, 2006).

In the fourteenth century, most tribute trade came from states lining the South China Sea, but in 1403 Emperor Yongle ascended the throne and began to promote a very active foreign policy that dispatched large-scale fleets and envoys across the Indian Ocean. A number of particularly massive fleets were led by Zheng He that greatly stimulated the tribute trade. At its peak, all the important ports along the main sailing routes in the Indian Ocean had joined into tribute trade including Malacca, Sumatra, Ceylon, Cochin, Calicut, Ormuz and Aden (Nakajima, 2018).

Voyages of Zheng He

Emperor Yongle would dispatch the great naval commander Zheng He on seven missions to the Indian Ocean to display China's power and gain tribute. These were massive missions composed of hundreds of ships and tens of thousands of soldiers. Zheng He also carried a wide array of luxury goods in 'treasure ships' to impress foreign leaders. These ships were packed with silk, tea, painted scrolls, gold and silver objects, textiles, carved and manufactured goods, and fine Ming porcelain. The treasure ships were huge, said to be some 140 metres (450 feet) long by 57 metres (185 feet) wide, carrying nine masts and as many as 50 cabins (Lunde, 2005b). This compares with European ships at the same time that were often much smaller than 30 metres (100 feet). Zheng He's ships had watertight bulwark compartments and were steered with balanced rudders, technologies far in advance of the Europeans.

The first three voyages in 1405, 1408 and 1409 followed the established trade routes down the coast of Vietnam, with stops at Sumatra and Java, a passage through the Strait of Malacca, and then to Ceylon and India. Wherever he landed, Zheng He would seek out the local ruler, present messages of goodwill from Emperor Yongle and then bestow the potentate with a number of gifts. He would invite that ruler to visit China himself or send an ambassador to the royal court. If they were interested, Zheng He would take them on his ship to return with him at his journey's end.

In general the rulers were passive, but some gave him trouble. One was the king of Ceylon, Alagakkonara, who tried to rob his ships. In response, Zheng He simply abducted the king and took him back to China as a hostage, where he readily agreed to pay an annual tribute for his freedom. Two Sumatran rulers were later handled in a similar manner when they gave him trouble. Zheng He often helped resolve local disputes. In one particularly notable instance, he captured the pirate Chen Zuyi, who had been terrorizing the seas around the Strait of Malacca.

In Zheng He's fourth voyage in 1413 CE, he visited Ceylon and India and then crossed over to the Arabian Sea and sailed to Ormuz and Aden. From there he went up the Red Sea to Jeddah and sent a party to Mecca. In response to these contacts, 19 foreign rulers sent diplomatic missions with tributes to the emperor (Cartwright, 2019b).

In his last three voyages in 1417, 1421 and 1431, he travelled even further and became the first Chinese to visit the coast of East Africa, making stops at Mogadishu, Malindi, Mombasa and Zanzibar. The ruler of Mogadishu sent an embassy to Yongle along with a shipload of exotica including lions, leopards, camels, ostriches, rhinos, zebras and giraffes. The giraffes in particular made a huge splash in China, being considered living evidence of the *qilin*, a mythical Chinese unicorn.

Zheng He died in Calicut in the middle of his last expedition in 1433 CE and his body was returned to China. Before he died, he summarized his own accomplishments in a tablet he erected in 1432 CE in Fujian, China. On it he related:

> We have traversed more than one hundred thousand *li* (27,000 nautical miles) of immense waterspaces and have beheld in the ocean huge waves like mountains rising sky high, and we have set eyes on barbarian regions far away hidden in a blue transparency of light vapors, while our sails, loftily unfurled like clouds day and night, continued their course (as rapidly as) a star, traversing those savage waves as if we were treading a public thoroughfare.
> (Cartwright, 2019b)

Chinese Return to Isolationism

There would be no more great maritime expeditions after Zheng He, as Yongle's successor, Xuande, decided to stop the expensive missions and returned China back to its isolationist policy. He even banned the construction of new ocean-going ships and prohibited voyages outside Chinese coastal waters. He may have been spurred by the cost of rebuilding the Great Wall of China to hold back the Mongols or the expense of ongoing battles with Japanese pirates. Had the Ming maintained their military presence in the Indian Ocean, the Portuguese likely would have had a much more formidable task in the 1500s in their conquest of the Indian Ocean.

The Ming also became concerned about the cost of hosting the tribute trade missions and began to restrict the number of them. Their return on investment was poor and many foreign states had grown powerful enough that they no longer felt that China was at the centre of the world. Korea, Ryukyu (Okinawa) and Vietnam continued active tribute trade, but missions from the Indian Ocean zone were decreased dramatically. The total number of tribute missions dropped from more than 400 between 1403 and 1435, to only about 100 between 1510 and 1539 (Nakajima, 2018). The Chinese overseas residents who had been the core of the tributary system began to develop their own regional bases of trade and became locally affiliated. They took local names, converted to Islam and married local wives.

As the fifteenth century progressed, the enforcement of the maritime exclusion policy gradually slackened, and more and more Chinese smugglers began to make voyages from Fujian and Guangdong to South East Asia. In addition, many unofficial trading vessels from South East Asia began to arrive at Guangzhou Bay and engaged in what was called 'mutual trade' with the Chinese merchants. It became a 'wink, wink, nod, nod' system, where the local authorities of Guangdong overlooked the unofficial nature of these ships and collected customs duties from them (Nakajima, 2018).

References

Boxer, C.R. (1977) *The Portuguese Seaborne Empire: 1415–1825*. Hutchinson & Co. (Publishers) Ltd, London.
Cartwright, M. (2018) 1453: The fall of Constantinople. *Ancient History Encyclopedia*. Available at: https://www.ancient.eu/article/1180/ (accessed 28 January 2021).
Cartwright, M. (2019a) Kilwa. *Ancient History Encyclopedia*. Available at: https://www.ancient.eu/Kilwa/ (accessed 28 January 2021).
Cartwright, M. (2019b) The seven voyages of Zheng He. *Ancient History Encyclopedia*. Available at: https://www.ancient.eu/article/1334/ (accessed 28 January 2021).
Cortesão, A. (tr.) (1944) *The Suma Oriental of Tomé Pires and The Book of Francisco Rodriques*. Hakluyt Society, Second Series No. LXXXIX. Digitized version from McGill University Library. Available at: https://mcgill.on.worldcat.org/search?databaseList=&queryString=The+Suma+Oriental+of+Tom%C3%A9+Pires+and+The+Book+of+Francisco+Rodriques#/oclc/10870283 (accessed 12 April 2021).
Crowley, R. (2012) *City of Fortune: How Venice Ruled the Seas*. Random House, New York.
Dale, S. (2012) Three Muslim empires. *Historically Speaking* 13(2), 31–32.
Fernandes, A.P. (1987) Goa's role in the international trade in 16th and 17th centuries. *Proceedings of the Indian History Congress* 48, 284–293.
Hall, K. (2006) Multi-dimensional networking: fifteenth-century Indian Ocean maritime diaspora in Southeast Asian perspective. *Journal of the Economic and Social History of the Orient* 49(4), 454–481.
Lunde, P. (2005a) The navigator Ahmad ibn Majid. *Aramco World: Arab and Islamic Cultures and Connections* 56(4). Available at: https://archive.aramcoworld.com/issue/200504/the.navigator.ahmad.ibn.majid.htm (accessed 28 January 2021).

Lunde, P. (2005b) The admiral Zheng He. *Aramco World: Arab and Islamic Cultures and Connections* 56(4). Available at: https://archive.aramcoworld.com/issue/200504/the.admiral.zheng.he.htm (accessed 28 January 2021).

Nakajima, G. (2018) The structure and transformation of the Ming tribute trade system. In: Garcia, M.P. and De Sousa, L. (eds) *Global History and New Polycentric Approaches: Europe, Asia and the Americas in a World Network System*. Palgrave Studies in Comparative Global History. Palgrave Macmillan, Singapore, pp. 137–162.

Ptak, R. (1992) The northern trade route to the Spice Islands: South China Sea–Sulu Zone–North Moluccas (14th to early 16th century). *Archipel* 43, 27–56.

Schoff, W.H. (1912) *The Periplus of the Erythraean Sea: Travel and Trade in the Indian Ocean by a Merchant of the First Century*. Longmans, Green, and Co., New York.

Sudharmawathei, J.M. (2017) Foreign trade relations in Sri Lanka in the ancient period: with special reference to the period from 6th century BC to 16th century AD. *Humanities and Social Sciences Review* 7, 191–200.

16 Portuguese Discovery and Conquest

Setting the Stage – Medieval European Knowledge of the Spice Trade

To most medieval Europeans the spices came from some sort of distant paradise, likely the Garden of Eden, which 'had miraculously survived the cataclysmic biblical flood that wiped out everything but Noah and his ark' (Shaffer, 2013, p. 22). Travellers like Marco Polo had instructed the West about many manners of the Asians and Chinese, but 'accounts of the wonders and marvels of the East were either disbelieved by their compatriots or else they were too highly colored and fragmentary to give an accurate idea of Asia to the western world' (Boxer, 1977, p. 16). In the European mind, spices were thought to likely be in great abundance and would be easy to obtain, if only they could find their source. This belief stood at the root of the European Age of Discovery. It enticed explorers like Columbus and da Gama to embark on their great journeys.

As the Europeans saw it, Islam was the great obstacle that had to be removed to freely obtain these riches. Holy Wars were fought for the purity of souls, but always on the minds of the Crusaders was a desire for greater access to the exotic luxuries of Paradise. The Europeans believed firmly that there must be Christian enclaves within the Muslim world that still existed and could aid them in their cause. At the forefront of that list was a great mythical leader named Prester John (Priest John) who ruled a powerful Christian empire first thought to be somewhere in India, then Asia and later Africa. This belief had a long and profound effect on the history of European exploration and discovery.

The myth of Prester John was perpetuated by many sources (Crowley, 2015; Lamb, 2018). The legend may have begun in 1122, when Patriarch John of the Indian Church of St Thomas came to a papal meeting in Rome and bragged of a powerful Indian state of which he was the temporal and spiritual leader. Such a myth was reinforced in 1144 when an emissary from Syria brought Pope Eugenius the news that Edessa had fallen to the Muslims. In his report, he tried to provide hope by mentioning possible help from a John, king and priest, who lived in the distant Orient and had reportedly defeated the Persians and taken their capital.

In 1165, the Byzantine Emperor Manuel Comnenos, the Holy Roman Emperor Frederick Barbarossa and Pope Alexander III all received a letter from Prester John. It was likely devised by an unknown author as propaganda for a third Crusade. In this letter he told of his desire 'to visit the sepulchre of our Lord with a very large army, in accordance with the glory of our majesty, to humble and chastise the enemies of the cross of Christ and to exalt His blessed name' (Lamb, 2018). This letter periodically passed through Europe in various forms for hundreds of years.

A fourteenth-century English knight, Sir John Mandeville, wrote a popular travelogue where he described his travels to Venice, Constantinople, Egypt, the Holy Land and the Far East. In it, he tells of many encounters with Christian kingdoms and gives a lengthy account of the glorious spices of India. Much of his account was fanciful, but his book was widely read, and well outsold the more factual publication relating Marco Polo's travels (Crowley, 2015).

Emergence of Portugal

In the fifteenth century, a new power emerged dramatically in Europe: Portugal (Jacobs, 1973; Boorstin, 1983). It burst on the scene when King João I ('John the Bastard') invaded and took over the Moorish port of Ceuta in 1415. Ceuta was a valuable acquisition as it stood at the head of the trade networks caravanning gold across the Sahara from the Senegal River and served as the western outpost of the Muslim spice trade. Located adjacent to the Strait of Gibraltar, the port town also served as a base for Barbary pirates who periodically raided the Portuguese coast.

On the far edge of Europe facing the Atlantic Ocean, separated from the lucrative trade of the Mediterranean, Portugal lusted mightily for the wealth of the great mercantile states of Venice and Genoa. Its path to that wealth would ultimately come by trailblazing a way to India down the African coast. King João I's son Henrique ('Henry the Navigator') would start that process in motion, sending a series of missions that each jumped a little further along the African coast in search of rivers that led into the heart of Africa. His dream was to directly connect with the lucrative trade routes of Africa and perhaps find a riverine link to the Nile. The prospect of finding a sea route to India was particularly attractive to him as a way to leapfrog over the Muslim dominance in the spice trade. He also hoped to find the fabled kingdom of Prestor John.

And so Henrique's navigators began down the African coast. They first discovered Madeira, the Azores and the Canary Islands. As Henrique's expeditions continued, the goal of each was simply to go further south, towards the ultimate goal to round the tip of Africa and sail to India. In 1455, Alvise Cadamosto reached the Gambia River and explored the Cape Verde Islands. Next, Diego Gomez went further south, exploring what is today the Senegal coast. As the Portuguese caravels moved down the coast of Africa, gold, jewels and spices flowed into Lisbon making it a busy commercial centre.

Henrique died in 1460 before his ultimate goal was reached, but his dream was carried on by his great nephew King João II, who took the throne in 1481. Under João's direction, Diego Cam in 1482 crossed the equator and explored the Congo River. In another mission, Cam reached Cape Cross in today's Namibia, about three-quarters of the way down the continent. In 1487, João sent another sailor, Bartolomeu Dias, down the coast, but this time rather than making incremental gains in distance by hugging down the coastline he decided to swing due west at mid-continent and see where the winds would take him. It is not known why he undertook this strategy, but it proved to be the breakthrough that made future travel possible from Western Europe to the Indian Ocean. After a long scary journey west out into the uncharted Atlantic Ocean he was rewarded by return winds that flung his ship around the Cape of Good Hope to the other side of the continent in stormy cold seas.

The crew of Dias spotted land on 3 February 1488, about 300 miles east of the Cape of Good Hope and they settled in a bay they named São Bras (today Mossel Bay). Here they met and tussled with the local Khoikhoi people but were able to take on fresh water and some provisions. Dias continued up the coast for about another 200 miles until his increasingly anxious crew forced him to turn back and head home. He returned on 12 March 1488, knowing full well that he had discovered the route to the Indian Ocean!

While these explorations were going on, Columbus was making a pitch to King João to fund a mission that would head due west in search of the Spice Islands. Columbus was actually in the court when Dias returned and announced his startling discovery. Based on this report, the king decided to stick with Henrique's vision of the southern route, forcing Columbus to turn to King Ferdinand and Queen Isabella of Spain for backing. Columbus did eventually make his journey under the flag of Spain and for a while had everyone convinced that he had found the Spice Islands, but time would tell that King João had backed the right horse.

Treaty of Tordesillas

To reward King Henrique's crusading efforts against the Moors in Africa, Pope Nicholas V had issued a bull in 1455, *Romanus Pontifex*, that authorized the king to convert all Muslims between his Kingdom of Portugal and India, and gave the Portuguese throne a trading monopoly across that portion of the world. In effect, the Pope gave King Henrique control not only of the discovered but also the undiscovered lands of Africa and the Indian Ocean.

In 1492, soon after Columbus had returned from his first exploration, the Spanish Pope Alexander VI issued a new bull that awarded Spain all the newly discovered territories west of a longitudinal line drawn 100 leagues (~350 miles) west of the Cape Verde Islands. This essentially gave Spain territories across the Atlantic. In 1493, he issued another vaguely worded bull that reversed Pope Nicholas V's *Romanus Pontifex*, granting the Spanish not only the lands across the Atlantic, but also 'all islands and mainlands whatsoever, found and to be found, discovered and to be discovered, that are or may be or may seem to be in the route of navigation or travel towards the west or south, whether they be in western parts, or in the regions of the south and east and of India' (Anonymous, n.d.).

This bull infuriated João II, as it gave Spain claim to all that the Portuguese had already discovered along the western coast of Africa and the territories in the Indian Ocean that they were forging a pathway towards. He went directly to Ferdinand and Isabella who were happy to compromise with him, being quite satisfied with their new western discoveries and respectful of Portugal's growing power. In a small town in central Spain, Tordesillas, a treaty was soon negotiated and signed in June 1494 that divided the world into halves along a line drawn 70 leagues (~240 miles) west of that placed by Pope Nicholas V. All to the west of this line was to be Spain's and all to the east was Portugal's. The power of these two nations was so great that 'they could partition the entire planet like two schoolchildren swapping marbles at recess' (Bernstein, 2008, p. 168).

Vasco da Gama Discovers an Unknown World

With ownership of the world now split into halves by treaty, it was high time for Portugal to finish its quest for the sea route to India. The nobleman Vasco da Gama was chosen by João II to complete this task. João II's goals for him were straightforward, albeit challenging: find the route to India, wage holy war on the Moors and locate Prestor John. Da Gama was given a squadron of four well-armed ships, three years of supplies and a store of cheap goods to trade with what was assumed would be unsophisticated natives. The trinkets they carried 'to delight a West African chief were: brass bells and basins, coral, hats and modest garments' (Crowley, 2015, p. 53).

Da Gama set off on 8 July 1497, and after passing down the western coast of Africa to the Canary and Cape Verde Islands, he followed the route of Dias into the Atlantic Ocean. After 95 days without sight of land, he boomeranged back to the coast of Africa some 125 miles north of the Cape of Good Hope (Fig. 16.1). He had not been as fortunate as Dias to be hurled around the Cape.

When da Gama and his crew reached land, they were in desperate shape with scurvy, their hands and feet grotesquely swollen, and their bloody gums distended over their teeth. Scurvy was to become the scourge of all the future European voyages to India and led to countless deaths. No one escaped the symptoms after a couple of months at sea without fresh fruit and vitamin C.

At St Helena Bay, the healthiest sailors were able to mend sails, collect water and search for fresh meat. They also had their first interactions with the Khoikhoi, which again turned sour when a misunderstanding led to a skirmish and da Gama received a minor spear wound. The Portuguese resolved to never

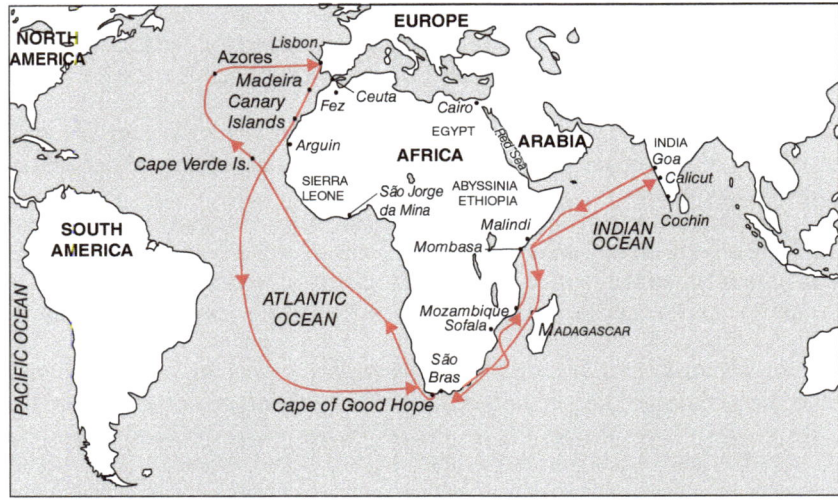

Fig. 16.1. Route of Vasco da Gama. (Redrawn from Walrasiad, CC BY 3.0 <https://creativecommons.org/licenses/by/3.0>, via Wikimedia Commons. File: https://commons.wikimedia.org/wiki/File:Map_of_Portuguese_Carreira_da_India.gif)

again approach land without being heavily armed and ready to 'fight at the slightest provocation' (Crowley, 2015, p. 47).

Da Gama then headed down the coast, travelled through stormy seas around the Cape and on 22 November sailed into São Bras. There he was forced to abandon one of his most damaged ships and had another bad altercation with the Khoikhoi. This one he chose to end with a couple of blasts from his ships' cannons, a power play the Portuguese would deploy frequently henceforth.

Vasco da Gama's first landing on the eastern coast of Africa was on Mozambique, in early March of 1498. At first, he had pleasant interactions with the local natives who were interested in trading for the cheap goods he was carrying, but this atmosphere changed dramatically when he encountered his first settlement of Muslim traders in Mozambique City. The local sultan was insulted by the poor quality of what was being offered and after a series of altercations, da Gama fled the city and continued north.

As da Gama moved up the coast, he was astonished to find a series of rich, sophisticated city states lining the coast of Africa. What he had stumbled upon was the south-western periphery of the prosperous trade network that stretched all the way from Africa to India, down to Malaysia, and through the islands of Indonesia to China. He was moving into a largely Muslim world that was far more powerful than the Portuguese had anticipated; a world outside his and the Portuguese experience that 'was richer, more deeply layered and complex than the Portuguese could initially grasp' (Crowley, 2015, p. 52).

Da Gama also made the startling discovery that Muslim trading vessels were unarmed, a situation totally alien in the Mediterranean Sea. Genoa and Venice had long been waging commercial wars at sea along with the Catalans, Spaniards and Franks. Da Gama found that he was easily able to prey on any Muslim trading vessels he came upon, taking gold, silver, foodstuffs and hostages from the unarmed ships. He felt justified in becoming a pirate as, after all, part of his mission was supposed to be a holy war against the Moors.

De Gama made a brief stop at Mombasa where the locals proved hostile. Then on 14 April he arrived at what proved to be the friendlier port of Malindi, whose sultan was at war with Mombasa and desired Portuguese help. Da Gama was able to acquire some trade goods there and most importantly, was provided with a Gujarati pilot who showed the Portuguese the way to India on the monsoon winds – knowledge that had been lost by the Europeans a thousand years before when the mighty Roman trading empire had collapsed.

While in Malindi, da Gama grew quite excited when four Hindu trading vessels arrived from Calicut, India. He could not fully understand what these sailors were trying to tell him, but they were clearly not Muslims and he became convinced that they practised some sort of aberrant version of Christianity. The Portuguese would later learn that they were Hindu not Christian, but at the time da Gama was convinced that Prester John and a pocket of Christianity must exist in this sea of Muslims.

Vasco da Gama left Malindi for India on 24 April 1498 and arrived at Calicut less than a month later. Here he found a Hindu kingdom, ruled by the king of Calicut, the *zamorin* or *samoothiri*. When Da Gama showed the Zamorin the gifts that he had brought, they were again rejected, causing his highness to seriously doubt whether da Gama represented a powerful king. On his side, da Gama was sure that the Zamorin was under the influence of rival Muslim merchants. After a scary, confusing audience with the Zamorin and shows of force on both sides, da Gama was finally allowed to do some trading for spices with the poor goods that he had. He was not, however, given any trade concessions. The Zamorin insisted that da Gama pay customs duty, preferably in gold, just like any other merchant.

A frustrated da Gama left Calicut and India on 20 September 1498, ignoring local warnings that the monsoons had not yet turned. His ships subsequently became trapped at sea by still winds and when they did finally reach Malindi 132 days later, his sailors were again in terrible shape with scurvy and half had died en route. So many had died that da Gama could no longer man all three of his remaining ships and the leakiest was scuttled. The two ships rounded the Cape of Good Hope with little incident and reached the western coast of Africa on 25 April 1499. Here the two ships got separated and headed to Portugal by different routes. Vasco da Gama stayed behind with his dying brother on Cape Verde for about a month, being the last to arrive in Lisbon in early September.

Upon his arrival home, da Gama was given a hero's welcome and showered with honours for opening up the sea route to India. His success had come at a great cost in human lives, but the small quantities of spices he had brought back signalled great future profits for the crown. He also had discovered the total expanse of the spice trade and its structure in the Indian Ocean. He had found evidence that Prester John existed (albeit incorrect) and that trading vessels in the Indian Ocean were unarmed. The route was set for Portugal to blast its way into dominance of Indian Ocean trade.

The Portuguese Conquest of India Begins

Upon da Gama's return, the Portuguese began a concerted naval strategy 'aimed at capturing – by force or by treaty – the major Indian Ocean ports with the goal of shattering the Muslim monopoly of the carrying trade in the western Indian Ocean' (Lunde, 2005). The Portuguese would use intimidation, hostage-taking and booming cannons to access the mercantile riches of the Indian Ocean world.

Pedro Álvares Cabral was chosen to lead the next expedition to India. His first order of business was to persuade the Zamorin of Calicut to fall in line with all other 'Catholic' kings and wage war on the Muslim enemies of the faith. He was told to take possession of any 'Moor' merchant ships he encountered and bring back as much spice as he could. He was also tasked to deliver a group of Franciscan missionaries to Calicut to help the Hindus get their religion back in line with Catholic orthodoxy.

Cabral was given a fleet of 13 vessels and he sailed out on 9 March 1500. With him was Bartolomeu Dias, the first explorer to sail around the Cape of Good Hope in 1487. Following the now standard western swing into the Atlantic Ocean, Cabral added discovery to his mission when he stumbled upon Brazil and claimed it for Portugal. After lingering for a while to explore a bit and take on provisions, he resumed his journey and headed east. While en route in the southern Atlantic, his fleet was hit by a nasty storm and five ships were lost. However, the remaining seven managed to boomerang around the tip of Africa and rendezvous in the Mozambique Channel. From there they headed north along the coast of Africa until they arrived at Malindi, where they procured fresh produce and water, and then continued on towards India.

After a two-week stay on Anjediva Island south of Goa, where he futilely waited to ambush Arab shipping, Cabral reached Calicut on 13 September. He arrived carrying much more luxurious gifts for the Zamorin and proper letters of introduction from King Manuel. This time a commercial treaty was successfully negotiated, and he was allowed to set up a factory to process spices.

All was going well in Calicut, until Cabral decided to seize his first Arab merchant ship loaded with spices. In response, the Arabs rioted and killed over 50 Portuguese soldiers and most of the Franciscan friars. Cabral reacted by seizing another ten Muslim ships, pirating their cargoes, killing their crews and setting the ships ablaze. He topped it all off with a full day of shore bombardment of Calicut and another nearby port. Thus began a full war between the

Kingdom of Portugal and the Zamorin of Calicut; a war that would last for over ten years and totally disrupt the political order of the Malabar Coast of India.

On 24 December, Cabral left smouldering Calicut and sailed south along the coast towards Cochin, a small Hindu city state at the outlet of the Vembanad lagoon. Cabral found a much more receptive environment here than in Calicut, was able to trade for enough pepper and other goods to fill the holds of his ships and was allowed to establish a small trading post. He also came to seriously doubt that the Hindus were Christian, after coming in contact with communities of Malabari Jews and Syrian Christians who knew better.

Cabral was forced to leave Cochin hurriedly on 16 January 1501, when he learned that the Zamorin had sent a flotilla of some 80 warships to engage him. These he managed to elude and after a final friendly stop in Cannanore arrived back in Lisbon five months later on 23 June 1501 as another full-blown hero. He had not been able to control the Zamorin, but he had brought back profitable quantities of spices and even more importantly, had taken the first steps to disrupt the long-established patterns of trade between the Arabs, Mamluks and Venetians.

The Pace Quickens

Over the next five years Manuel would:
> dispatch a volley of fleets of increasing size, eighty-one ships in all, to ensure success in a life-and-death struggle for a permanent position in the Indian Ocean. It was a supreme national effort that called on all the available resources of manpower, ship building, material provision and strategic vision to exploit a window of opportunity before Spain could react. In the process the Portuguese took both Europe and the peoples of the Indies by complete surprise.
> (Crowley, 2015, p. 101)

Da Gama was chosen in 1502 to lead the next India armada. He was given command of ten ships, which were supported by an additional two flotillas of five ships each commanded by his uncle and nephew. His explicit orders were to take control of Indian trade by forcing the East African sultans and the Zamorin of Calicut to bow down to Portuguese might. This he set out to accomplish in what became a gruesome rampage of piracy and massacre.

After his push around the Cape of Good Hope, da Gama began his onslaught by terrorizing the Muslim ports up and down the African east coast, obtaining the total submission of the Sultan of Kilwa after threatening to burn down his city. Next, he ambushed an Arab ship filled with pilgrims, seized its cargo and then set it afire, burning to death hundreds of passengers including women and children. He then headed to Calicut to force a trade agreement with the Zamorin, who initially proved open to negotiation but baulked when da Gama asked for the expulsion of all Muslims in the city. In an attempt to intimidate the Zamorin, da Gama took one of his high priests hostage and had him tortured by cutting off his lips and ears and sewing a pair of dog's ears to his head. He then bombarded unfortified Calicut for two days, grabbed more captives from ships in the harbour, cut off their noses and ears and had them sent to the Zamorin.

The enraged Zamorin responded by engaging a huge fleet to ambush da Gama who was forewarned and decided to flee like Cabral. He loaded up his ships with spices at the friendly ports of Cochin and Cannanore and left India in a fury, attacking and looting several Muslim ships along the Malabar Coast. Da Gama arrived in Lisbon in September 1503 to a mostly cool reception from Manuel I. He had brought back much spice and had temporarily disrupted Indian shipping but had failed again to bring the Zamorin into submission, even with his extreme barbarity.

The First Viceroy of Portuguese India

In 1505, King Manuel I decided that it was time to establish a permanent presence in the Indian Ocean. The man given that job was Francisco de Almeida, who was appointed the first viceroy of Portuguese India. Almeida was charged with building forts and trading posts all along the East African and Indian coasts, and to bottle up Muslim trade with Calicut. He was given a force of 22 vessels, 1000 sailors and 1500 soldiers to do the job.

After rounding the Cape of Good Hope, Almeida began his mission by ravaging East Africa, storming and sacking Kilwa, torching Mombasa and Mozambique, and in essence taking full control of a 1000-mile stretch of the African east coast. He then set off for India, leaving a garrison of 550 men behind in Kilwa. On 13 September, Almeida stopped at Anjadip (Anjediva) Island, where he built his first fort, and then travelled to Cannanore, where he built another fort and left a force of 150 men and two ships.

Almeida next sailed to Cochin to strengthen the existing Portuguese fortifications. Upon arriving, he was told to his horror that the factors left behind by da Gama had all been killed. He retaliated by sending his son Lourenço with six ships to Quilon harbour where he destroyed 27 ships. Upon leaving, Lourenço encountered a huge flotilla sent by the Zamorin to engage him and roundly defeated them in the battle, 'or rather slaughter, in which between 3,000 and 4,000 of the enemy were killed or drowned, most of the vessels (from 200 to 300) being sunk, and only a few of the larger ones captured' (Ferguson, 1907, p. 360).

Lourenço then decided to go to the Maldives to pursue and destroy more Arab ships. The Moors had begun altering their normal spice route from Malacca and Sumatra to Cambay through the Maldives to escape capture. Lourenço got lost on the way and instead wound up in Ceylon, where he made the most of the situation and took on a load of pepper. Surprisingly he did not take on much cinnamon which had become the primary export of Ceylon, perhaps because 'the bales of cinnamon had to be handled and stowed carefully, pepper was one of the easiest cargoes to load, being simply poured into every available space of the ship, the spaces being then closed' (Ferguson, 1907, p. 346).

Socotra and Ormuz

As Almeida was busily establishing his presence in India, King Manuel dispatched another large fleet from Portugal in his all-out attempt to monopolize Indian Ocean trade. Placed under command of Tristão da Cunha, this fleet was to establish a base on the island of Socotra near the entrance of the Red Sea, then sail to Ormuz to take control of the mouth of the Persian Gulf. This attack was to be led by da Cunha's cousin Afonso de Albuquerque, who carried with him a secret letter from Manuel making him viceroy in three years' time. The overall goal of the Portuguese was to pinch the Muslims at three pressure points: in India itself where the spices were obtained; and in the two shipping lanes north to Europe, the Ottoman-controlled Persian Gulf and the Mamluk's Red Sea.

The fleet landed at Suq on Socotra in April 1507 and captured the port after a stiff but brief battle. Tristão da Cunha then headed to India, leaving Albuquerque on the island with about 500 starving men and six decaying ships. Socotra turned out to be much harsher and remote than the Portuguese had anticipated, so Albuquerque sailed directly to Ormuz to wage war and hopefully acquire supplies.

Albuquerque's force, even in its weakened condition, was able to take the city and surroundings with relative ease. At one point during the last battle, the Portuguese were surrounded by 50 armed merchant ships and as many 200 light oared craft, but the density of ships actually worked in Albuquerque's favour as his artillery fire had greater effect. Persian bowman on the light oared ships gave him some problems, but the Portuguese were able to capture or burn most ships, including a great carrack from Gujarat. Fearing a bloody assault on land after witnessing the sea battles, the Vizier of Ormuz surrendered by raising a white flag over the royal palace. To sue for peace, he agreed to pay a hefty tribute, allow the Portuguese to trade without customs duties and to build a fortress on the island. Ormuz was now a vassal state of Portugal.

The other Portuguese invasion of Socotra proved to be a total failure. The building of a fortress at Suq was begun but the stark infertility of the land led to famine and sickness in the garrison. The harbour proved poor for wintering and many moored Portuguese ships were destroyed. Socotra also proved too distant from the Red Sea to effectively interfere with the Gujarat spice trade. The Portuguese abandoned the island after a four-year occupation, and it passed into Muslim hands.

The Portuguese–Mamluk Naval War

As more and more Arab ships were being destroyed by the Portuguese in Indian harbours, desperation grew among the Mamluks and Venetians on how to save their lucrative trade network. As Paul Lunde relates:

When Vasco da Gama set out from Lisbon in 1498, the Mamluks of Egypt and Syria dominated the Middle East. They controlled the key Mediterranean ports of Alexandria, Beirut and Tripoli as well as all of the Red Sea and the overland routes to Makkah. The spices of India and the East Indies reached Venice through Mamluk territory, and taxes on the trade provided a significant part of Mamluk revenues. For more than 200 years the Venetians and the Mamluks had maintained diplomatic and commercial relations, and their prosperity was interdependent. The Portuguese intrusion into the Indian Ocean was a direct blow to the commercial interests of Venice, and the Venetians did everything they could to encourage the Mamluks to repel the Portuguese.

(Lunde, 2005)

The Venetians began pushing the Mamluks to rise up and fight the Portuguese intrusions, which would protect their own trade ambitions. They pleaded with the Mamluks to try and stop trade between the Indian city states and the Portuguese and to support the fight of the Zamorin of Calicut, who had also sent an ambassador to Egypt asking for help. Finally in 1505, after final diplomatic attempts with the Pope had failed, the Mamluks decided they must take action and Sultan Qansuh al-Ghuri ordered the building of a fleet to fight the Portuguese. This endeavour became a joint effort combining the knowledge of Venetian skilled shipwrights and timber and weapons brought overland on camel by the Mamluks from the Ottomans' realm. The fleet was completed in November and under the leadership of Amir Husain Al-Kurdi, left Suez and travelled to Jidda (Jeddah), where they built a fortress. They then headed to Aden, where they intended to slip off to India and engage the Portuguese.

The fleet arrived in Diu in 1507, where it joined up with local Indian forces and surprised the much smaller fleet of Lourenço in the Battle of Chaul, sinking six of eight ships and killing the viceroy's son. Fearing retribution from Almeida, the governor of the city and naval officer Malik Ayyaz (called Meliqueaz by the Portuguese) wrote to the viceroy trying to appease him by saying his son had fought bravely, but also warning him that Diu was well defended and he had hostages. Furious, Almeida wrote back:

> I the Viceroy say to you, honored Meliqueaz captain of Diu, that I go with my men to this city of yours, taking the people who were welcomed there, that in Chaul fought my people, and killed a man who was called my son, and I come with hope in God of Heaven to take revenge on them and on those who assist them, and if I don't find them I will take your city, to pay for everything, and you, for the help you have done at Chaul, this I tell you, so that you are well aware that I go, as I am now on this island of Bombay, as it will tell you the one who this letter brings.
>
> (Crowley, 2015, p. 227)

The viceroy rallied his forces and on 2 February 1509 crushed the Egyptian–Indian coalition and had most of the Egyptian captives hanged, burned alive or dismembered. Meliqueaz was forced to surrender, pay a hefty tribute and to return his hostages in good health. He also felt compelled to offer the Portuguese the opportunity to build a fortress in Diu which the viceroy refused, not wanting to overstretch his forces. The remnants of the Egyptian fleet returned to Egypt, and the Mamluks never seriously challenged the Portuguese again. It would not be until 1538 that another Muslim fleet would be sent against the Portuguese, this time by the Ottoman Turks.

Almeida's successors came to regret his refusal of a fortress at Diu but were not themselves able to build one for decades. Meliqueaz was able to resist both peaceful negotiations and conquests. It wasn't until 1535 when Bahadur Shah, the sultan of Gujarat, was threatened by the Mughal emperor Humayun and needed Portuguese support that the Portuguese were allowed to station a garrison in the city. Bahadur Shah soon changed his mind and tried to throw the Portuguese back out, but they now had a firm toehold and refused to go. They built a fortress that was strong enough to survive a siege in 1538 by the combined forces of Coja Sofar, the lord of Cambay, and the Ottoman Suleiman Pasha, and another assault by Sofar alone in 1546. The Portuguese were able to control Diu until 1961.

Albuquerque Takes Over

As the first viceroy of India, Almeida had successfully established Portuguese sea power in the Indian Ocean, but a secure naval base had not been established to protect Portugal's precarious position. Portugal still faced the determined opposition of the Zamorin and Muslim merchants of Calicut. A central naval base was needed where the Portuguese could maintain a permanent fleet to control all the Indian Ocean sea lanes. This job fell to Alfonso de Albuquerque, who replaced Almeida as viceroy in 1509 after a brief power struggle. He decided that the place to build this base was Goa on the Malabar Coast. It had a good harbour, an active shipbuilding industry and was an active trade centre.

But before Albuquerque could get started on this grand plan, another fleet arrived in the Indian Ocean under Marshal Fernão Coutinho, with specific instructions to overthrow the Zamorin of Calicut. Albuquerque reluctantly joined in this endeavour with disastrous consequences. In a direct assault on the city, the combined forces of Coutinho and Albuquerque took over and looted the Zamorin's palace, but the locals rallied in bloody street fighting, killing Coutinho and seriously wounding Albuquerque. Only a small group of surviving Portuguese were able to flee the city, with the Zamorin still in power.

Miraculously recovering from his wounds, Albuquerque returned to his original plan a few months later and joined forces with the local Hindus to take Goa in 1510 from the Sultanate of Bijapur. He was supported by Timoji, a Hindu pirate who commanded a force of 2000 men. They were able to take the city in a short, sharp fight with the support of the local Hindu community. Goa would remain the capital of the Portuguese seaborne empire, the Estado da Índia, until 1961 (Lunde, 2005).

Albuquerque next set out to take control of Aden and Malacca in addition to Ormuz. Malacca was the gateway to the Bay of Bengal, the Spice Islands and

China. Aden was closer to than Socotra to the Bab-el-Mandeb Strait entering the Red Sea and the lucrative trade routes into Egypt, North Africa and the Mediterranean.

Albuquerque decided he would start with Malacca and in 1511, headed there with a fleet of 16 ships and an army of 1000 Portuguese and Malabari soldiers. Manuel had earlier sent a small fleet to take it in 1509 but they had failed and been forced to flee. In a battle that raged for 40 days, Albuquerque finally overwhelmed a force of 4000 and expelled all the Gujarati, Bengali and Muslim Tamil merchants from the city. He also constructed a fortress on the former site of the sultan's palace to defend against future attacks.

Albuquerque returned to Goa in September of 1512, just in time to relieve the city from a siege and to suppress an internal revolt. He then departed to Aden, which proved to be much more difficult to defeat than Malacca, having high, strong walls and much more fervent defenders who had nowhere else to go.

Albuquerque sailed from Goa on 18 February 1513 to try and capture Aden and control the mouth of the Red Sea. His fleet consisted of 20 ships carrying 1700 Portuguese and 800 allied Malabari soldiers. From Socotra Island at dawn on Easter Sunday (26 March), the Portuguese mounted their attack on the fortress at Aden without warning. They were able to capture an outwork and a considerable amount of ordinance but were later repulsed with many casualties. They attempted to use scaling ladders to enter the fortress, but most collapsed under the weight of the soldiers, leaving many of them trapped in isolation atop the wall. Albuquerque was forced to withdraw to his ships and retreat to Kamaran Island north of the Gulf of Aden. There they harassed Red Sea shipping for a while until so many of his men became ill that he was forced withdraw and return to Goa.

In February 1515 Albuquerque sailed on his last expedition to Ormuz to re-establish control of that city. Violent court intrigue had left a leader there who had stopped paying the Portuguese tribute. Albuquerque assembled a fleet of 27 vessels and over 2000 soldiers and arrived in Ormuz, with trumpets blaring and artillery booming. The stunned Shah decided it was not worth a fight and agreed to return to vassalage. He was stripped of his army, required to pay the annual tribute again and had to allow the Portuguese to greatly strengthen their fortress. Albuquerque headed home that September a victor again but died onboard within sight of Goa.

The Portuguese in the Bandas and Moluccas

Once established in Malacca, Albuquerque learned of the location of the Spice Islands, which had long been held secret by the Muslim traders (Pinto, 2016). Malay guides were hired, and three ships were sent on an exploratory mission under the command of António de Abreu and his lieutenant Francisco Serrão. They arrived in the Bandas in 1512 and were successful in filling the hulls of their ships with nutmeg, mace and cloves. Abreu successfully returned to Malacca with his booty, but the overstuffed ship commanded by Serrão floundered and broke up on a reef.

Serrão and his shipwrecked crew were brought to the nearby island of Ternate by the local sultan, Bayan Sirrullah. The sultan of Ternate hoped to use Serrão to ally himself with a powerful foreign nation and tip the balance of power in the area. At this time, Ternate was in fierce competition with the nearby island of Tidore. Serrão came to be Sirrullah's personal adviser, led a band of mercenaries and remained on the island until his death.

The sultan of Ternate allowed the Portuguese to build a fort in 1522 and establish a trading colony, but the relationship between the Ternatans and Portuguese quickly grew strained (Bernstein, 2008). Over the next half-century a series of Portuguese governors were sent to this distant outpost who grew increasingly greedy and brutal. Perhaps the worst was Jorge de Menezes, who is credited with the European discovery of New Guinea and was governor of the Moluccas from 1527 until 1530. Among his mean deeds were the plundering of a Spanish fort in Tidore and the poisoning of the sultan of Ternate. He also allowed his soldiers to pillage Ternate when a Portuguese supply ship was late. When the natives had the audacity to object and killed a few of his soldiers, he had their leader taken hostage, cut his hands off and set his dogs upon him. The man somehow escaped into the water and used his teeth to drown the dogs one by one before he drowned himself. De Menezes' atrocities were so great, even by Portuguese standards, that he was arrested and banished to Brazil.

The sultans of Ternate continued to serve at the whims of the Portuguese for decades, but by mid-century began to assert themselves. One major figure was Sultan Hairun who reigned from 1535 to 1570. He started as a Portuguese puppet and at one point even converted to Christianity, however he became increasingly agitated by the brutality of the Portuguese and eventually aligned himself with the Muslims of Ternate. He was assassinated in 1570 and was replaced by his son Babullah who led an uprising supported by the Sultanate of Tidore and Muslims from as far away as Aceh and Turkey. He and his followers laid siege to the Portuguese fort and after four years finally overran it in 1574 and rid themselves of the Portuguese. From then on Ternate was a strong, fiercely Islamic and anti-Portuguese state.

Focusing on Ternate and Malacca, the Portuguese never put much effort into controlling the nutmeg and mace trade of the Banda Islands. It was not until 1529 that a Portuguese trader Captain Garcia Henriques tried to build a fort on the main island of Neira and was repelled by the local Bandanese (Milton, 1995). From that point on, the Portuguese left the Bandanese pretty much alone and chose to purchase their nutmeg in Malacca.

Portuguese Move into Ceylon

After Almeida had blown into Ceylon in 1505, the Portuguese maintained regular contact with the king of Kotte who controlled much of the south-western quarter of Ceylon and the coastal city of Colombo. In 1518, they were allowed to build a fort in Colombo and given favourable trade concessions.

In 1521, the three sons of King Vijayabahu VI of Kotte had him assassinated, portioned the kingdom and then commenced fighting each other. One, Bhuvanaikabahu, took control of the north-western half of Kotte; another, Pararajasinghe, became the ruler of Raigama in the southern quarter of the old kingdom; and the third, Mayadunne, became king of Sitawaka in the east (Pieris and Naish, 1999). Bhuvanaikabahu sought help from the Portuguese to maintain and build his kingdom, while Mayadunne allied himself with the Indian Zamorin of Calicut. Mayadunne became a fierce opponent of the Portuguese and dedicated his life to overturn Bhuvanaikabahu, to preserve the independence of Ceylon. Pararajasinghe remained independent.

Bhuvanaikabahu became more and more dependent on the Portuguese for his defence and in 1556 his heir – Dharmapala – was converted to Christianity by the Franciscan order of the Roman Catholic Church. When his conversion to Christianity was announced there was a public outcry and he became even more easily controlled by the Portuguese. In 1580 the Portuguese convinced Dharmapala to deed his kingdom to them and after he died, they took formal possession of it. When Mayadunne died, his son Rajasinha continued to fight successfully on land, but was held at bay by Portuguese sea power. His kingdom was taken over by the Portuguese in 1593 when he died without a clear successor.

In the late sixteenth century, the Portuguese turned their eyes towards the Jaffna Kingdom which ruled the northern part of Ceylon. Mostly Hindu, Jaffna had fought mightily against Catholic conversion and through most of the sixteenth century, the influence of the Portuguese was minimal until they invaded in 1591 and installed a puppet. Great unrest and royal dispute continued unremittingly until another invasion by the Portuguese in 1619 allowed them to fully annex the kingdom. The king of Jaffna and every surviving member of his royal family was shipped to Goa, where the king was hanged, others were beheaded and the remainder forced to become monks or nuns. As the seventeenth century progressed, the Portuguese expanded their control to the lower reaches of the central highlands and the east coast ports, Trincomalee and Batticaloa. When it was all said and done, the Portuguese had a monopoly in cinnamon and elephants that generated good profits, along with pepper and betel nuts.

References

Anonymous (n.d.) The Papal Bull *Dudum siquidem* of September 26, 1493. *Reformation.org*. Available at: http://www.reformation.org/dudum-siquidem.html (accessed 12 April 2021).

Bernstein, W.J. (2008) *A Splendid Exchange: How Trade Shaped the World*. Grove Press, New York.

Boorstin, D.J. (1983) *The Discoverers*. Random House, New York.

Boxer, C.R. (1977) *The Portuguese Seaborne Empire: 1415–1825*. Hutchinson & Co. (Publishers) Ltd, London.

Crowley, R. (2015) *Conquerors: How Portugal Forged the First Global Empire*. Random House, New York.
Ferguson, D. (1907) The discovery of Ceylon by the Portuguese in 1506. *The Journal of the Ceylon Branch of the Royal Asiatic Society of Great Britain & Ireland* 19(59), 284–385.
Jacobs, W.J. (1973) *Prince Henry the Navigator*. Franklin Watts, New York.
Lamb, A. (2018) The search for Prester John. *History Today*. Available at: https://www.historytoday.com/miscellanies/search-prester-john (accessed 28 January 2021).
Lunde, P. (2005) The coming of the Portuguese. *Aramco World: Arab and Islamic Cultures and Connections* 56(4). Available at: https://archive.aramcoworld.com/issue/200504/the.coming.of.the.portuguese.htm (accessed 28 January 2021).
Milton, G. (1999) *Nathaniel's Nutmeg: The True and Incredible Adventures of the Spice Trader Who Changed the Course of History*. Penguin Books, New York.
Pieris, P.E. and Naish, R.B. (1999) *Ceylon and the Portuguese, 1505–1658*. Asian Educational Services, New Delhi.
Pinto, P.J.d.S. (2015) Share and strife – the Strait of Melaka and the Portuguese (16th and 17th centuries). *Orientierungen – Zeitschrift zur Kultur Asiens* 2013, 64–85.
Shaffer, M. (2013) *Pepper: A History of the World's Most Influential Spice*. St Martin's Press, New York.

The Portuguese Build an Empire 17

Setting the Stage – Long-Term Impacts of Albuquerque's Battles

Albuquerque's victory in Malacca gave Portugal a major foothold in the Far Eastern pepper trade, but the Portuguese were never able to fully dominate it. Rival states were established at Johor on the south end of the Malay Peninsula by the former sultan of Malacca, Mahmud Shah, and at Aceh on the northern end of Sumatra by the displaced merchants. Supported by the Ottoman Turks, these renegade states and those that followed were able to continually harass the Portuguese and became adept at running pepper through their blockades into the Red Sea.

The Malay sultans were able to cause the Portuguese severe hardship but were never able to rid themselves of them (Boxer, 1977). The sultans besieged Malacca repeatedly in the first half of the sixteenth century but could not take the city. Malay and Javanese forces did finally team up in 1550 to capture most of Malacca, but the Portuguese recouped, drove the invaders to the sea and butchered them. Massive, coordinated attacks in 1567 and 1574 were also repulsed by the Portuguese.

The constant state of battle for Malacca encouraged most Asian traders to bypass it and its status as an entrepôt steadily declined over the years. Rather than achieving their goal of dominating South East Asian trade, the Portuguese had instead only disrupted and dispersed it. Trade was now scattered across a number of ports across the region and was no longer centralized in one place. Far from dominating the region, the Portuguese had only grabbed a piece of the action and at a great cost.

The successors of Albuquerque kept trying to bottle up the Bab-el-Mandeb Strait but never succeeded. The last straw came in 1538, when the Ottomans sent 100 ships and 20,000 men down the Red Sea, captured a number of Portuguese ships and occupied Aden. This action ended any serious Portuguese threat to the Red Sea routes and from that point on, Gujarati ships carrying Sumatran pepper passed easily through the strait. By mid-century

Alexandria was supplying the Venetians with as much spice as they had before the Portuguese discovered the Cape route to India.

Ormuz became the jewel in the Portuguese spice trade. Everyone, from everywhere, sought out its port – Arabs, Persians, Indians, Jews and Christians. All seagoing trade between India and the Middle East passed through Ormuz. As described in 1606 by the Flemish merchant Jacques de Coutre:

> At the trading season during the monsoons, more than 200 large ships come filled with merchandise. Some are loaded with cinnamon from Ceylon, others from Cochin with cloth from Bengal and Coromandel and cloves, nutmeg, mace, sugar and other goods that come from south of Cochin. Other ships come from Makkah and Aden to buy; still others come from Mangalore and Barcelore, loaded with rice.
>
> (Lunde, 2005)

The Cartaz System

While Albuquerque was not able to completely bottle up all the major spice ports of the Indian Ocean, he was able to set up a 'cartaz system', where the Portuguese claimed suzerainty over the Indian Ocean and no one else was allowed to sail unless they purchased a safe conduct pass (Lunde, 2005; Maloni, 2011). The cartaz obliged Asian ships to call at a Portuguese-controlled port and pay customs duties before proceeding on their voyage. Ships without this document were considered fair game and their goods could be confiscated. It was, pure and simple, a protection racket. The cartaz system, plus customs duties and outright piracy, provided most of the funds defraying the costs of the Portuguese navy and its garrisons.

As Krishna described the cartaz system:

> no Asiatics except the subjects of the allied princes could stir out of their ports without obtaining passports from the Portuguese. Those who defied these sovereigns of the seas did it at the peril of their persons, ships and goods. The Malabars, Arabs and the Gujerats did sometimes steal out without permits, but they were always liable to be captured by the Portuguese ships. Within these restrictions all encouragement was given to Indian and other Asiatic merchants to resort to various ports with their goods, and Portuguese boats were on the sea to defend them against the depredations of the Malabars and other pirates, and to convoy to their destination.

> Large fleets for the security of the sea in India and Europe were a permanent feature of the political organization of the Portuguese. They equipped two armadas at Goa, one of which, called Armada del Nord, went as far as Ormus; the other, Armada del Sud, sailed as far as Comorin. Each was composed of fifty to sixty war galliots, without counting the merchantmen called Names de Ohatie, which were convoyed by Names de Armada to the various ports. One or two grand galleys like those of Spain were also added to the fleet. These ships departed in the month of October and remained on their cruising duty for six months.

Galleys were rowed by prisoners and convicts, but galliots by the Canarine, natives of Salsette, Colombo, etc. ... The ships of war were well armed. The great galleys had from two to three hundred soldiers called Lascarits, others had a hundred each, while, smaller frigates carried forty to fifty men-at-arms ... Besides these two regular armadas, others went to Malacca, Sunda, Mozambique and other places where they were required.

(Krishna, 1924, p. 49)

By the mid-sixteenth century, Asian merchants began shipping their goods on Portuguese ships and vice versa. Portuguese ships came to be crewed by a hodgepodge of sailors from Arabia, Malabar, Gujarat, Malaysia and Indonesia, with a Portuguese officer or two. 'Pidgin Portuguese became the lingua franca of the Indian Ocean ports' (Lunde, 2005).

The India Run

Throughout most of the sixteenth century, the headquarters of the Portuguese trading empire was Goa. It was the residence of the Portuguese viceroy and headquarters of the Jesuit order that operated all across India and South East Asia. Most Portuguese shipments of pepper and ginger originated from the Malabar Coast of India. The other centres of Portuguese trade were Malacca, which controlled trade and shipping from India to Indonesia and China, Jaffna in Ceylon for trade in cinnamon, and Ternate in the Moluccas for trade in cloves, nutmeg and mace.

Every year in April or March an average of 12 or 13 ships would leave on a round-trip mission between Lisbon and India (Newitt, 2005; Anonymous, 2020). In a large part, they sailed the route pioneered by Vasco de Gama (see Fig. 16.1 in Chapter 16, this book).

These trade missions, called *Carreira da Índia* by the Portuguese, were scheduled to coincide with the annual cycles of the monsoon winds. The goal was to pass around the Cape of Good Hope in June to July and arrive in East Africa by August to resupply and then catch the summer monsoon winds to India, for an early September arrival. Ships that failed to reach the African coast by late August would be stuck there and have to wait until next spring to make the Indian Ocean crossing.

The scheduled stop in Africa after crossing the Cape of Good Hope was on Mozambique Island, which had an excellent harbour. The downside to Mozambique Island was that it was a dry and desolate coral island, and drinkable water and provisions had to be brought by boat from the eastern coast of Africa by Portuguese factors before the arrival of the armada. Having adequate stocks on hand was critical as the condition of ships and crews upon arrival was generally quite woeful. They had been at sea for months without repairs and resupply and by now the crew members were beginning to die from dysentery and scurvy. The passage around the turbulent, stormy Cape had also been a veritable horror for all aboard.

As Krondl describes it:

> By the time the ships reached the Cape of Good Hope, the hardtack was likely to be crawling with worms, and the water in the barrels was fetid ... The smell emulating from the ship, where five hundred unwashed men had been eating, sleeping, and relieving themselves of every bodily fluid (seasickness was all too common) in extremely close quarters for at least four months must have been astonishingly putrid as it sailed up the African east coast and into Mozambique harbor to restock. If lucky, a ship would lose no more than 10 percent of its human load on the outboard voyage.
>
> (Krondl, 2007, pp. 143–144)

The ships generally sailed directly to India from Mozambique, but if they had time, some would continue up the south-eastern coast to Malindi, an ally of the Portuguese since Vasco da Gama's arrival in 1498. Malindi had many fertile cultivated fields and lots of groves of oranges and lemons which effectively combatted scurvy. Malindi was also at the perfect latitude for a ride on the south-westerly monsoon for an Indian Ocean crossing, and there were generally many experienced Indian Ocean pilots (Swahili, Arab or Gujarati) available for hire.

Once in India, the ships loaded up and followed the reverse course back to Portugal. The returning ships left Cochin and Goa at the beginning of the following year, riding the winter monsoon back to Africa and then sailing the rest of the way home, carrying between 7000 and 10,000 hundredweight of spices. They arrived in Lisbon in the summer (June–August) after a round trip that took a little over one year. Pepper was their primary commodity, but cinnamon, cloves, nutmeg and mace were also important.

The long trip to India and back was a very dangerous one. The loss of ships on the Carreira da Índia ('India Run') was frightful. Out of 806 that left for India, 425 or just over half returned to Portugal successfully. Of those not returning, 20 turned back before reaching India, 66 were lost, four were captured by enemies, six were voluntarily scuttled when they were no longer seaworthy and 285 stayed in India for other purpose. The Mozambique Channel was the area where the most ships went down, averaging almost 40% of the losses from 1497 to 1550 (Guinote, 1999).

Importance of Indian Cotton in the Spice Trade

For centuries before the Portuguese arrived in India, cotton textiles were at the very centre of Indian Ocean trade. 'Cotton textiles were central to the dynamics of interaction created within the space of the Indian Ocean ... a global system in which cotton cloth was the commodity traded and India was its manufacturing core' (Riello, 2013, p. 25). In his *Suma Oriental*, Tomé Pires estimated that before the Portuguese arrived, Malacca was annually receiving at least five vessels from Gujarat and southern India plus several from Bengal, all laden with cloth.

The Portuguese soon discovered that textiles were an essential ingredient in the quest for spices. The spices that the Portuguese wanted could only be obtained either in India for gold or in Indonesia and the Spice Islands for textiles. They got into the cotton trade in a big way and in fact the profits in their inter-Asiatic trade came to surpass those of their trade to Portugal. Each year the captain of Malacca sent eight or nine ships to the Coromandel Coast to buy cloth for the spice trade. Only one of the ships was for the Crown, the others were for private entrepreneurs (Laarhoven, 1994).

Sixteenth-Century Shifts in European Spice Trade

The presence of the Portuguese in eastern India and Malacca led to a major shift in the routes spices travelled into Europe. Before the Portuguese began carrying spices around the Cape, most of the pepper that found its way to Europe came from the Malabar Coast of India on Gujarati and Arab ships. It was moved through the Red Sea to Cairo and Alexandria, where the Venetians purchased it and then sold it throughout Europe.

In the early years of the sixteenth century, the Portuguese conquest of India and their takeover of Malacca dried up much of the Venetians' ability to obtain spices. Venetian spice imports fell from around 1600 tons a year at the end of the fifteenth century to less than 500 tons a decade and a half later (Lane, 1940). However, it did not take the Gujarati and Arab merchants long to shift to other locations in Sumatra and Java for their pepper to keep supplying the Venetians. Muslim Aceh became a pepper trading powerhouse, producing 7 million pounds (3.18 million kilograms) annually (Boxer, 1969). The Venetians continued delivering pepper to most of Europe through the Mediterranean, while the Portuguese focused on Northern Europe through the Dutch.

In 1517, the Ottomans took over Egypt and stopped most of the Venetian trade in spices, but again this embargo did not dissuade the Venetians for long (Lane, 1940; O'Rourke and Williamson, 2009). Even though the Ottomans and Venetians were bitter enemies and fought territorial wars, it was mutually beneficial to keep their trade links open, much as had been the case with the Romans and Persians battling over the Middle East long ago. The Ottomans sold wheat, spices, silk, cotton and ash (for glass) to the Venetians, and in return received finished goods such as soap, paper and textiles. Luxury objects including carpets, metalwork, finished glass and art objects also passed hands. By the mid-1500s as much or more spice was passing through Venetian hands into Europe as before the Portuguese conquests (Bernstein, 2008).

The Portuguese Distribution Network in Europe

A royal trading house called the *Casa da Índia* ('House of India') was established around 1500 by King Manuel I to monitor and organize Portuguese trade activities with the Far East. Some ships were owned and outfitted by private consortiums, but generally the cost was too prohibitive for the private sector with the risks involved. It was more common for the Crown to support a ship's construction and activities and allow Portuguese merchants to contract for cargo.

Once the spices had arrived in Lisbon, most were transported to a factory in Antwerp which served as a distribution centre for Northern Europe. While the Portuguese had temporarily disrupted Venice's trade in spices in the early sixteenth century, by the second decade Venetians' trade with the Muslims was again flourishing as the Portuguese were never able to completely bottle up the Red Sea portal. By focusing their spice trade on Northern Europe, the Portuguese could avoid competing directly with the established Venetian Mediterranean trade, and it provided them with a source of the silver they needed to buy the spices in the Far East. This silver came from Central Europe, under the control of the major German trading families.

The international trade that was now flowing into Antwerp made it the leading financial and commercial city of Europe in the sixteenth century. The bulk of the spice trade between Portugal and Flanders was directly contracted between the Casa da Índia and private foreign consortiums and shipped by Dutch, Hanseatic and Breton ships. Thus, most of the spice profits came to be garnered by the private European merchants. These profits would ultimately give the Dutch the financial resources and desire to forcefully take direct control of the Indonesian spice market themselves.

Portuguese Spice Profits

Portuguese imports of pepper held strong over most the sixteenth century. The total weight of the spice cargoes averaged 40,000 to 50,000 quintals (1 quintal = 130 pounds or 59 kilograms) annually in the first half of the century and 60,000–70,000 quintals later on (Boxer, 1977). During the early decades of the *Estado da Índia* ('State of India', or Portuguese India), 95% of the cargoes were composed of pepper. Records have been left of one cargo in 1518 that totalled almost 5 million pounds (2.27 million kilograms), of which 4.7 million pounds (2.13 million kilograms) was pepper, 12,000 pounds (5443 kilograms) cloves, 3000 pounds (1360 kilograms) cinnamon and 2000 pounds (907 kilograms) mace (Krondl, 2007). Most of the pepper and other spices were purchased in Malabar on the open market. Portuguese profits on the pepper trade could run as high as 500% (Lunde, 2005).

By the end of sixteenth century, the Portuguese share of the pepper exports began to slip and dropped to about 10,000 quintals a year. By then, the bulk of the pepper reaching Europe was travelling overland across the Levant from the Red Sea rather than by the Atlantic route. This trade was fuelled by the Achinese on Gujarati ships who were moving some 50,000 quintals to Jeddah in the Red Sea each year.

Overall, spice production in Asia actually doubled in the second half of the sixteenth century and prices increased two- to threefold. A considerable quantity of pepper was produced in Sumatra and western Java, and the bulk of it was purchased by the Chinese. This pepper was cheaper than that from Malabar, but the Portuguese were unable to break into this market and reduce their costs.

Portuguese imports of mace, nutmeg and cloves increased somewhat in the second half of the century and their value tripled, but the profit margins were slim due to the cost of sending boats to the distant Molucca and Banda Islands and maintaining the forts on the islands. Most of the spices the Portuguese bought in Indonesia was sold to Asian and Chinese traders in Malacca, Goa and Ormuz, and very little was shipped home.

The Mughals and Portuguese in India

As the Portuguese were pushing hard to take over Indian Ocean trade, another land-based empire was also invading India. The Mughal Empire, which was founded in 1526, took over most of northern India during the reign of Akbar (r. 1556–1605). The Mughals were largely ignored by the Portuguese (and vice versa) until they reached the sea, first in Gujarat in 1572/3 and then Bengal in 1575/6. Before that, the greatest worries of the Portuguese lay with the previous territorial powers like Vijayanagara, Bijapur, Gujarat, Ottoman Turkey and Safavid Iran. For the Mughals, their concerns had been primarily with Iran and Turkey.

Where the two powers touched, they largely maintained an uneasy peace as their commerce was based on very different sources. The Mughals derived most of their revenue from their extensive landholdings and maritime matters were of little concern to them; while for the Portuguese, the sea was their main source of revenue. The Estado da Índia obtained 15 times more of its revenue from the sea than the land, and about 60% of its total revenue came from customs duties. The Portuguese coastal trading route from Goa to Cambay was even more lucrative to them than the India Run to Portugal itself (Pearson, 1998).

By the late 1570s, the previous hostile relationships between the Gujarati and Portuguese had been replaced by acquisition and accommodation. However,

> the Portuguese rightly feared that if Akbar wanted to, he could stop their trade, and probably also take their forts, which they could not succor from the sea during the three months (June to September) of the southwest monsoon. And even if they could have held out, they were still vulnerable, as a Jesuit letter of 1600 gloomily noted: we remain surrounded by him, and although it is not possible for him to take our cities and forts by arms, nevertheless if he takes the places from where we get our provisions it would be our certain end, which God not permit, because then we would have no other refuge except Ceylon.
>
> (Pearson, 1998, p. 413)

Akbar did start a war with the Portuguese but pulled back and chose peace. He only asked that he be able to trade freely to the Red Sea, and his mother and other elites could conduct an annual pilgrimage to visit the tomb of Muhammad. To keep the peace, Akbar was given one duty-free cartaz a year by the Portuguese on which the Gujarat merchants took their most valuable trade goods and gained as much profit as possible.

Akbar decided to concentrate his military actions on the land, where he felt prestige was greatest. 'Glory was to be won by campaigning on land, leading one's contingent of cavalry, galloping across the plains. To courtiers, including the emperor, the sea was a marvel, a curiosity, a freak. This was not an arena where power and glory were to be won' (Pearson, 1998, p. 413).

The Last Frontier: Portugal and China

As Portugal's position in the spice trade gradually slipped in the sixteenth century, there was one major area of growth: China and Japan. Once the Portuguese were positioned in Malacca, they almost immediately began trading with China, mostly illegally. Developing diplomatic relations with China came hard, as Malacca was one of China's tributary vassal states and conservative factions in the Chinese court were dead set against allowing formal diplomatic ties with Portugal. Tomé Pires, sent to establish an embassy in China, ultimately died there in prison after China demanded that the Portuguese relinquished their control of Malacca and return it to the deposed sultan (Chang, 1962). The Portuguese persisted for decades in illegal trading activities, even taking Chinese children into slavery, with periodic violent Chinese reactions. It wasn't until 1557 that diplomatic relations were normalized, when the Ming Court let the Portuguese establish a trading factory at Macau in southern China, not far downstream from Canton.

The Portuguese discovered Japan in 1543 when a ship was blown off course and landed on the island of Tanegashima. In 1571 they were allowed to establish permanent port facilities in Nagasaki. This proved to be a major boon to the Portuguese as for nearly a century, trade between China and Japan had been outlawed by the Ming. The Portuguese became the trade intermediary between these two empires, dealing largely in silver and silk thread (Lunde, 2005). A Jesuit community of missionaries was given control of the Portuguese interests in Nagasaki.

In 1526, silver mines were discovered in Iwami Ginzan, Japan that produced 1.0 to 1.2 million troy ounces of silver a year, making Japan the second-largest silver producer in the world next to Spanish Peru. As Lunde tells it:

> With the grant of Macao [and access to Japanese silver], the Portuguese were able to greatly extend the geographical and commercial scope of the *Estado da India*. They brought silver, ivory, ebony and sandalwood from Goa and Malacca to Macao, took Chinese silks and other goods to Nagasaki in exchange for silver,

exchanged the silver for gold with Chinese merchants in Macao, and returned to Malacca and Goa with gold, silk, camphor, copper, mercury and porcelain, all in great demand in the Indian Ocean trading network. This trade was especially lucrative because the value of silver was greater in China than in Japan.

(Lunde, 2005)

Portuguese trade with China and Japan was highly regulated. Only a limited number of ships were allowed to make the trip and the merchants paid tens of thousands of ducats for the privilege; however, profits could be 200,000 ducats (one ducat was worth US$80 in today's dollars) (Bernstein, 2008). Huge ships were built specifically for this trade, increasing from the standard 500 tons to a mammoth 2000 tons.

Dutch and English Look for Alternative Routes to the Spices

Throughout the sixteenth century, the Dutch and English lusted for their own piece of the East Asian action. To do this, they would have to move into a colonial world now claimed and jealously guarded by the Portuguese. It wouldn't be until the end of the century that the Dutch and English felt strong enough to try and take over that empire. Until then, they contented themselves with privateering, but they also sought an alternative, northern route to the riches of South East Asia (Fig. 17.1). Some expeditions were sent west to North America (John Cabot, Jacques Cartier, Martin Frobisher and John Davis), some sailed east along northern Russia (Richard Chancellor, Oliver Brunel and Willem Barents) and one sailed both ways (Henry Hudson) (Milton, 1999).

The first European to explore coastal North America was Venetian John Cabot commissioned by Henry VII of England. He made two trips in 1497 and 1498 of which only sketchy details exist. In his first trip he landed in southern Labrador, the island of Newfoundland or Cape Breton Island, which he was sure was the north-eastern coast of Asia. On his second trip, he and his crew likely died at sea trying to advance further in what he thought was the direction of Japan.

Frenchman Cartier was next to search for the Northwest Passage in three expeditions between 1534 and 1542. In his first voyage, he travelled along the west coast of Newfoundland, discovered Prince Edward Island and explored the Gulf of St Lawrence. He returned with great tales of potential riches and two captive Iroquois, being rewarded with another mission with more ships and a bigger crew. Guided by the two Indians, he sailed this time up the St Lawrence River as far as the island of Montreal, where rapids prevented him from going any further. There he spent a difficult, cold winter and many of his men died of scurvy before it was safe to return. In 1541 he was sent on a third mission, this time to support the nobleman, Jean-François de La Rocque de Roberval, in establishing a French colony. He left a year ahead of the colonists to continue his quest for riches, found minerals that he thought contained gold and diamonds and overwintered in Quebec, again greatly suffering from

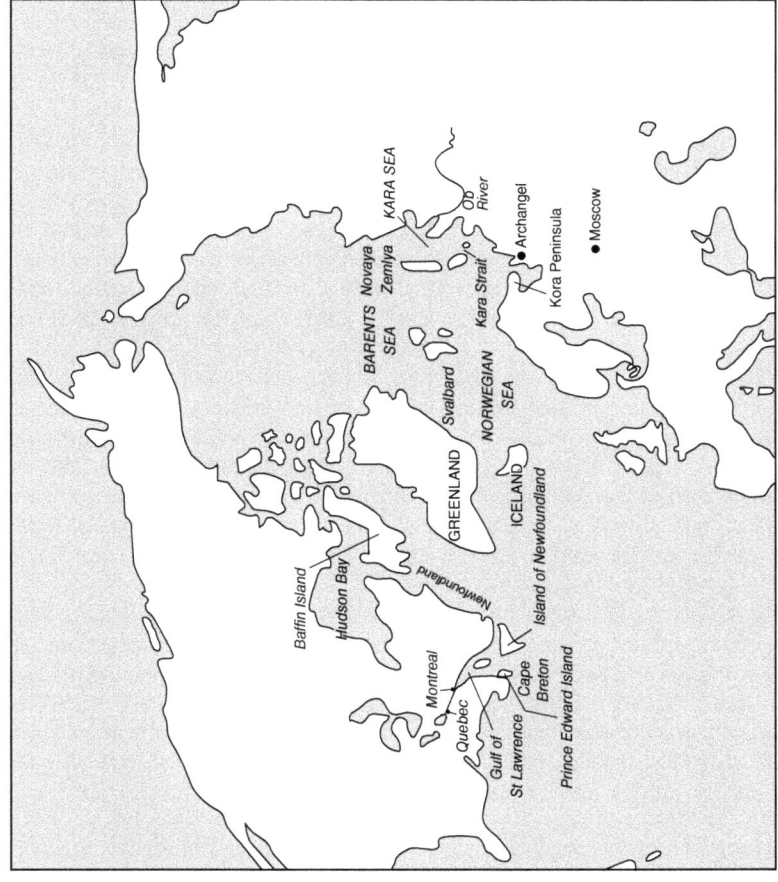

Fig. 17.1. Polar regions where explorers searched for Northwest and Northeast Passages.

the cold. In the spring he fled home without waiting for the colonists, only to find that his stash of gold and diamonds was worthless 'fool's gold'.

In 1576, Englishman Frobisher set sail for North America in the first of three expeditions to find the fabled Northwest Passage. During his first trip, he reached Labrador and Baffin Island and returned home convinced that North America was a land rich in gold. Based on that promise he got royal backing for two more missions in 1577 and 1578. On these he spent most of his time looking for precious metals and attempted to start one ill-fated settlement. When he returned empty-handed for the third time, he lost his financial backing.

Englishman John Davis was next to get royal support to search for the Northwest Passage, leading three missions between 1585 and 1587. In his first he bumped into the icebound east shore of Greenland, headed south, rounded Cape Farewell, and then sailed northwards along the coast of western Greenland. He then turned west in what he thought was the direction of China and sailed for some distance up Cumberland Sound before turning back. In his other two missions he spent a lot of time exploring the coast of Greenland and never got much further than Baffin Bay in his search for a route to Cathay.

Englishman Chancellor sailed in 1553 on the first mission sent west towards China sponsored by an organization called the 'Company of Merchant Adventurers to New Lands' ('The Mystery, Company and Fellowship of Merchant Adventurers for the Discovery of Unknown Lands', in full) (Milton, 1999). Three ships embarked on the voyage with Chancellor commanding and Sir Hugh Willoughby as his second. The three ships got only as far as northern Norway before high winds and churning seas separated them. Chancellor was never able to find the others, while Sir Hugh and the third ship were able to reconnect. These two ships wandered about the Arctic Circle for 300 miles until the icy seas forced them to land near the current border of Norway and Russia. Here they eked out an existence for several months only to freeze to death during the polar night when they could no longer find food. The two battered ships were found five years later, filled with ghostly bodies frozen in place in the act of doing normal daily activities such as writing, eating and opening lockers.

Chancellor and his crew fared much better than Sir Hugh and his. When Chancellor deemed the pack ice was too dangerous to continue battling, he landed close to present-day Archangel and then hiked to Moscow. Here he and his crew enjoyed several months of the barbaric pleasures of the Court of Ivan the Terrible and negotiated a treaty for English trade with Russia. In the summer of 1554, they rejoined their ship and were able to return home. To take advantage of the new trade treaty, the Company of Merchant Adventurers to New Lands was re-chartered by Mary I in 1555 and became known as the 'Muscovy Company'.

Flemish Oliver Brunel made a number of expeditions to the Arctic from 1565 to 1584 and was instrumental in establishing trade between the Dutch and Russians (Baron, 1985). In his first expedition, he explored the northern coast of Russia and established a trading post at the mouth of the Dvina River (now Arkhangelsk, Russia). He was thrown into prison after complaints from the Muscovy Company who had exclusive trade rights with Russia but got out when a wealthy Russian merchant family, the Strogonovs, intervened. As their agent he helped set up regular trade with the Dutch. In 1576, Brunel made an overland trip from Moscow to the Ob River in western Siberia, which he was convinced was the pathway to China. He never actually travelled down the river but sent a description back to Holland that raised huge excitement.

In 1581, the Strogonovs had two ships built by a Swede to make their own search for a route to Cathay and dispatched Brunel to Antwerp to hire skilled crews. When he got there, he betrayed his benefactors and enlisted the support of Dutch merchants for his own expeditions. Brunel made two attempts to find the way to the Ob but never got any further than the Kara Strait, and in his last trip he drowned when his ship capsized. There is no record of a Strogonov expedition ever taking place.

Still charged by Brunel's promise of the Northeast Passage, the Dutch next tapped Willem Barents in 1594 to continue the search. The government offered a 25,000 guilders award if he could find the way. He made it as far as the west coast of Novaya Zemlya before being forced to turn back when confronted by a sea filled with large icebergs. In a second expedition in 1595 he got only as far as the Kara Sea which had frozen solid. In 1596, he reached Svalbard and explored Spitzbergen, before getting locked in the ice to the east of Novaya Zemlya. He and the crew spent the winter there in a shelter they called *Behouden Huys* ('house of safety') and in the spring made their escape by rowing southwards in two open boats. Barents died on this journey, but most of the crew survived to reach the Kola Peninsula, where they met a Dutch vessel that took them back home to Holland.

Henry Hudson Searches Both Ways

Henry Hudson undertook four major expeditions in search of a passage from Europe to the Orient (Milton, 1999). In 1607, he was sent by the English Muscovy Company in search of an ice-free sea that would lead west from Europe to China. He was stopped by ice near the Svalbard archipelago. A year later, he got as far as the islands of Novaya Zemlya before being set upon by ice fields. In 1609, the Dutch East India Company sent him on another mission east but this time when he was stopped by ice, he decided to sail west towards the New World rather than return to Amsterdam. He found a large waterway that we now call the Hudson River, which he followed as far as Albany before realizing it was not his ticket to the Pacific.

In Hudson's last expedition in 1610, he was commissioned by the British East India Company to try going further north in another search for the fabled Northwest Passage. He sailed through the strait that now bears his name and into the Hudson Bay where he explored its southern regions and then spent months sailing aimlessly through its great expanse. As winter approached, with no clear route to the Pacific in sight, Hudson's starving crew mutinied and put the explorer, his son and all hands sick with scurvy into a small lifeboat and set them adrift. The crew made it back to England, but Hudson was never heard from again.

References

Anonymous (2020) *Portuguese India armadas*. Wikipedia. Available at: https://en.wikipedia.org/wiki/Portuguese_India_Armadas (accessed 24 February 2021).

Baron, S.H. (1985) Muscovy and the English quest for a northeastern passage to Cathay (1553–1584). *Acta Slavica Iaponica* 3, 1–17.

Bernstein, W.J. (2008) *A Splendid Exchange: How Trade Shaped the World*. Grove Press, New York.

Boxer, C.R. (1977) *The Portuguese Seaborne Empire: 1415–1825*. Hutchinson & Co. (Publishers) Ltd, London.

Chang, T.-T. (1962) Malacca and the failure of the first Portuguese embassy to Peking. *Journal of Southeast Asian History* 3, 45–64.

Guinote, P.J. (1999) Ascensão e declínio da *Carreira da Índia* (Séculos XV–XVIII). In: *Vasco da Gama e a Índia*, Vol. II. Fundação Calouste Gulbenkian, Lisbon, pp. 7–39.

Krishna, B. (1924) *Commercial Relations Between India and England (1601 to 1757)*. George Routledge & Sons Ltd, London.

Krondl, M. (2007) *The Taste of Conquest: The Rise and Fall of the Great Cities of Spice*. Ballantine Books, New York.

Lane, F. (1940) The Mediterranean spice trade: further evidence of its revival in the sixteenth century. *The American Historical Review* 45(3), 581–590.

Laarhoven, R. (1994) The power of cloth: the textile trade of the Dutch East India Company (VOC) 1600–1780. PhD thesis, Australian National University, Canberra.

Lunde, P. (2005) The coming of the Portuguese. *Aramco World: Arab and Islamic Cultures and Connections* 56(4). Available at: https://archive.aramcoworld.com/issue/200504/the.coming.of.the.portuguese.htm (accessed 28 January 2021).

Maloni, R. (2011) Control of the seas: the historical exegesis of the Portuguese 'cartaz'. *Proceedings of the Indian History Congress* 72, 476–484.

Milton, G. (1999) *Nathaniel's Nutmeg: The True and Incredible Adventures of the Spice Trader Who Changed the Course of History*. Penguin Books, New York.

Newitt, M.D. (2005) *A History of Portuguese Overseas Expansion, 1400–1668*. Routledge, London.

O'Rourke, K. and Williamson, J. (2009) Did Vasco da Gama matter for European markets? *The Economic History Review* 62(3), 655–684.

Pearson, M.N. (1998) Portuguese India and the Mughals. *Proceedings of the Indian History Congress* 59, 407–426.

Riello, G. (2013) *Cotton, The Fabric That Made the Modern World*. Cambridge University Press, New York.

The Spanish Build Their Empire   18

Setting the Stage – Spain Stretches Its Muscles Across the Atlantic

The Spanish sent their first group of colonists to Hispaniola in 1493. Santo Domingo was founded on the south-eastern coast of Hispaniola in 1496 and secondary cities were established all across the island wherever gold could be found. The Spaniards exploited the island's gold mines using the local Taíno as slaves and within 25 years of Columbus' arrival, most of the Taíno had died from enslavement, massacre or disease. These mines soon ran out and in the second decade of the sixteenth century the Spaniards were forced to move on to the other large islands, repeating the same cycle over and over again. In only a generation they largely exhausted the demographic and mineral potential of the Greater Antilles and headed off in three directions – first to central Mexico and the Isthmus of Panama and then later to Peru.

The first permanent European settlement on the Isthmus of Panama was established by Vasco Núñez de Balboa in September 1510 at Santa María la Antigua del Darién near the mouth of the Tarena River on the Atlantic Ocean (Sarcina, 2019). He arrived there with all that was left of an expedition from Hispaniola that had built a fortress at nearby Necoclí but had been forced out by the local inhabitants. The locals refused to trade food with the Spanish and frequently attacked them. Of the 300 men who had initially arrived with Balboa only 42 survivors remained. Balboa was in the first group of Europeans to see the Pacific Ocean on 25 September 1515 when he led an expedition across the south-west corner of the isthmus.

In 1519, Hernán Cortés landed on the central, Atlantic coast of Mexico with a small band of Spanish soldiers and within two years had marched across the highlands and conquered the powerful Aztec Empire, giving the Spanish King Charles V his first major territory on the American mainland. The peoples they first encountered offered the Spaniards little resistance and they were able to almost immediately establish the city of Veracruz. The conquistadors then moved inland and encountered the Tlaxcalans, who engaged them briefly in battle, but decided after suffering heavy losses to ally with the Spaniards against their traditional enemy, the Aztec. As the Spaniards marched towards the Aztec capital of Tenochtitlán, most of the local subordinate states also joined with them.

Once in Tenochtitlán the Spaniards managed to seize the emperor Montezuma, but they were forced from the city with severe casualties. They retired to nearby Tlaxcala, where they were reinforced and then began a full-scale siege. After four months the Spaniards were able to capture the Aztec capital and made it their headquarters as Mexico City. The rest of central Mexico capitulated more easily, and several more Spanish cities were established in the region.

Magellan Finds Another Route to the Pacific

About the same time that Cortés was establishing a Spanish Empire in Mexico, Ferdinand Magellan was beginning his great adventure around the world. On 20 September 1519, Magellan set sail from Spain with five ships and 270 men in his attempt to find a western sea route to the Spice Islands (Bergreen, 2003). After years of trying to convince Manuel I of Portugal to support his navigational dream, he had finally gained support from the young King Charles I of Spain (and future Holy Roman Emperor).

Magellan had considerable Indian Ocean experience, being part of the Portuguese expedition to Goa in 1505 and Malacca in 1511. He had served with Francisco Serrão, the shipwrecked mercenary who became the military advisor to the sultan of Ternate. Serrão's letters to Magellan about the Spice Islands would help him persuade the King of Spain to finance his circumnavigation. Both men would be dead when Magellan's ships limped into the Spice Islands in 1521.

Magellan headed to West Africa and then to Brazil, where he searched along the coast for a passage to the Pacific. He tried the Río de la Plata, a large estuary south of Brazil, which ended in failure. He overwintered in Patagonia at Port St Julian (today's Puerto San Julián, Argentina), where his Spanish captains mutinied against him, but he was able to crush the revolt, brutally executing one of the conspirators and leaving another behind with a group of his co-conspirators.

Finally, on 21 October 1520, he discovered what came to be known as the 'Strait of Magellan' and after a 38-day struggle to navigate this treacherous strait arrived blissfully at the Pacific Ocean. Magellan named it *Mar Pacifico* because its waters appeared so calm in comparison to the churning waters of the strait. The Pacific proved remarkably placid, but Magellan had greatly underestimated its breadth and had to travel another 99 days until he stumbled upon land, the Island of Guam, on 6 March 1521. By then his crew's resources were completely depleted and the sailors were sustaining themselves by soaking and chewing the leather parts of their gear. The local people fed and supplied Magellan and his crew in intermittent periods of civility and fighting.

Magellan then continued his quest and ten days later, he arrived at the Philippine island of Cebu, only 400 miles from his goal of the Spice Islands. He was now in territory that had been frequented by Arab and Chinese traders. Magellan befriended the local chief of Cebu who agreed to convert to Christianity, in hopes that the Europeans would assist him in conquering a rival tribe on the neighbouring island of Mactan. Magellan led the attack, assuming his European weapons would quickly and easily overwhelm the natives. The locals, however, fought fiercely, ambushing his small force and killing Magellan on 27 April 1521 with a poisoned arrow.

The Fate of Magellan's Ships and Crew

At Magellan's death, there were only two ships left of his original five. One had been wrecked during a storm off the coast of South America, one had been purposely destroyed when there were not enough men left to operate it, and one had deserted during the crossing of the Strait of Magellan. The remaining two ships, the *Victoria* and *Trinidad*, were now barely seaworthy but sailed on, following a rather haphazard route to the Moluccas and arrived there on 5 November 1521.

With the arrival of these two Spanish ships in Tidore, Portugal no longer had the Spice Islands all to itself (Bergreen, 2003). Spain had forged a western entry into the area by travelling across the Atlantic Ocean, rounding the tip of South America and then crossing the Pacific Ocean. A long and tortuous journey to say the least, but now proven possible. Any ship and crew that survived this journey, filled its hulls with spices and then made it the rest of the way home to Spain would be richly rewarded.

Of course the Portuguese were distressed at this development, but the legal ownership of the Moluccas was very much up in the air. The Treaty of Tordesillas, signed in 1494, divided any newly discovered lands between Spain and Portugal along a meridian west of the Cape Verde Islands, but no line of demarcation had been set on the other side of the world. This meant that both countries could lay claim to the Spice Islands, as long as Portugal travelled there from the east and Spain from the west. Thus began a period of conflict that didn't end until the Spanish and Portuguese royal courts were unified in 1580 through intermarriage.

Victoria left the island of Tidore to go home on 21 December 1521, while the *Trinidad* lingered longer for repairs. The *Victoria* headed home under the command of Juan Sebastián Elcano, travelling west across the Indian Ocean. The *Trinidad*, led by Gonzalo Gómez de Espinosa, waited until 16 April 1522 to leave and headed in the opposite direction. He and his crew feared the route across the Strait of Magellan and disastrously decided to follow a heading they thought would take them more easily across the Pacific to the Spanish settlements at Panama.

The *Trinidad*, loaded with spices, made its way north searching for easterlies until the battering of frigid storms, starvation and scurvy forced Espinosa to turn back to Tidore seven months later. As he approached the island, he learned to his horror that a fleet of seven Portuguese ships led by António de Brito had arrived, looking for Magellan and the *Armada de Molucca*. Now in desperate straits, Espinosa could only send a note begging for mercy and supplies. Brito sent an armed party to capture *Trinidad* and instead of meeting any resistance, they found almost a ghost ship. As Bergreen (2003, p. 180) tells it: 'His soldiers boarded *Trinidad* expecting to overwhelm the crew, but were repelled by the grievous spectacle of men near death, a foul and unhealthy stench that no one dared to brave, and a ship on the verge of sinking.'

In spite of their condition, Brito punished Espinosa and his debilitated crew by making them build a fortress at Ternate with lumber from the *Trinidad*, and then sent them as forced labourers all across the Indian Ocean, including Banda, Java, Malacca and Cochin. After years of captivity, only commander Espinosa and two other crew members ever got back to Spain.

Victoria's crew also suffered through major struggles before she got home. The ship left Tidore in dismal condition, with her sails torn and her hull leaking so badly that water had to be pumped out continuously to keep her afloat. She barely survived the passage around the stormy Strait of Magellan and half the crew died of scurvy and starvation. *Victoria* never would have made it home at all except Elcano was able to get some provisions at Portuguese Cape Verde by pretending to be a ship returning from the Atlantic Americas. The Portuguese caught on quickly to the ploy, but the *Victoria* managed to flee before being captured, although 13 crew members were left behind.

On 6 September 1522, Elcano and his remaining crew arrived in Sanlúcar de Barrameda in Spain, almost three years after they had departed. Only 18 of the original 270 men were on board. The other 13 men held captive in Cape Verde got back to Spain a few weeks later, but the overall toll in lives on Magellan's voyage around the world had been horrific. Still, the *Victoria* held 50 hundredweight of spices in her hold which proved highly profitable and more than paid for the expedition. The Crown would be encouraged to make several more such forays.

Spanish Struggle to Keep a Foothold in the Spice Islands

Soon after the arrival of *Victoria* in Spain, Charles V sent a second expedition, led by García Jofre de Loaísa with Elcano as his chief pilot (Nowell, 1936; Kelsey, 1986). They were given three goals, to: (i) seek and rescue the *Trinidad* which had not yet returned; (ii) colonize the Spice Islands; and (iii) find the location of the mythical land of Ophir, mentioned in the Bible as the source of the silver, gold and gems used to decorate Solomon's great Temple. Spanish scholars had of late become obsessed with this possibility and their best guess was that Ophir was somewhere near China.

Loaísa was given seven vessels in 1525 for this mission and 450 men including tradesmen and administrators for the Spice Islands settlement. He sailed south along the African coastline and then west to Brazil. He then made his way down the coastline of South America to Patagonia and there decided to give up the search for the *Trinidad* and head to the Spice Islands. The fleet suffered extreme weather in the Strait of Magellan and two ships were lost. The raging storms continued in the Pacific and the four remaining ships got separated, never to see each other again (Wagner, 1929). One sailed north and after a 10,000-kilometre (6200-mile) journey reached the Pacific coast of Mexico. Another disappeared entirely. A third sailed across the Pacific to Celebes (Sulawesi), where its crew was either killed or enslaved.

Only one galleon reached the Spice Islands in September 1526, and by this time both Loaísa and Elcano had died. The survivors were able to establish a fortress on Tidore, but were forced to abandon their now unseaworthy ship, leaving them no way to get news back to Spain.

With no news of the Loaísa disaster, Charles sent another fleet to the Moluccas in 1526 commanded by Sebastian Cabot, who got no further than the west coast of South America. By now Charles was getting very anxious for any news about his fleets in the vicinity of the Spice Islands and ordered Hernán Cortés in New Spain (Mexico) to send a rescue mission west to find out about all these lost Spaniards. Cortés sent a three-ship fleet across the Pacific under the command of Álvaro de Saavedra Cerón. Only one of these ships made it to the Moluccas and while it did find the survivors, it could not find favourable winds to get back across the Pacific and its captain also perished.

With several armadas lost or unreported in the Moluccas, the Spanish sovereign finally reached the conclusion that the Moluccas were not worth his effort, and in 1529 decided to treat with the Portuguese. In the resulting Treaty of Zaragoza, the Spaniards sold their rights to the Moluccas to the Portuguese for 350,000 ducats and the dividing line between territories was fixed at 17° east of the Maluku Islands. There was an escape clause that allowed the Spanish to have the islands back, if they reimbursed the Portuguese, but this never happened. The Spanish left the Moluccas largely alone except for an invasion of Ternate in 1575, a few years before Philip II united the two Iberian powers and they became one.

The Spanish Explore the Pacific

The Treaty of Zaragoza in 1529 technically gave Spain control of most of the Pacific including the Philippines, but for the next 20 years the Spaniards were not able to exploit this opportunity commercially. They could reach the area relatively easy by sailing westwards across the Pacific from Mexico, but they did not have a route back to New Spain. Until an eastward return route was discovered, they could only return to Spain by completing the circumnavigation of the globe and crossing waters controlled by their jealous Portuguese rivals.

The Spanish sent several expeditions into the Pacific in the 1530s and 1540s (Nowell, 1936; Kelsey, 1986). Hernando de Grijalva sailed west from Peru in 1537 in search of the land of Ophir but was stabbed to death by one of his men when he refused to go to the Moluccas. The mutineers wound up wrecking the ship on the northern coast of New Guinea. A two-pronged expedition was sent from New Spain in 1542, with one armada commanded by Ruy López de Villalobos heading due west towards the Philippians and hopefully Ophir, and the other led by Juan Rodríguez Cabrillo travelling north in hopes of finding China or a north-west passage back to Europe. Villalobos did reach the Philippine islands and named them after King Philip, and likely discovered the Hawaiian Islands, but he and his crew were imprisoned by the Portuguese when they sought refuge in the Moluccas. Cabrillo worked his way up the coast to California, where he died of an infection. His crew continued on, perhaps as far north as Oregon, before nasty winter storms forced them back to Mexico.

In 1565, the first permanent Spanish settlement was established in the Philippines by Miguel López de Legazpi at Cebu and in 1571 his forces conquered Manila. Thus began a long and protracted struggle between the Spanish and Malay people that lasted 250 years. The Spanish were every bit as ruthless as the Portuguese, being quick to kill, maim and torture, and demanding that the local Muslims convert.

Andrés de Urdaneta Finds the Way to New Spain

Among the members of Legazpi's expedition was Andrés de Urdaneta, who served as his navigator and spiritual advisor. Urdaneta's background was truly remarkable. He was one of the few survivors of the Loaísa expedition that had reached the Spice Islands in 1526, only to be taken captive by the Portuguese. Urdaneta spent eight-and-a-half years as a prisoner in South East Asia and did not get back to Spain until 1536 to complete a world circumnavigation that took him 11 years.

While a captive in the Indian Ocean, Urdaneta spent his time well, studying the stars, ocean currents and the winds, and he put down all he learned in an extensive series of diaries, memoirs and maps. These were discovered by the Portuguese and taken from him, but he carried that information in his head and reconstructed it when he finally escaped to Spain. This provided the Spanish Crown with an invaluable wealth of information on Indian and Pacific Ocean navigation.

Urdaneta spent the next 30 years in Spain, becoming an Augustinian friar, studying the accounts of other Spanish voyages in the Pacific and developing a strategy to cross the Pacific to New Spain. His big opportunity finally came on 1 June 1565, when Legazpi sent him from Cebu on the *San Pedro* to test his theories. He had to sail north as far as 38° before he found the favourable winds he was sure existed and arrived in Acapulco on 8 October 1565. In an epic journey of 20,000 kilometres (12,000 miles) that took 130 days, 14 of the crew died of scurvy and only Urdaneta and Felipe de Salcedo, nephew of Legazpi, had enough strength to cast the anchors.

Manila Galleons

The charting of 'Urdaneta's Route' made possible a trans-Pacific galleon trade and the profitable colonization of the Philippines. Soon ships were travelling regularly from Manila to New Spain (Bernstein, 2008). A complex trade network evolved that was truly global in nature (Fig. 18.1). Into Manila would flow spices from the Moluccas and silk and porcelain from China. These would be shipped across the Pacific by the Spanish to Acapulco, a journey of four to six months. There the galleons would be loaded with silver and sent back to purchase more exotic goods. The products brought from Asia to Acapulco were then transported overland across the peninsula to Veracruz, and then shipped along with silver to Spain. These goods were traded all across Europe.

The amount of silver that flowed into China for silk and other exotic products was simply breathtaking. As described by Debin:

> China became a huge suction pump, drawing silver first from Japan, then from Mexico and Peru. According to conservative estimates, fully 75 percent of the 400 million pesos of silver bound for the Philippines during the period 1565–1820 ended up in China. On average, roughly two million pesos of silver were shipped through Manila in the seventeenth century. However, it is important to note that the strength of the suction power from China was sustained by silk threads – Chinese silk was the single most important export item to both Japan

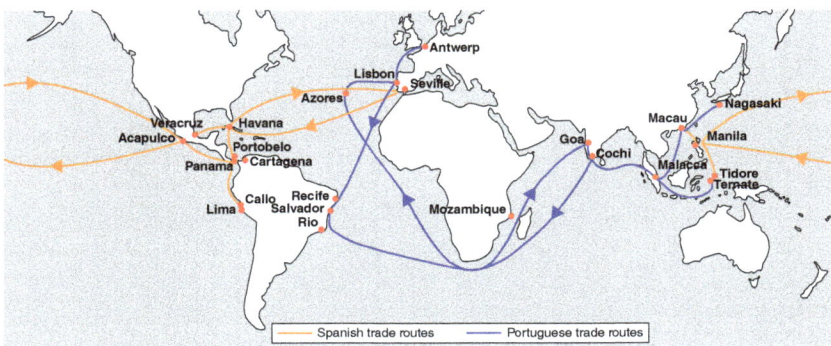

Fig. 18.1. Portuguese and Spanish trade routes. (Redrawn from World_Topography.jpg: NASA/JPL/NIMAderivative work: Uxbona, Public domain, via Wikimedia Commons. File: https://en.wikipedia.org/wiki/Spanish_treasure_fleet#/media/File:16th_century_Portuguese_Spanish_trade_routes.png)

and Spanish America. In the high stage of the trade, China sent three- or even four-million pesos worth of silk goods a year to New Spain.

(Debin, 1998, p. 51)

The silver came from Potosí, Bolivia where hundreds of thousands of enslaved Incan lives were sacrificed by the Spanish to extract that silver from the bowels of the earth. Potosí became a major commercial hub and grew to be one of the largest cities in the world with as many as 200,000 people. The mines became the centre of Spanish wealth and were the reason Spain remained powerful during the colonial period. From 1556 to 1783, they extracted some 45,000 tons of silver from these mines.

The Spanish Crown used a system of convoys of ships (the *flota*) to limit attacks by English and Dutch pirates and privateers. At its peak, these fleets consisted of more than 50 larger boats. Many attacks were made in the Gulf of Mexico and Caribbean Sea by the likes of Francis Drake and Thomas Cavendish (see below), who not only went after ships but also sacked port cities. For over three centuries the primary role of the Spanish navy was to escort the galleon convoys that sailed around the world.

This lucrative trade network was not without its detractors at home in Spain. By then the modest sericulture industry that had emerged in Spain was threatened by the superior Chinese silks now flooding into Spain (Bernstein, 2008). In 1582, the Spanish Crown actually prohibited trade between Manila and Peru at the behest of the Spanish silk industry. This edict was largely ignored, as were similar edicts made in 1593, 1595 and 1604. In 1611, the Crown even prohibited trade between its two colonies in the Americas, Peru and New Spain. This unenforceable edict was also largely ignored, as were another five reissues. The first global trade network persisted on.

Spanish Establish a Silk Industry in New Spain

After the conquest of Mexico and the looting of all the gold and silver they could find, the Spanish needed to develop some local industries to generate revenue (Grace, 2004). One possibility was silk production as New Spain had a native mulberry tree called the *Morera criolla*. The Spanish finished their conquest by 1521 and by 1523, the first silkworm eggs had been exported to Mexico.

During the next 20 years, much experimentation was conducted and experienced growers from Spain were imported to teach the native workers sericulture. By 1540, silk was being produced from the state of Michoacán to the Mayan area of Yucatán, becoming concentrated in Oaxaca and to some extent Zapotec. The Dominican friars established missions across this area and taught the natives sericulture. Many thousands of hectares of mulberry trees were planted.

To process the raw silk, the Crown organized weaving guilds in the cities. Exclusive franchises were given to Mexico City in 1543, Puebla in 1547 and

Oaxaca in 1552 to weave silk velvets, taffetas and satins for the Spanish gentry and Church. Some of the silk was exported to Guatemala and Peru, but the great majority stayed in Mexico.

The Mexican silk industry remained strong until about 1590, but internal and external pressures led to its demise. The native workforce was decimated by at least six plagues caused by introduced diseases from Europe and suffered from the direct importation of higher-quality Chinese damasks, satins and thread from the Manila galleons.

English Sea Dogs Growl at Spanish

During the reign of Elizabeth I of England (1558–1603) a number of 'privateers' brought great wealth to England by plundering Spanish settlements in the Americas and capturing ships and their cargoes (Chatterton, 1914; Milton, 1999; Cartwright, 2020a). These men were in reality pirates, but with license from Elizabeth they were given the credibility to wreak havoc on the Spanish fleets. The English could not trade legitimately with Philip II's New World colonies, so the queen resorted to robbery to weaken the Spanish *flotas* and try to force a change in policy. It also was a way to wage a religious war between Protestant England and Catholic Spain. Most of the English activity was in the Atlantic and Pacific Oceans, but some spilled into the Indian Ocean as well.

Sir Francis Drake ($c.$1540–1596) made a career out of prowling in the Atlantic and Caribbean attacking Spanish settlements and seizing ships loaded with treasure. He ranged far and wide raiding across the Cape Verde Islands, Cuba, San Domingo, Colombia, Florida and Hispaniola. One of his most audacious feats was to ransack and burn the Spanish settlement of Nombre de Dios in Panama in 1573 and capture a Spanish caravan loaded with silver.

Between 1577 and 1580, Drake made a circumnavigation of the globe in his 150-ton *Golden Hind*, after starting out on a mission to plunder Spanish settlements in the Americas. He began by attacking ships in the Cape Verde Islands, then sailed across the Atlantic and worked his way down the coast of South America and up the other side, stopping periodically to raid and loot. In March of 1579 he grabbed his biggest prize off the coast of Peru, the *Cacafuego*, filled with so much gold that it took his crew six days to offload it. He then worked his way further north along the Pacific coast all the way to Oregon in search of the Northwest Passage. Failing this, he turned back south to what is today San Francisco, then headed across the Pacific to Indonesia and took spices on board in the Moluccas. From there he crossed the Indian Ocean, went around the Cape of Good Hope, and arrived back home after travelling for two years and nine months. The value of his cargo was estimated to be well over £600,000, an amount more than double the annual revenue of England (Cartwright, 2020a).

Thomas Cavendish (1560–1592) was the next English privateer to find fame and glory, but unlike Drake, he specifically set out to sail across the world

(Cartwright, 2020b). Cavendish started his journey on 21 July 1586 with a fleet of three ships, his flagship the *Desire*, the *Content* and the *Hugh Gallant*. Cavendish sailed down the coast of Africa, crossed the Atlantic Ocean and reached the coast of Brazil by the end of October. He then sailed to the Strait of Magellan, and after a difficult 46 days arrived at the Pacific coast and began raiding Spanish settlements and capturing any ships he came upon.

Cavendish continued up the coast to North America where he stumbled on his greatest prize, the carrack *Santa Ana*, which was on its way from the Philippines and almost bursting with 22,000 gold pesos and 600 tons of silks and spices. Its capture was particularly gratifying to Cavendish as it belonged directly to Philip II. Cavendish then crossed the Pacific, pausing in the Mariana Islands and the Philippines, before passing the Moluccas and Java, then sailing around the Cape of Good Hope in May 1588. Cavendish paused again to explore the remote island of St Helena in the South Atlantic and then headed home. He arrived triumphantly in Plymouth on 9 September 1588 after a journey of 780 days.

English Sea Dog Activity Against Portuguese

The first Englishman to sail directly to the Indian Ocean in search of plunder was James Lancaster (c.1554–1618), who left on 10 April 1591 in the *Edward Bonaventure* accompanied by the *Penelope* and *Merchant Royal*. It proved to be a particularly disastrous voyage. Lancaster lost so many men to scurvy before he even reached the Cape of Good Hope that the *Merchant Royal* was sent home to allow adequate manning of the other two ships. Soon after rounding the Cape the *Penelope* and her crew were lost in a storm. The *Edward Bonaventure* made her way across the Arabian Sea to the Malay Peninsula, where Lancaster laid in wait and plundered many Portuguese vessels. He then headed to Ceylon where he hoped to capture more Portuguese ships but was forced to head home when his crew threatened to mutiny.

In a long and meandering return trip, where captain and crew suffered through becalmed seas, starvation, scurvy and a forced detour to the West Indies, only 25 officers and men ever reached England, and they arrived back on French vessels. The *Edward Bonaventure* was lost to the Spanish in Santo Domingo, when a carpenter cut its cable while Lancaster and most of the crew were on Mona Island gathering food and water. All told, no profit was gained from this mission, but the English had learned much about the spice trade and the vulnerability of the Portuguese monopoly.

Sir Walter Raleigh (1552–1618) was more of a courtier and explorer than privateer but was responsible for the single biggest prize of all Elizabeth's privateers. After three failed expeditions to found a colony on the coast of North America, he was the organizer of an armada that captured the Portuguese carrack, *Madre de Deus*, off the Azores in August 1592 (Cartwright, 2020c).

The massive ship was filled with over 500 tons of cargo, including jewels, pearls, gold, silver coins, ambergris, cloth, tapestries, pepper, cloves, cinnamon, nutmeg, frankincense, red dye, cochineal and ebony. Raleigh, once a favourite of Queen Elizabeth, had intended to command this fleet but had been imprisoned at the last minute by the jealous queen when she discovered he had secretly married one of her ladies-in-waiting. He was released from prison when the ship arrived in Plymouth, to supervise its unloading and to make sure that Elizabeth got her cut of the profits.

Lancaster sailed off again in 1594 with three ships in the other direction to plunder the Portuguese colony of Recife in Pernambuco, north-west Brazil. This expedition went extremely well. After picking up a chance-met squadron led by John Venner at the Cape Verde Islands, they assaulted and easily took the city, seizing huge quantities of sugar and Brazil-wood (the source of red dye used to colour wool). They also got extremely lucky and captured the spice cargo of a damaged East India carrack that had just been unloaded. Lancaster hired another three Dutch and four French ships to help transport the booty and headed home with 15 ships, all well laden. All but one made it home safely, making Lancaster a very wealthy man.

References

Bergreen, L. (2003) *Over the Edge of the World: Magellan's Terrifying Circumnavigation of the Globe*. HarperCollins, New York.

Bernstein, W.J. (2008) *A Splendid Exchange: How Trade Shaped the World*. Grove Press, New York.

Cartwright, M. (2020a) The sea dogs – Queen Elizabeth's privateers. *Ancient History Encyclopedia*. Available at: https://www.ancient.eu/article/1576/ (accessed 28 January 2021).

Cartwright, M. (2020b) Thomas Cavendish. *Ancient History Encyclopedia*. Available at: https://www.ancient.eu/Thomas_Cavendish/ (accessed 28 January 2021).

Cartwright, M. (2020c) The capture of the treasure ship *Madre de Deus*. *Ancient History Encyclopedia*. Available at: https://www.ancient.eu/article/1572/ (accessed 28 January 2021).

Chatterton, E.K. (1914) *The Old East Indiamen*. T. Werner Laurie Ltd, London.

Debin, M.A. (1998) The great silk exchange: how the world was connected and developed. In: Flynn, D.O., Frost, L. and Latham, A.J.H. (eds) *Pacific Centuries: Pacific and Pacific Rim History Since the 16th Century*. Routledge, Abingdon, UK, pp. 38–69.

Grace, L. (2004) 460 Years of silk in Oaxaca, Mexico. In: *Textile Society of America Symposium Proceedings 482*. Available at: https://digitalcommons.unl.edu/tsaconf/482 (accessed 28 January 2021).

Kelsey, H. (1986) Finding the way home: Spanish exploration of the round-trip route across the Pacific Ocean. *Western Historical Quarterly* 17(2), 145–164.

Milton, G. (1999) *Nathaniel's Nutmeg: The True and Incredible Adventures of the Spice Trader Who Changed the Course of History*. Penguin Books, New York.

Nowell, C.E. (1936) The Loaisa expedition and the ownership of the Moluccas. *Pacific Historical Review* 5, 325–336.

Sarcina, A. (2019) Santa María de la Antigua del Darién: the aftermath of colonial settlement. In: Sarcina, A. (ed.) *Material Encounters and Indigenous Transformations in the Early Colonial Americas*. Brill, Leiden, the Netherlands, pp. 175–196.

Wagner, H.R. (1929) *Spanish Voyages to the Northwest Coast of America in the Sixteenth Century*. Special Publication No. 4. California Historical Society, San Francisco, California.

19 The Dutch and English Conquest of South East Asia

Setting the Stage – The Rise of the Dutch and English Empires

Even though the Dutch did not have their own direct link to the spices of South East Asia in the 1500s, their merchants still became very wealthy selling pepper, cloves, cinnamon and nutmeg across Europe. In the first half of the sixteenth century, the great majority of spices flowing into Lisbon were purchased by Dutch merchants from Antwerp for distribution across northern Europe. By this time Antwerp had grown into one of Europe's largest and richest cities through European trade. Many foreign merchants lived and did business there, and its port was the busiest in Europe. The city attracted both Catholic and Protestant merchants from England, Germany and the northern provinces of the Netherlands. It also had a booming textile industry, was the major salt importer in Europe and became the sugar capital, importing the raw commodity from Portuguese and Spanish plantations on both sides of the Atlantic.

Antwerp's fortune's fell precipitously in the second half of the 1500s, when it became the centre of the religious-political struggle between Catholic Spain and the Dutch Protestants. In 1568, the Netherlands revolted against Philip II because of his high taxes and persecution of Protestants, leading to the Eighty Years' War. Philip was a staunch Catholic filled with religious fervour and happily used bloody force to control the rebellion. During the 'Spanish Fury', over 6000 people were slaughtered in Antwerp and close to 1000 homes were burned to the ground. The king's captain-general Alba, in his 'Council of Blood', brutally put more than 1000 political opponents and religious reformers to death, often by fire.

In 1579, seven northern provinces in the Netherlands formed an alliance against Spain (the Union of Utrecht) and in 1581 these United Provinces declared their independence (the Act of Abjuration) (Anonymous, 2020a). They were able to avoid the brunt of the Spanish Fury by being largely Protestant and becoming an economic powerhouse controlling a worldwide network of seafaring trade routes. As the situation worsened in Antwerp all the intellectuals, merchants and Protestants fled to Amsterdam in the United Provinces. Between 1585 and 1622, the population of the city grew from 30,000 to 105,000 and became the most powerful city in Europe, taking over from Antwerp.

The Dutch path to world power was aided greatly in 1588, when a huge armada sent by Philip II of Spain to invade Protestant Elizabeth I's England was roundly defeated. Composed of 130 ships with 2500 guns, 8000 seamen and almost 20,000 soldiers it appeared invincible, but as the seven-mile-long line of Spanish ships approached the British Isles, the English navy was able to bombard them from a safe distance, taking full advantage of their long-range heavy guns. When the battered armada anchored too closely together in an exposed position off Calais, the English sent eight burning ships into their midst and the panicked crews of the Spanish ships cut their anchors and fled out to sea to avoid catching fire. This Spanish fleet, now in complete disarray, was chased by the English and cut to pieces as they picked off all stragglers one by one. The devastated armada was forced to retreat north and return home in a hard journey around Scotland and Ireland. When the last of the surviving fleet got home in October, half of the original armada was lost and some 15,000 men had perished.

This decisive defeat of the Spanish Armada greatly bolstered the confidence of the English and Dutch and encouraged them to forge their own routes to the riches of India and South East Asia. However, the way forward was still clouded in mystery. Outside Portugal there was very little knowledge about where the spices were procured and how to get there. The early voyages east of the Englishmen Drake, Cavendish and Lancaster had provided some sketchy details, but overall little was known about the full extent of the Indian Ocean trade network. Two remarkable men were to provide this information: Englishman Ralph Fitch and Dutchman Jan Huygen van Linschoten.

Ralph Fitch

In 1591, the first English worldwide traveller Ralph Fitch returned home after an astonishing eight-year journey to Mesopotamia, the Persian Gulf, the Indian Ocean, India and South East Asia (Ryley, 1899). He returned with incredible first-hand knowledge of the inner workings of the spice trade and its opportunities.

His journey began in February 1583 when he embarked for Aleppo, Syria with two merchants, John Newberry and John Eldred, a jeweller, William Leedes, and a painter, James Story. Their trip was financed by the Levant Company, chartered by Queen Elizabeth I, to generate trade and political alliances with the Ottoman Empire. They travelled down the Euphrates, crossed southern Mesopotamia to Baghdad, and then followed the Tigris to Basra where Eldred stayed behind to trade.

Fitch and the others found passage down the Persian Gulf to the Portuguese fortress at Ormuz, where they were arrested as spies and sent across the Indian

Ocean to Goa. There they were held captive until two English Jesuits helped them escape. Story chose to join their sect and remain, while the other two continued across India until they reached the court of the great Mogul leader Akbar at Agra. Leedes got a post from Akbar and stayed put, while Fitch and Newberry continued on to Allahabad, after joining a huge convoy carrying salt, opium, lead and carpets. Newberry decided to head home at this point but was never heard from again.

Fitch kept travelling, sailing down the Yamuna and Ganges Rivers from 1585 to 1586, and then across the sea to Pegu and Burma, where he further travelled deep into the interior of the continent on the Irrawaddy River. Early in 1588, he visited Malacca, and tried and failed to get passage into the South China Sea. At this point, he decided it was time to head towards home, travelling first to Bengal, round the Indian peninsula to Cochin and Goa, then on to Ormuz, up the Persian Gulf to Basra and then down the Tigris and Euphrates to Aleppo and Tripoli.

Fitch arrived back in London on 29 April 1591, eight years after he had left, having visited the whole extent of the spice route spanning from the Levant to South East Asia. He appeared in London after being long presumed dead and until his real demise 20 years later, his eyewitness accounts served as a firebrand and blueprint for the early English missions to the Indian Ocean.

Jan Huygen van Linschoten

To a large extent, the Dutch path to the East Indies was paved by the adventurer Jan Huygen van Linschoten, who in 1595 published his *Itinerario*, almost an exposé of the Portuguese spice trade (Saldanha, 2011; Anonymous, 2020b). Huygen was born in Haarlem, Holland in 1563, the son of an innkeeper. At the age of 16 he left for Spain to work with his brother in Seville for five years and then headed to Lisbon where he was employed as a merchant. In 1583 he landed a job as an accountant for the Archbishop of Goa, João Vicente da Fonseca, and sailed off to India. Once there, Huygen devoted himself to learning about Goa and its inhabitants, assimilating a great deal of nautical and mercantile information by visiting the docks and talking to sailors. He obtained much first-hand knowledge of the 'secret' Portuguese trade routes and practices, and even befriended the Portuguese viceroy who let him see top-secret charts which he deviously copied.

Huygen loved his life in Goa but was forced to leave in 1588 when his sponsor died. On the way home, his ship was wrecked near the Azores and he stayed for two years on the island of Terceira working to salvage and sell the spices that were aboard the ship. He also began writing his future publications. He finally got home in 1592 but his wanderlust was not extinguished, and he sailed to the Arctic with Willem Barents on the first two of his failed missions to find the Northeast Passage to China.

It was not until 1595 that Huygen finally settled down and finished the *Itinerario*, describing his experiences in the East and offering detailed information on the route to India, the nature of the spice markets and the condition of the Portuguese Estado da Índia. The book clearly documented the growing European suspicion that the Estado da Índia had become decadent and overstretched. It showed that the Portuguese did not really have control of the markets in Java, which would ultimately become the centre of Dutch Indonesia. Most importantly, the book gave almost complete nautical and economic information about the Indian Ocean that until its publication had been a closely guarded secret of the Portuguese.

The First Dutch Expeditions to the East Indies

With the maps and knowledge provided by Huygen, the Dutch merchants sprang into action. In 1595, nine Amsterdam merchants joined together and organized the first Dutch expedition to the East Indies. They chose their fellow merchant Cornelis de Houtman to lead it and provided him with four ships: the *Mauritius*, *Amsterdam*, *Hollandia* and *Duifje* (Masselman, 1963; Milton, 1999). The plan was to follow the traditional Portuguese route around the Cape of Good Hope and then head to Bantam, the main pepper port of west Java, and fill up with spices.

The whole affair proved to be a fiasco, filled with bloody infighting, senseless carnage and needless death. By the time the expedition got to the Cape, 71 sailors had already died of scurvy. Discipline on the ships had broken down completely and open warfare had erupted among several discontented factions. A brief truce was called when the expedition reached Sumatra so they could obtain a few spices, but the bitter infighting soon returned after loading. There was an attempt on Houtman's life and the alleged perpetrator was poisoned to death.

When the ships did finally arrive at Bantam, they found that the Portuguese traders in the area had convinced the Bantams to raise their prices to absurdly high levels. In response, Houtman went on a rampage, bombarding the city, damaging the royal palace and capturing innocent victims for torture. At the nearby port of Sidayu, a group of locals managed to board one of the ships and kill 12 of the crew only to be chased back to shore and murdered. When the Dutch ships passed the island of Madura, which did not know of their rampages, a small flotilla of greeting natives was blown out of the water, killing their prince.

At this point, the crew and ships were in desperate shape. Hundreds of men had died on the journey and the ships were greatly weathered and leaking badly. One ship was abandoned altogether and burned. After more violent arguments over whether to continue on to the Spice Islands, agreement was finally reached to head home, even though few spices had been purchased. The expedition arrived home after two years at sea with two-thirds of the crew dead and only the pitiful amount of spices they had obtained in Sumatra. Miraculously, a profit was still turned, as the price of spices had risen so dramatically while the ships were at sea.

A second expedition was sent out in 1598 with six ships led by Jacob van Neck, with Wybrand van Warwijck and Jacob van Heemskerck each commanding a ship. Heemskerck was one of the survivors of the last Barents expedition to the Arctic. Van Neck's ship and the others were separated after rounding the Cape of Good Hope but were reunited some months later at Bantam. Unlike Houtman, van Neck was a shrewd negotiator and accepted the inflated prices of the natives to forge future relationships. He returned to Holland in July 1599 with four ships loaded with spices that netted huge profits, including nearly a million pounds of pepper and cloves, and another half-ship of nutmeg, mace and cinnamon (Masselman, 1963). He and crew were met by an ecstatic Amsterdam, with parades in the street and all the church bells ringing.

When van Neck headed home with four of the expedition's ships, Warwijck's set off for Ternate while Heemskerck's sailed to the Banda Islands (Masselman, 1963; Milton, 1999). Warwijck reached Ternate without incident and was received well by the king, who was anxious for support in his war with neighbouring Tidore. Warwijck loaded his ship with spice and headed home, reaching Amsterdam in September 1600, also to a hero's welcome.

Heemskerck got a much cooler reception at Great Banda Island, as the natives were wary of foreigners after their 90 years of experiences with the Portuguese. However, with lavish gifts and diplomacy, Heemskerck finally won their trust and was able to fill his ship with nutmeg. While on the island, he was treated to a huge pyrotechnic display when the volcano Gunung Api erupted, and he was witness to a bloody invasion from the natives of the neighbouring island of Neira. Heemskerck and crew arrived safely home in late 1600 also to an ecstatic public.

Jacob van Heemskerck and the Law of Prize and Booty

In 1602, Admiral Jacob van Heemskerck was sent to the Kingdom of Johor at the tip of Malaysia to form an alliance with Sultan Alauddin Riayat Shah III, to help the Dutch fight their common enemy, the Portuguese (Anonymous, 2018). To test the Portuguese capabilities, the sultan alerted the admiral that a large Portuguese carrack from Macao was headed towards Johor and suggested the Dutch should capture it. The admiral agreed and in a daring attack captured the ship *Santa Catarina* and obtained an immense booty of silk, musk and porcelain, along with nearly 1000 soldiers, sailors and captive natives intended to be sold as slaves. The capture of the *Santa Catarina* reshaped the culture of the Dutch nation and yielded one of the most important international legal doctrines regarding the law of the sea. It also marked the start of the Dutch–Portuguese War and the beginning of the end of the Portuguese monopoly on trade in the East Indies.

Upon the ship's arrival in Holland, the Dutch East India Company and Heemskerck filed a lawsuit before the Admiralty Board to secure legal rights to the ship and its contents. The Admiralty Board declared the seizure to be a 'good prize' and ordered it to auction. When the thousands of pieces of fine porcelain from the *Santa Catarina* were auctioned, it triggered a wild craze for *Kraakporselein*, or 'carrack-porcelain'. It was the first time that treasures from the Far East had been offered to the public. The subsequent public demand for fine porcelain ultimately led the Dutch to produce their own world-renowned Royal Delft Blue.

The whole incident surrounding the seizure of the *Santa Catarina* was quite controversial from a legal point of view. Heemskerck had attacked the ship without a privateering commission from the Dutch Republic, and as such it could be considered an act of piracy. In fact, many shareholders were appalled that their enterprise was supporting piracy. To head off a public outcry, the directors of the Dutch East India Company commissioned a young lawyer and humanist, Hugo de Groot (Hugo Grotius), to write a thesis justifying the incident (he was the perfect person for the job as he was a cousin of Heemskerck). What Grotius produced was a 500-page thesis with the title *De Jure Praedae commentarius* (*On the Law of Prize and Booty*) in which he defined the Laws of War and defended free access to the ocean for all nations. The whole thing was not published until 1868, except for one seminal chapter with the title '*Mare Liberum*' ('The Freedom of the Seas'). This was widely circulated and oft reprinted.

English East India Company

As the Dutch began to corner the spice market in Europe, the English could not sit idly by and had to get serious. In 1600 the 'Company of Merchants of London Trading with the East Indies' was formed, which came to be called the 'English East India Company' (EIC). It was established by a royal charter, as a joint stock company giving it a monopoly over all trade east of the Cape of Good Hope and west of the Strait of Magellan. It was run by a governor and a court of 24 directors, who were elected by a general court of investors or subscribers. The original general court was made up of over 200 merchants, gentry and aristocrats.

James Lancaster of privateering fame in the 1590s was made the first director of the company. He was also given command of the company's first expedition of four ships in 1601, composed of the *Hector*, *Susan*, *Ascension*, and the flag ship *Red Dragon*. The *Red Dragon* already had a rich history at sea for under the name *Scourge of Malice*, the 38-gun ship had served George Clifford, 3rd Earl of Cumberland, another of the privateers who had made multiple raids on the Spanish Main and Spanish colonies in Brazil and the West Indies.

After leaving England the fleet was becalmed in the doldrums and when they finally arrived at Table Mountain at the southern tip of Africa, a total of 105 men had died from scurvy. Lancaster's flagship fared better than the others as he had provided his crew with lemon juice during the voyage and he was able to send members of his own crew to help man the other ships into the harbour. The expedition headed along the coast of Africa and after adverse winds limited their progress, they stopped in Antongil Bay in Madagascar for several months to recover from yet another bout of scurvy.

They eventually sailed south-east to Aceh on the tip of Sumatra where the sultan was willing to sell them pepper without customs duties. The problem was that their cargo of woollen cloth was not particularly desired in a tropical country, so Lancaster captured a Portuguese carrack filled with a rich cargo of gold, silver and Indian textiles, which the Sumatrans were happy to take in trade.

There was not enough pepper in Aceh to fill the hulls of more than one ship, the *Ascension*, so Lancaster sent the *Susan* further south to Pariaman in central Sumatra to purchase pepper and took the *Hector* and *Red Dragon* to Bantam on the western side of Java, where he had heard that pepper was cheaper than on Sumatra. Using a letter from Queen Elizabeth I, he negotiated beneficial terms for trade there and a factory was established. A pinnace was also purchased there from the Dutch that Lancaster sent to the Moluccas to obtain cloves and nutmegs.

Lancaster left Bantam for home on 20 February 1602. The *Red Dragon* was almost abandoned at the Cape of Good Hope after her rudder was damaged in a storm, but she was able to limp home with the *Hector* after a confrontation between Lancaster and the crew about abandoning her. They arrived back in London in September 1603 to a city ravaged by the plague. The crowds that had cheered them at the beginning of their voyage were now cowering in their homes afraid to go outside. Despite the prevailing gloom, Lancaster was duly received by the new King James and knighted for his achievements. Lancaster continued on as the chief director of the EIC for another 16 years but made no more voyages himself.

Dutch East India Company (Verenigde Oost-Indische Compagnie)

Intense competition for East India trade arose in the Dutch Republic after van Neck's and Heemskerck's successful expeditions. Dozens made the trip to Bantam, the Banda Islands and Ternate and established trading stations. From 1598 to 1602, 51 ships in eight fleets sailed to the Indian Ocean and beyond (Crump, 2006).

In 1602, the government formed the Dutch East India Company (*Verenigde Oost-Indische Compagnie* or VOC) in an effort to stabilize profits. It was given the power to essentially govern the East and was allowed to run its own shipyards, build forts, keep armies and make treaties. It was the first public company in the world to issue negotiable shares and became immensely powerful, dominating Eastern trade for two centuries. The VOC was governed by 17 representatives of the investors, called the *Heeren XVII* or the Lords Seventeen.

Soon after its formation, the VOC began to identify and single out a number of focal points in South and South East Asia that it considered front-line polities (Borschberg, 2009) These were deemed by the VOC to be important points of commerce, but politically weak and/or unstable. There were two regions identified at the forefront: the Malay Peninsula and the Maluku and Banda island groups. The VOC set out to arrange military alliances with the rulers of these front-line polities and develop agreements whereby they supplied spices exclusively to the Dutch. All these polities were already in armed conflict with the Spanish/Portuguese or irritated with them and easily co-opted to fight with the Dutch. Entry into the Portuguese spheres of India and China was put on hold by them until the Dutch had established a foothold in Indonesia.

When the VOC began its grab for control of the spice trade, the balance of power in the western Indonesian archipelago rested on the shoulders of four major pepper producers: Aceh and Melaka on the Malay Peninsula; Johor in western Java; and Bantam on north-west Java. Melaka was under control of the Portuguese, while the others were ruled as independent sultanates. In the eastern Indonesian archipelago, the clove trade was centred on Ambon Island and the Molucca islands of Ternate and Tidore. Most of Ambon was under the control of the Spanish, although the Dutch had recently made trade agreements with the Hitu in the northern Muslim portion of Ambon. Ternate and Tidore were now independent of Iberian control but were still under constant Spanish threat and in continuous battle with each other. The nutmeg and mace trade was centred in the Banda Islands which were ruled by a group of rich men that had remained largely free of European control.

The Dutch Establish an East Asian Trading Network

The first VOC fleet sailed on 18 December 1603 and within a few years, the company had established a network of hundreds of bases across Asia, ranging from simple offices and warehouses to huge commercial centres (Fig. 19.1). By 1605 there were factories on Java, Sumatra, Borneo, the Spice Islands, the Malay Peninsula and the mainland of India. Its headquarters were established in Batavia (Jakarta, Java), where a great fort was built that was surrounded by workshops, warehouses and homes for the resident staff. The whole operation was ruled by a governor general who chaired the Council of the Indies, the company's executive body. The governor general operated essentially as a head of state.

Each year two or three fleets would leave the Dutch Republic headed for the East Indies, a trip that took about eight months. Death rates were astronomical among the Europeans both en route and in the tropics, and the manpower there had to be constantly replenished. The VOC built special ships for the long voyage to East Asia which were a combination of battleship and cargo carrier. They were massive ships from 500- to 1000-tons burden, and had huge holds to carry produce, food and ammunition. The ships carried hundreds of passengers with little thought about hygiene. Officers and passengers occupied cabins on the quarterdeck, while sailors and soldiers slept tightly packed like sardines

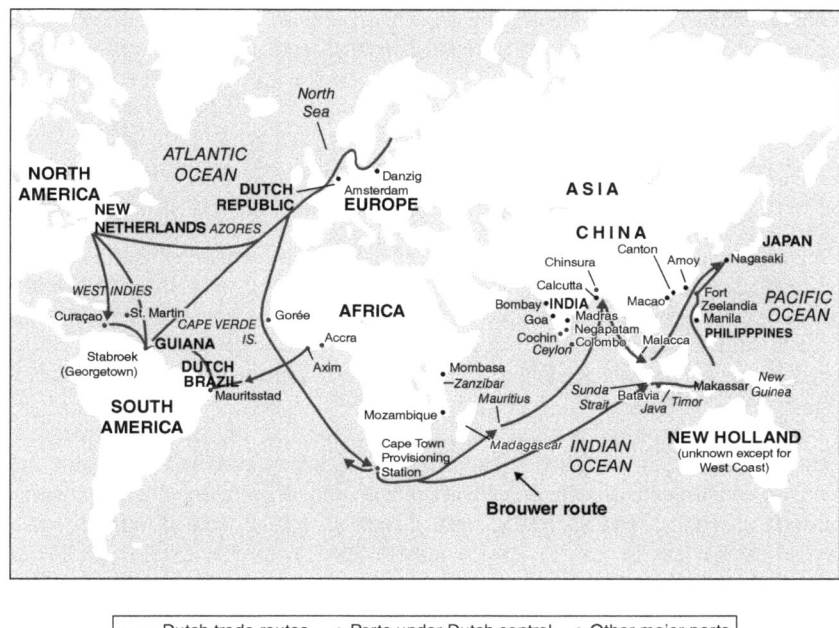

Fig. 19.1. Dutch trade routes to the East Indies and beyond. (Redrawn from https://voertaal.nu/wp-content/uploads/2018/12/Dutch_trade_routes-600.jpg)

below decks. Tight discipline had to be maintained to keep these masses under control. All and all, the trip east was a difficult and tedious experience.

First English–Dutch Confrontations

The next Englishman to lead an expedition east was Henry Middleton, who had sailed with Lancaster on the first voyage and was given the same four ships. He departed on 25 March 1604 and after an arduous, scurvy-filled journey, arrived at Bantam on 21 December. Along the way they successfully harpooned a whale in Table Bay and rendered its oil for their lamps.

Soon after he arrived, Middleton presented the sultan of Bantam with a letter from the English king and a hodgepodge of gifts including muskets and spoons (Milton, 1999). He then was allowed to load up two ships, the *Hector* and *Susan*, with the pepper that had been stockpiled by the surviving English merchants left behind by Lancaster. One can only imagine the Englishmen's relief to be reconnected with their countrymen after months of harsh treatment by the natives.

Middleton and the remaining two ships then headed to the Spice Islands and the island of Ambon. Along the way, they suffered an awful spate of the 'bloody flux', a particularly deadly type of dysentery. Upon arrival Middleton was able to get permission to trade by convincing the local Portuguese commander that King James I and King Philip II had signed a peace treaty. This was indeed true, but wily Middleton could not have known this, as the treaty was signed five months after he had gone to sea.

Just as they were about to begin loading the ships, an ominous sight appeared on the horizon: a VOC armada of nine ships along with a supporting group of pinnaces and sloops. The Dutch had sent a fleet to challenge the Portuguese control of the island and the English had stumbled into the middle. Fearing the coming onslaught the English beat a hasty retreat, while the Portuguese suffered a full-scale invasion, where their fortress at Victoria was taken. Middleton decided to separate his fleet and go elsewhere, sending the *Red Dragon* to the Moluccas for cloves and the *Ascension* to the Banda Islands for nutmeg and mace.

Little did Middleton know that he was about to fall into the middle of another Dutch–Portuguese confrontation. As he approached the islands of Tidore and Ternate, he came across two Ternatan ships being chased by seven Tidore war boats. The lead boat carried their king and three Dutchmen, who pleaded with him to rescue the second boat filled with more Dutchmen. It was to no avail as the second boat was caught and all were killed except the three who had managed to swim to the *Red Dragon*.

The Tidorans had aligned themselves with the Portuguese who were battling the Dutch and their ally, the sultan of Ternate. The Spaniards tried to block Dutch ambitions with the help of Tidore, by building fortresses in 1605 on the western and southern coasts of Ternate and another one on Tidore. To defend themselves and their Ternatan allies, the Dutch constructed a fortress in 1607 on the east coast of Ternate.

Middleton arrived at Tidore on 27 March just as the Dutch and their allies were about to attack and told the Portuguese that if they did not trade with him, he would join the Dutch in their war against them. This pressure succeeded and he was allowed to fill his hulls with cloves before heading home. A few days later, the Dutch did successfully take over the Spanish fort on Tidore and began the process of gradually pushing the Spanish out of Ternate. They succeeded by making the costs to maintain the Spanish strongholds in the Moluccas exceed the benefits, and Spain voluntarily withdrew from the Moluccas in 1663.

The Ill-Fated Missions of Admiral Pieter Willemszoon Verhoeff

In 1607, Admiral Pieter Willemszoon Verhoeff was sent with a fleet of 13 ships to Indonesia to capture Malacca from the Portuguese and then build a fort on the Banda island of Neira. Both of these missions ended in failure with long-standing impacts, although the Dutch had a great party sandwiched between these events.

When the fleet arrived at Malacca, Verhoeff first sent word to the Sultan of Johor, Alauddin Riayat Shah III, requesting help, which was refused. Verhoeff attacked Malacca anyway, but could not take the heavily fortified city, and immediately set a course south to make sure that the sultanate remained loyal to the Dutch. Upon arrival at the capital of Johor, he and his crew were treated with exceptional warmth and hospitality (De Witt, 2016). They were carried by elephants to the sultan's palace and treated to a banquet in honour of them, with exotic local foods, musical performances and dancing girls. They were also given a boat ride down the Johor River. It turned out that it was Aidilfitri, the last day of Ramadan, and the Muslims had not been able to help them at Malacca because their assault had coincided with the holy fasting month. The festivities went on long and hard, with the Dutch organizing a mock battle to entertain the Malays and Verhoeff receiving a gold Malay dagger from the prime minister in return for the admiral's personal sword.

In April 1609, Admiral Verhoeff arrived at the Banda island of Neira to build the fort and ultimately suffered a much greater tragedy than at Malacca. Many of the rulers of Banda, the *orang kaya*, had been persuaded by the Dutch to sign a treaty granting a monopoly on spice purchases, but were bitterly opposed to the Dutch having a physical presence on the island. On 22 May, the *orang kaya* called a meeting with the Dutch admiral to supposedly negotiate prices and instead ambushed him and two of his officers, decapitated them, and then murdered another 46 of the Dutch soldiers. The rest of the Dutch sailors and soldiers escaped, but this event greatly soured their attitude towards the Bandanese forever. The ambush and broken treaty would be used repeatedly to justify Dutch oppression and ultimately the extermination of the Bandanese.

Brouwer Route

In 1610 Dutchman Hendrik Brouwer found a way to cut the voyage from Africa to Java in half (Fig. 19.1). Instead of taking the previous monsoon route, which followed the coast of East Africa northwards, through the Mozambique Channel round Madagascar and then across the Indian Ocean, he headed directly from the Cape to Java taking advantage of the strong westerly winds in the Roaring Forties – latitudes between 40° and 50° south. By 1617, the VOC was requiring all its ships to take this route. The Brouwer route proved to be a great success for the Dutch, but the English were more hesitant to use it (Lee, 1934). Captain Humphrey Fitzherbert on *Royal Exchange* tried the route in 1620 with success, but the second English ship to use it sailed too far east before turning north and was smashed off the Pilbara coast of Australia. This spooked the English for two decades. The Dutch also lost a few ships off the coast of Australia, but they proved to be more fearless.

Jan Pieterszoon Coen

One of the men who survived the Banda massacre was Jan Pieterszoon Coen, who carried at that time the modest title of assistant merchant. Upon his return to Holland in 1610, Coen gave the VOC's directors a report on trade possibilities in South East Asia that greatly caught their attention. He was rewarded with the higher rank of chief merchant and sent again overseas in 1612 to the Moluccas, and when he returned, he was appointed head of the company's post at Bantam, in Java. In November 1614, he was made director general of all the company's commerce in Asia.

Coen, a Calvinist, was a ruthless man who demanded strict adherence to contracts forged with Asian rulers. He gained commercial monopolies by supporting these rulers against their indigenous rivals or other European powers, and using an iron fist the Dutch gained control of a large portion of spice trade through heavy military and naval investment. Coen justified his liberal use of force as 'there's no trade without war; there's no war without trade' (quoted by Goodman, 2010, p. 61).

As Coen and the VOC worked to control the spice trade, the English did everything in their power to keep a piece of the action for themselves. The first major confrontation between the Dutch and English under Coen's leadership came in Java. Cohen had begun his reign as VOC director by transferring the VOC headquarters from Bantam to Jacatra (present-day Jakarta) when the sultan of Bantam resisted his attempts in 1610 to control the pepper trade. Also not trusting the local Jacatra ruler, he had converted the warehouse there into a fort.

In 1618, Sir Thomas Dale was sent from England to break up Coen's growing monopoly. In a pitched battle, the undermanned fleet of Coen was able to hold its own for a while but was forced to flee to Ambon, where there was a larger Dutch fleet. While he was away, the sultan of Bantam also decided to enter the fray, forcing the English to withdraw and laying siege to the Dutch fort. When Coen returned in 1619, he was able to force the Bantams back out of Jacatra and burned the city to the ground. On its ashes he built the new Dutch city of Batavia, which became the headquarters of the VOC in South East Asia.

While the English and Dutch were battling over Jacatra, another battle was ranging between them on the tiny Banda island of Run. There, Captain Nathaniel Courthope, 39 English defenders and their native allies were trapped in a small fort under siege from the Dutch. Courthope had reached Run in 1616, signed a contract with the locals accepting James I as their sovereign and was now holed up in a fortress he had built. Courthope and his supporters held out for 1540 days before he was killed and the remaining English retreated. The Dutch subsequently went on a rampage – killing or enslaving all adult Bandanese men, exiling their dependants and chopping down every nutmeg tree they could find.

Ironically while Courthope and crew were under siege, Coen received word that the Dutch and English leaders had reached an agreement in 1620 to cooperate in the East Indies rather than fight. They were to leave each other's existing trade settlements alone and cooperate together against common enemies. A frustrated Coen let the battle over Run play out and named a huge swath of Java as the 'Jacatra Kingdom' to keep it out of English hands.

In January 1621, Coen decided to make a full-scale conquest of the Banda Islands under the guise that the Bandanese had been disregarding previous commercial agreements. Using Japanese mercenaries, the Dutch took the island of Lonthor by force but suffered fierce local resistance that was supported by English-supplied cannon. In response, Coen and the Dutch massacred thousands of inhabitants, replacing them with slaves from other islands and deporting some 800 inhabitants to Batavia. Only about 1000 of the original 15,000 on the island survived. The violence was so extensive, and the international condemnation so great, that Coen even drew a reprimand from the VOC's directors.

At this point, Coen thought the best future strategy would be to colonize the East Indies with Dutch settlers, who would handle the East India trade. To sell his plan to the VOC directors, he travelled back home in 1623. Everything was looking good until he was hit with another scandal, the so-called 'Amboyna massacre'. In 1623, a group of Englishmen from a small settlement at Cambello in Ambon were taken into custody by the Dutch factors, questioned, grievously tortured and sentenced to death. They were suspected of working with the locals in a plot to take over the Dutch settlement. While Coen was not directly involved, he was held morally responsible by the English as the leader of the VOC and roundly condemned. Since the English and Dutch were now on friendly terms, he was temporally forbidden to return to Batavia and it wasn't until 1627 that he was able to go back, travelling incognito. His colonization plan never took hold and he died probably of dysentery in 1629. It took until 1654, after years of many fruitless negotiations, that the Dutch finally paid 300,000 guilders in compensation to the descendants of the Amboyna massacre.

The Dutch actions in Ambon and the Banda Islands were clear indications that the VOC would do anything to safeguard its interests. All cultivation in Ambon and the Bandas became regulated under the strict supervision of the VOC, which could then control the world supply of cloves. In the Bandas, the VOC divided up the conquered land into parcels that were given in hereditary tenure to Dutchmen, who exploited the land with slave labour. In Ambon, the Dutch ruled by agreement where clove production was regulated by village chiefs who got 10% of the price paid to the producers.

Dutch Finally Take Portuguese Malacca

Once the VOC had pushed the English out of most of the Indonesian Archipelago, they turned their eyes towards Portuguese shipping in an attempt to destroy Malacca as a trade centre. As described by Sluiter:

> After 1636 Dutch fleets annually blockaded Goa and the entire Malabar Coast ... On August 2, 1640, its military and naval forces began the siege of Malacca for

the third time. One hundred and sixty-five days later this key stronghold fell and with it virtually the last Portuguese hope of stemming the Dutch ascendancy in the Far East. This was unmistakably the end of an epoch, not only in Far Eastern but in world history. A half century of vigorous, aggressive Dutch expansion had done more than anything else to shatter an ancient status quo, the Iberian colonial monopoly.

(Sluiter, 1942, p. 41)

The VOC in the North Moluccas

During the course of the seventeenth century, Ternate become a formidable military power after throwing off the yoke of the Iberian powers and began collecting a number of vassal states throughout the eastern archipelago. Tidore remained independent but operated largely in the shadow of Ternate. Makassar on South Sulawesi became an important entrepôt for European and Asian traders seeking to purchase spices and high-value goods away from the Dutch.

Over time, both Tidore and Ternate fell increasing under Dutch control. In 1652 they both agreed to a spice eradication programme (*extirpate*) where, for an annual payment, all clove trees not owned by the VOC would be destroyed. This monopoly was to be upheld by making regular expeditions to eliminate any spice trees outside Dutch-controlled territories. These expeditions became notorious for their brutality and continued until the nineteenth century.

Makassar proved long to be resilient to the VOC's 'monopolistic aspirations'. Its success:

> stemmed from a combination of factors. Makassar was not merely a trade entrepôt; it was also the main political centre of South Sulawesi. The port city was the seat of government of the kingdom of Gowa, which, jointly with the neighbouring kingdom of Tallo, stood at the head of a wider federation of principalities ... Gowa fielded formidable armies and was defended by extensive fortifications. It also had an expansive diplomatic reach. The diplomatic connections of Gowa and Tallo reached from the Moluccas to Mecca, including ties with the English and Danish East India Companies and the Portuguese.

(Mostert, 2018, p. 26)

As a result, the VOC and Gowa–Tallo were in a state of conflict for more than 50 years and had outright wars in 1633 and the 1650s and 1660s. It all came to a head in the Makassar War of 1666–1669 when the VOC and a host of local allies led by Arung Palakka of the Bugis ethnic group in South Sulawesi broke the Gowan's political power. A VOC fleet, under Cornelis Speelman, fought the Gowa fleet, while Palakka led a 'difficult' land campaign to win South Sulawesi. Palakka became the most powerful man in South Sulawesi and governed with authoritarian rule until his death in 1696.

East India Companies' Effect on European Trade

The trade goods from India coming into Europe from the East India companies had major impacts on the previous balance of trade. As succinctly and eloquently told by Krishna:

> Many far-reaching consequences of the direct trade carried on by the Dutch and English with the Indies become visible at a very early period in Europe as well as in Asia ... The Hanseatics, who carried on a very profitable sea-borne trade as far as Venice, were so completely wormed out of it by the Dutch and English that they had to sell their large ships and return home overland. Then the English Levant Company made complaints, as early as 1604, on the decay of their trade into the Levant, alleging that all spices, silks, indigo and goods of the Indies, which used to be brought through Persia into Turkey, and home thence to England, were being brought direct from the Indies. The share of the Turks, Egyptians and Venetians was, however, more considerably reduced. The wealth, revenues, prosperity and population of the old commercial cities like Cairo, Alexandria, Aleppo, Constantinople, Mocha and Ormus were given a serious blow by the diversion of Oriental trade into new routes.
>
> But most of all, the direct and enthusiastic participation of the Dutch and English in the Eastern trade spelled a great disaster to the Portuguese trade and prosperity. The sinking condition of the Portuguese and the capture of their trade and possessions by the new adventurers form the prominent features of the period. Even in 1607 the Portuguese are reported to have sustained so great a loss in the East Indies that it was thought in these places a wound almost incurable. Then the trade to Lisbon for spices was overthrown through the East India trade with England and Holland, and so was also the sale of cloths and jerseys, which were the chief commodities before transported by the Portuguese from Europe ... In fact, the trade of the Dutch and English had so shortened in their return that half their galleons did not come, and those that came from Portugal brought new supplies for the garrisons, but returned so empty that the charge was but defraud ... 'Never were such opportunities,' writes Sir Thomas Roe, 'to discharge the Portugal from all these coasts. He is declining on all sides, and a little weight on his head now laid would sink him.'
>
> (Krishna, 1924, p. 84)

Atlantic Trade of Europeans

When the Twelve Years' Truce expired, the Dutch–Iberian conflict resumed all across the world. Tempted by the great wealth being shipped from the Americas by the Spanish in the treasure fleets, the Dutch West Indies Company (*West-Indische Compagnie* or WIC) was chartered in 1621. It began as a company of privateers supporting the war with Spain and evolved into a trading company moving goods from North and South America.

In the first few years after the incorporation of the WIC, the Dutch and Iberians battled mightily across the Americas, including the salt flats of

Punta de Araya in Argentina, northern Brazil, Peru, Manila, Formosa and the Caribbean (Sluiter, 1942). With 14 ships, Admiral Piet Heyn in 1626/27 sailed right into Bahia, Brazil and captured 23 loaded sugar vessels. He also captured the entire Spanish treasure fleet in Matanzas Bay in Cuba in 1628 with an armada of 31 heavily armed ships.

The booty was worth over 11.5 million guilders and it took eight days to unload all the silver. In 1630, the Dutch captured the towns of Recife and Olinda in Brazil and took control of the burgeoning Portuguese sugar industry. They controlled these cities from 1637 to 1644 and expanded the territory across a 2000-kilometre (1250-mile) strip of the north-east coast of Brazil. The Dutch also captured the Portuguese fort at Elmina in West Africa from which they traded gold and transported slaves to the Brazilian sugar plantations.

The Portuguese did manage to take back their Brazilian colonies in 1654. When the Dutch were forced to leave, they took their money, experience and shipping to the Caribbean, which was closer to Europe and being actively colonized by the English and French (Hancock, 2017). Within 30 years, the Caribbean took over dominance in worldwide sugar production, driving the price of Brazilian sugar down significantly and greatly reducing its export. Starting with the tiny island of Barbados, a hodgepodge of islands controlled by the English, French and Dutch fuelled a dramatic growth in European sugar use as tea, coffee and chocolate consumption grew by leaps and bounds in the seventeenth and eighteenth centuries. As sugar production spread across the Caribbean, it spawned a massive growth in slavery, stimulated almost constant warfare and piracy, and made many people extremely rich.

References

Anonymous (2018) The capture of the *Santa Catarina* (1603) Peace Palace Library, Den Haag, the Netherlands. Available at: https://www.peacepalacelibrary.nl/2018/09/the-capture-of-the-santa-catarina-1603/ (accessed 28 January 2021).

Anonymous (2020a) Timeline Dutch History. Rijksmuseum, Amsterdam. Available at: https://www.rijksmuseum.nl/en/rijksstudio/timeline-dutch-history (accessed 28 January 2021).

Anonymous (2020b) Jan Huygen van Linschoten's 'Itinerario': Key to the East. *Utrecht University Library Special Collections*. Utrecht University, Utrecht, the Netherlands. Available at: https://www.uu.nl/en/utrecht-university-library-special-collections/collections/early-printed-books/geographical-descriptions/itinerario-by-jan-huygen-van-linschoten#:~:text=Key%20to%20the%20East&text=It%20was%20an%20age%20of,Jan%20Huygen%20van%20Linschoten%27s%20Itinerario (accessed 28 January 2021).

Borschberg, P. (2009) *The Johor–VOC Alliance and the Twelve Years' Truce: Factionalism, Intrigue and International Diplomacy, c.1606–1613*. International Law and Justice Working Paper No. 2009/8. Institute for International Law and Justice, New York University School of Law, New York.

Crump, T. (2006) The Dutch East Indies Company – the first 100 years. Gresham College, London. Available at: https://www.gresham.ac.uk/lectures-and-events/the-dutch-east-indies-company-the-first-100-years (accessed 28 January 2021).

De Witt, D. (2016) Diving into Johor's European connection. *Senses of Malaysia* (Nov–Dec 2016). Available at: http://www.expatgo.com/my/2016/11/23/diving-johors-european-connection/ (accessed 8 April 2021).

Goodman, B. (2010) The Dutch East India Company and the tea trade. *Emory Endeavors in History* 3, 60–68.

Hancock, J.F. (2017) *Plantation Crops: Power and Plunder, Evolution and Exploitation*. Routledge, London and New York.

Krishna, B. (1924) *Commercial Relations Between India and England (1601–1757)*. George Routledge & Sons, Ltd, London.

Lee, I. (1934) The first sighting of Australia by the English. *The Geographical Journal* LXXXIII(4), 317–321.

Masselman, G. (1963) *The Cradle of Colonialism*. Yale University Press, New Haven, Connecticut and London.

Milton, G. (1999) *Nathaniel's Nutmeg: The True and Incredible Adventures of the Spice Trader Who Changed the Course of History*. Penguin Books, New York.

Mostert, T. (2018) Scramble for the spices: Makassar's role in European and Asian competition in the Eastern Archipelago up to 1616. In: Clulow, A. and Mostert, T. (eds) *The Dutch and English East India Companies: Diplomacy, Trade and Violence in Early Modern Asia*. Amsterdam University Press, Amsterdam, pp. 25–54.

Ryley, J.H. (1899) *Ralph Fitch: England's Pioneer to India and Burma*. T. Fisher Unwin, London.

Saldanha, A. (2011) The itineraries of geography: Jan Huygen van Linschoten's *Itinerario* and Dutch expeditions to the Indian Ocean, 1594–1602. *Annals of the Association of American Geographers* 101(1), 149–177.

Sluiter, E. (1942) Dutch maritime power and the colonial status quo, 1585–1641. *The Pacific Historical Review* 11(1), 29–41.

Age of Expansion 20

Setting the Stage – The EIC and VOC Move into India

When the Dutch and English first entered the Indian Ocean, the primary goal of both nations was to gain a monopoly in the spice trade. To do this, they had to militarily push out the Portuguese and prevent the other from gaining a foothold. Ultimately, the VOC came out the big winner taking control of the clove, nutmeg and mace trade of the Moluccas. It also took a considerable portion of the Indonesian pepper trade by force, but not all.

With the loss of the Spice Islands, the British shifted their attention to India and its pepper, saltpetre, cotton and indigo. The VOC also turned its eyes to India, but with far less lasting impact. To gain their foothold in India the English and Dutch were faced with two significant challenges: they would need to gain the favour of the Mughals who now controlled most of North India and they would have to push back the Portuguese who were well entrenched along the west coast. The Mughals had left the Portuguese ports mostly alone, preferring to trade with them rather than fight.

The arrival of the East India companies in India marked a new era in European–Indian commercial exchange (van Meersbergen, 2018). As Nadri describes:

> The companies displayed a different approach to trade and trading from their predecessors, the Portuguese, whose relationship with Indian maritime merchants frequently ended in conflict. Unlike the Portuguese, the English and the Dutch had a strong mercantile tradition and tended to be more pragmatic in their dealings with Indian and other Asian merchants. They forged a close commercial relationship with Indian merchants and secured permission from Indian rulers to establish trading stations or factories in port cities and in the interior ... The companies' large-scale trading enterprise and the kind of commodities that they exported from India required a close interaction with Indian merchants, brokers, and bankers. There was a well-developed market structure in place with a hierarchy of merchants and intermediaries as well as banking and brokering

services. These merchants and intermediaries were willing to extend their commercial services to the EIC and VOC and to take the business opportunities that the companies presented to them.

(Nadri, 2018, p. 127)

Mughals of India in the Seventeenth Century

In 1600 the Mughals under Akbar the Great (r. 1556–1605) ruled most of the subcontinent of India (Richards, 1993). They had arrived on the subcontinent about the same time as the Portuguese. Akbar was a 'workaholic' who seldom slept more than three hours a night and personally oversaw the administration of his vast country. He built his empire by conciliating conquered rulers through marriage and diplomacy, winning him the support of his non-Muslim subjects. He assimilated Hindu chiefs into high government ranks and even celebrated their holidays.

When Akbar died, Jahangir became the fourth Mughal emperor, ruling from 1605 until 1627. He was a largely ineffective leader, addicted to opium and subject to court intrigues. Under his tenure 'the number of unproductive, timeserving officers mushroomed, as did corruption' (Heitzman and Worden, 1995, p. 23). Jahangir was far less impartial than Akbar and supported mass conversions to Islam. He married a Persian princess and his court became filled with Persian artists, scholars and writers who found asylum in the Mughal court.

Jahangir was replaced by his son, Shah Jahan (r. 1628–1658), who had a great passion for building, exemplified by the Taj Mahal. He also strongly supported literature, painting and calligraphy and he probably had the largest collection of jewels in the world. His court was run with great ceremony and splendid display. This opulent lifestyle was not without consequences as it greatly strapped the Mughal Empire's economy at a time when resources were shrinking.

The last of the great Mughal leaders was Aurangzeb (r. 1658–1707), who took power by killing all his brothers and imprisoning his father Shah Jahan. Under Aurangzeb's reign, the empire became the world's largest economy, holding almost a quarter of the world's gross domestic product (Broadberry *et al.*, 2013), although it was failing. 'The bureaucracy had grown bloated and excessively corrupt, and the huge and unwieldy army demonstrated outdated weaponry and tactics' (Heitzman and Worden, 1995, p. 24). When Aurangzeb died in 1707, the great Mughal Empire that had controlled most of India for 180 years quickly fell apart and dissolved into many smaller independent states. Ultimately the whole subcontinent would fall under British rule.

EIC Entry into India

The first English expedition to India was led by William Hawkyns, who landed the *Hector* at Surat on 24 August 1608. Hawkyns proceeded directly to

Jahangir's court at Agra, where he hoped to secure a trading agreement. This took some time, as the emperor so greatly enjoyed the company of Hawkyns that he detained him for three years before granting a *farman* (licence) to build a factory. While at court, Hawkyns was provided with a handsome salary and a wife.

In 1612, Thomas Best was sent to Surat from England with a fleet of four ships – the venerable *Red Dragon* along with the *Hosiander*, *James* and *Solomon*. Soon after they arrived, a squadron of four Portuguese galleons and 16 barks confronted them and a pitched battle ensued over a three-day period. Three of the galleons were grounded and one was sunk, forcing the Portuguese to withdraw. The whole affair was witnessed by thousands of people on the shore and so impressed the *sardar* (governor) of Gujarat that he convinced the emperor that henceforth he should favour the English over the Portuguese. This put English trade on a permanent footing in India.

In 1615, Thomas Roe was sent to India as ambassador of King James I, and through lavish gifts and flattery to Jahangir was able to obtain a *farman* to trade and establish factories all across the Mughal Empire. Jahangir was not willing to give the king exclusive trading rights but did allow the English the right to compete with the traditional traders. The EIC also convinced the Vijayanagara Empire to allow them to open a factory in Madras. The English then began to establish trading posts up and down the coasts of India, and large English communities were established in the three major trading towns of Calcutta (Kolkata), Madras (Chennai) and Bombay (Mumbai).

Early Dutch Spread into India

The Dutch probably first became aware of the importance of Indian trade in their contact with Gujaratis in Aceh as the seventeenth century dawned (Fischer, 1965; Ashfaque, 2006). A brisk trade between the sultan of Aceh and the ports of Gujarat had been going on long before the Dutch arrived. Early in 1602 two Dutch factors, de Wolff and Lafer, were sent to Gujarat to investigate the possibility of establishing trade. They sent back home favourable reports, but on their route back they were captured and killed by the Portuguese.

The next Dutch merchant sent to Surat was David Van Deynssen, who arrived in 1606. However, 'his effort turned out to be just as ill-fated as that of de Wolff and Lafer, as the Portuguese succeeded in setting the Mughal authorities against him. After being tortured by the latter and threatened in many ways he committed suicide, leaving his merchandise in their hands' (Fischer, 1965, p. 208).

The VOC spent the next ten years trying to get compensation for Van Deynssen's belongings worth about 20,000 guilders. An opportunity appeared in 1615, when the Mughals offered to return the goods if the Dutch would give them naval support in their war against the Portuguese. A factor named Ravensteyn was dispatched overland from Masulipatam (Machilipatnam) to Surat to receive the goods, but by the time he got there the Mughals and Portuguese had made peace and Ravensteyn returned empty-handed.

Disgruntled, he wrote to his superiors at Amsterdam that the company would be better served by concentrating on the Moluccas than go after the 'uncertain opportunities of the Surat' (Fischer, 1965, p. 208).

The Dutch director general Coen was undeterred and dispatched Pieter Van den Broecke to Surat in August 1616. He too was initially frustrated, but finally in 1618 with the support of the Gujarat merchants, Emperor Jahangir issued a generous *farman* allowing the Dutch to trade in Surat. This document contained the following provisions:

- The Dutch may come and go for their trade whenever they like, and nobody should forcibly interfere with them; and at the same time they (the Dutch) should not use force against anybody.
- On all imports and exports by the Dutch the normal customs of Surat would be taken and nothing more.
- Nobody should prevent the merchants of Gujarat from buying any merchandise from the Dutch, if they wished to, and they should not be penalized for that.
- Any rarities, brought for presentation and sale (i.e., to the King or big nobles) should not be opened at the customs-house, but be kept sealed and in this state brought before the throne.
- If any of the Dutch die in India, no Indian should claim his belongings, but these should be returned to the countrymen of the deceased.
- In all judicial matters regarding affairs of the Dutch they should be dealt with by their own authorities.
- We (the Mughal authorities) should not try to convert any of the Dutch to Islam by force.
- All the victuals required by the Dutch ships may be taken free of customs.

(Fischer, 1965, p. 209)

This agreement was renewed 28 times between 1618 and 1729.

East India Companies Spread Out from Surat

Surat held most of the EIC's early attention in India, but its importance declined dramatically when a protracted drought hit in the 1630s that affected all of western India. As an eyewitness described:

> When wee came into the cytty of Suratt, we hardly could see anie living persons, where heretofore was thousands; and ther is so great a stanch of dead persons that the sound people that came into the town were with the smell infected, and att the corners of the street the dead laye 20 together ... In these parts ther may not bee anie trade expected this three years.
>
> (Barrow, 2017, p. 13)

Surat did manage to recover, but the EIC gradually shifted its focus towards the west coast of India to Bombay, then Madras on the eastern Coromandel Coast and finally Calcutta in Bengal.

Bombay was received as a gift from the Portuguese to the English king, Charles II, as a part of the dowry arrangement in his marriage to Catherine of Braganza. The EIC moved its headquarters to Bombay in 1687, when local political pressure at Surat became too intense and required a move. The local authorities were becoming increasingly unhappy with the EIC who were seizing ships of their European competitors and damaging Mughal trade. The English also feared the other regional power, the Maratha Confederacy, which was becoming increasingly hostile towards the Mughals and had sacked Surat on two occasions.

The VOC began establishing trading posts along the Coromandel Coast in the early seventeenth century (Fig. 20.1). Factories were set up in 1600 at Palecatte (Pulicat) and in 1615 at Masulipatam. A factory at Negapatam (Nagapattinam) was captured from the Portuguese in 1658. The VOC tussled with the Portuguese to hold these settlements and had many internal disputes with local leaders. As a result, most of their trade became organized around fortresses and strongholds. The VOC built Fort Geldria in Pulicat which became their Coromandel Coast headquarters. It was destroyed in 1680 by a tsunami but was rebuilt and served as the home of the VOC governor for the Coromandel until 1690.

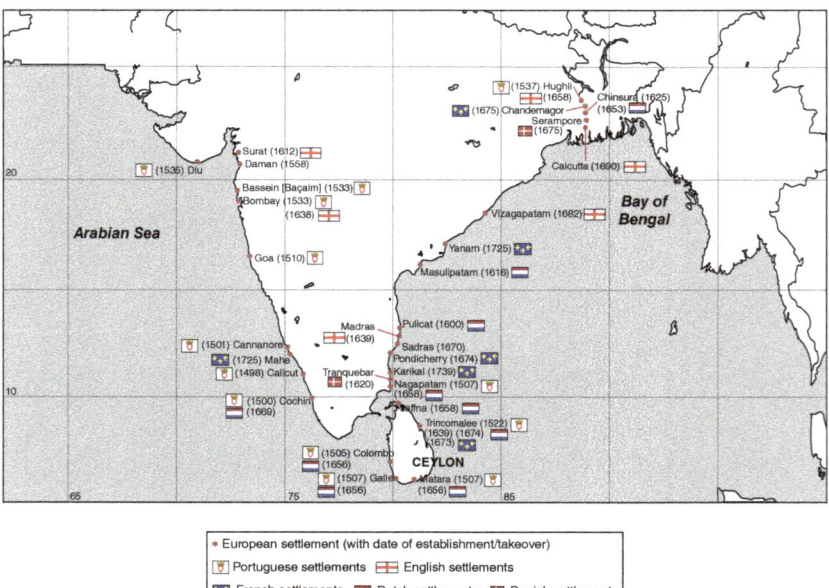

Fig. 20.1. European settlements in India. (Redrawn from Luis wiki, CC BY-SA 2.5 <https://creativecommons.org/licenses/by-sa/2.5>, via Wikimedia Commons. File: European settlements in India 1501-1739.png)

The first settlement of the EIC on the Coromandel Coast was at Masulipatam in 1611, but they moved their settlement south to Madras in 1639 and built Fort St George, to escape the ongoing warfare between the Qutb Shahi dynasty of Golconda and the Mughals. Madras was selected more for location than convenience, as it did not have a natural harbour and trade was conducted by catamarans between the land and anchored ships. In 1658 all other English settlements on the Coromandel Coast were made subordinate to Fort St George and by the 1670s, Madras had eclipsed the amount of trade in Surat. By 1700 the city had grown to over 150,000 inhabitants segregated into 'black' and 'white' towns.

Both the Dutch and English began trading in Bengal in the early 1600s, but it wasn't until the early 1630s that the local Mughal ruler granted the European companies full trade concessions. The Bengal governor Shah Shuja permitted the English and Dutch to trade in Bengal without any customs duties in lieu of annual payments to his government. A trading post was established by the Dutch first in Calcutta and in 1635 at Hughli (Hooghly) where a fortress was built in the 1680s. The latter had been a Portuguese colony before they were thrown out by Shah Shuja in 1631 for their slave trading among other tensions. The EIC established its first factories at Balasore in 1633, Kasim Bazar (Cossimbazar) in 1658, Hughli in 1658, Dhaka in 1668 and Calcutta in 1690.

Bengal would become the centre of both the VOC and EIC trade in India. The Bengal Subah province of the Mughal Empire was its wealthiest state and became the worldwide trade centre for muslin and silk (Prakash, 2006). In addition, much of India depended on Bengali products such as rice, silk and cotton. Saltpetre was also shipped to Europe from Bengal; opium was sold in Indonesia; raw silk to Japan and Europe; cotton and silk textiles were exported to Europe, Indonesia and Japan; cotton to the Americas and all across the Indian Ocean.

Bengal came to account for about 40% of the Dutch imports from Asia, including more than 50% of textiles and around 80% of silk (Richards, 1993). Both European companies imported much more Bengali than Chinese raw silk. Practically no silk was imported from China in the latter half of the seventeenth century, and between 1710 and 1760, nine to 19 times more raw silk was imported from Bengal than China (Krishna, 1924).

In the 1650s, the VOC established a way station at the Cape of Good Hope to serve its ships sailing to eastern Asia. Its trading network continued to expand, and factories were set up across such far-ranging places as Persia, Bengal, Malacca, Siam, Formosa (Taiwan) and Malabar. The Dutch imported Chinese and Persian silks around the Cape to Europe and could undersell the Manila galleons. 'These galleons continued to sail but their cargos were now distributed in the New World alone' (Lunde, 2005). By 1660 the VOC was by far the richest company in the world, with wealth exceeding the gross national product of most nations.

Country Trade

In their Indian Ocean trade, the East India companies had no choice but to use silver and gold bullion to purchase spices and silks, as there was little interest in Asia for English and Dutch woollens and other manufactured goods. In the first 23 years of the company, the EIC exported over £750,000 in bullion and less than half of that in manufactured goods (Marshall, 1998). This put considerable strain on the English economy.

The companies soon discovered that pepper and other spices could be secured in trade with goods procured from other regions of the Indian Ocean in what became known in England as the 'country trade'. They began to use their silver to buy cotton piece goods in India and then exchange these for spices in the Indonesian Archipelago (Chaudhuri, 1963). This strategy had long been practised by the Gujarat merchants of India. The Portuguese had also figured this out before the English and Dutch arrived, but of course had kept very quiet about it.

There was a massive amount of internal trade going on in the Indian Ocean in the seventeenth century. As Krishna relates:

> We have referred to the coasting fleets of three hundred to four hundred vessels plying between Cambay and Goa alone, and fleets of more than 250 ships sailing from Goa to the south, a fleet more of half that strength coming to the Coromandel Coast from the south, and to the numerous ships plying on the coasts of Orissa, Bengal and other kingdoms. The 40,000 boats on the Indus, the fleet of 180 boats from Agra to Satgaon, the Bengal flotilla of 4000 to 5000 armed boats, a fleet of 100 galliots of the Malabars, the numerous vessels of the various ports of India all testify to the existence of hundreds of thousands of boats and ships of all descriptions plying in the rivers and seas of India.
>
> (Krishna, 1924, p. 51)

Buried in the *Itinerario* of Jan Huygen van Linschoten was a single sentence that described the essential role that cotton played in the Portuguese spice trade: 'The wares that are there desired and exchanged for spices, are diuers sortes and colours of cotton linen, which come out of seuerall Prouinces; [Cambaya, Choramandel and bengala]' (Laarhoven, 1994, pp. 41–42). This demand for textiles in Indonesia is one of the most fundamental discoveries of the VOC, and the acquisition of Indian cottons for trade in South East Asia became a key tenant of the Dutch trading strategy. 'With their extensive intra-Asian trade, the Dutch could trade lead, carpets, textiles, and wool to other Asian nations through Batavia for tin, pepper, cotton, wax, and spices that were in demand in China. These goods accounted for a fifth of the cost of Chinese purchases' (Goodman, 2010, p. 62).

Laarhoven estimated that:

> at the beginning of the 17th century when the Dutch entered the competition for the Indian trade cloths, the import to the archipelago would have reached at least a value of 900,000 cruzados in the Singapore Straits or approximately 550,000 cruzados buying price in India ... [This] translated into a transfer of 11,000,000

yards or 10,120,000 meters of cloths being imported in the archipelago annually ... early in the 1600s at least 880,000 pieces of textiles had passed through the Singapore Straits annually ... Taken together, these facts indicate a considerable, or rather an immense consumption of cloth by the people in Southeast Asia.

(Laarhoven, 1994, p. 39)

By the mid-1600s the intra-Asian trade of the VOC contributed greatly to its overall profitability. The directors worked hard to keep this trade as self-supporting as possible. Nijman describes that:

By making sufficient profits on the intra-Asian trade, the VOC was able to keep exports of gold, silver and copper from the Republic to a minimum. The relative importance of the intra-Asian trade varied considerably from commodity to commodity, but ... between 1670 and 1680, 75 percent of the VOC-controlled harvest of cloves was brought back to Europe while 25 percent was sold on the Asian market. During the same period, 44 percent of the nutmeg harvest and 68 percent of the mace harvest were sold on the Asian markets.

(Nijman, 1994, p. 217)

The VOC was much more active in the intra-Asian trade than the EIC or Portugal had been and became the most important agency through which precious metals moved across the region. 'Goods were likely to be sold where the prices were highest, be it in Asia or in Europe, instead of making automatic recourse to the home market ... its major focus lay in the creation, redirection and intensification of the trading system itself' (Nijman, 1994, p. 219).

The Port of al-Makha (Mocha)

By the seventeenth century, al-Makha, lying on the Red Sea coast of Yemen, had largely taken over the role of Aden in Indian Ocean trade. It had become the primary supplier of coffee to the Muslim world and received a considerable amount of pepper from Indonesia on mostly Indian vessels, but also some Dutch.

By selling spices in Arabia, the Dutch sought to gain cash in order to purchase textiles and indigo in Surat. They sought precious metals so that less silver had to be sent from Europe to Asia. They were playing a global game. They converted pepper from the Indonesian Archipelago into hard cash on the market of al-Makha. This was used to purchase textiles in Indian cities and then these products were bartered in the archipelago for the spices which were sent to Amsterdam, for extremely profitable redistribution all across Europe and even the Levant. It was a tricky game that they played, as they did not want to supply al-Makha with so much pepper that it was redistributed to their competitors in the Levant.

Most of the spices received in al-Makha were sold within Arabia, and only a limited quantity found its way by sea and caravan to the Mediterranean ports. These ports were now being supplied in the other direction. Bulut reports that a Turkish contemporary writer named Ömer Talip wrote:

> Formerly the goods of India, Sind and China used to come to Suez, and were distributed by Muslims to all the world. But now these goods are carried on Portuguese, Dutch and English ships to Frangistan [Europe], and are spread all over the world from there. What they do not need themselves they bring to Istanbul and other Islamic lands, and sell it for five times the price, thus earning much money.
>
> (Bulut, 2020, p. 17)

Affairs at Home and the East India Trade

While the EIC and VOC were locked in an intense struggle over Indian Ocean trade, the nature and intensity of that trade were often greatly affected by conditions at home. There were the English Civil War from 1642 to 1651 and three major Anglo-Dutch Wars between 1652 and 1674. All these conflicts at home naturally diminished Indian Ocean trade.

The First Anglo-Dutch War (1652–1654) began when the English instituted the Navigation Act, aimed at barring the Dutch from involvement in English sea trade. It allowed goods from other European countries to be brought to England only on either English ships or ships of the country of origin, and goods brought from Asia, Africa or America could be brought to England only on ships from England or the particular colony. At first, the Dutch inflicted serious damage to the English fleet, but the tide turned and in 1653 the English successfully instituted a naval blockade of the Netherlands. The Dutch broke through and a ferocious fight occurred off the coast near Terheide. Both sides suffered huge losses, and no one was able to claim victory. The two exhausted countries signed a peace treaty in 1654.

By 1665, relations between England and the Dutch Republic had again deteriorated to the point of warfare. English merchants and their chartered companies decided that the English navy had become powerful enough to cripple Dutch shipping with a full-blown combination of naval battles and privateering missions. Initially the English had several victories, including taking the colony of New Netherland (present-day New York), but there were also Dutch victories such as the capture of the English flagship *Prince Royal* during the 'Four Days' Fight'. In 1667, a contingent of Dutch ships under Admiral Michiel de Ruyter crossed the North Sea and sailed up the River Medway to Chatham, where many English ships were docked. Six of the biggest ships were set on fire and the English flagship, the *Royal Charles*, was brought home to Holland in triumph. This hammer blow left King Charles II and the English reeling, they had no option but to sue for peace.

Admiral Michiel de Ruyter also played a pivotal role in the Third Anglo-Dutch War (1672–1674). This war centred on a French plan to invade the Dutch Republic, supported by various German princes and the English. Michiel de Ruyter was able to inflict enormous damage in a series of four engagements with the French and prevented them from landing an army on Dutch soil.

A month later this war was ended by the Treaty of Breda. A lack of trust and cooperation between the French and English contributed greatly to the Dutch success.

VOC Gets into Trade with Japan

In 1609, two ships from the VOC left for Japan on a diplomatic mission. The Dutch were attracted to the silver of Japan but also wanted to establish trade between themselves and China. They arrived safely on the island of Hirado and after an audience with the first Tokugawa shōgun Ieyasu, received permission to trade and build in Hirado. Here the Dutch laid the foundations for a long-lasting relationship with Japan. The English also received permission to establish a trading post in Hirado in 1613, but under intense competition from the Dutch they closed it in 1623.

From Hirado the Dutch could begin importing silk thread and textiles from their Pantani factory located in the northern part of the Malay Peninsula. They could also use Hirado as a base to attack and plunder Spanish, Portuguese and Chinese ships travelling between Manila and Macao. The VOC continued its piracy until 1621, when the shōgun told it to stop under pressure from the Iberians.

Hirado gave the VOC a foothold in Japan, but the Dutch still didn't have a trading outpost anywhere near the coast of China. They tried to seize Macao in 1622 and failed miserably, when the Dutch admiral William Reijersen and an invading force of 800 soldiers were chased out of the city by a motley assemblage of Portuguese soldiers, Macanese citizens, Dominican friars, Jesuit priests and black slaves. They were more successful in a subsequent invasion of Penghu, an archipelago in the Taiwan Strait inhabited by mostly fishermen. They used this location as a base to raid up and down the Fujian coast until the Chinese authorities got fed up, sending a fleet of 50 warships and 5000 troops to expel them in October 1624.

In 1635, the harsh *Sakoku* ('closed door' policy) was decreed by the shōgunate; and in 1641, the VOC was forced to transfer its operations to the small man-made island of Dejima in Nagasaki harbour, where the operations could be more closely watched. The island had been originally used by the Portuguese, but they were totally thrown out of Japan after the *Sakoku* edict. The Japanese had grown weary of the Portuguese Jesuits' efforts to convert the populace to Christianity. Only the Dutch were permitted to remain after all other Westerners had been excluded.

The Dutch were indeed closely monitored and controlled in Japan. Each year the VOC had to travel to the court of the Tokugawa shōgun Edo to offer tribute of exotic and elaborate gifts such as clocks, telescopes, fine medicines and rare animals. They were largely free to do as they pleased on the island; but they could not import religious items, bring females in from the outside and they could not bury their dead in Japan. They were also ordered to work on Sunday. The VOC was happy to follow these restrictions as the profits were very high; the single factory in Japan in 1636 brought in an estimated net profit of three million florins (Sluiter, 1942).

The Dutch and the Pirates

The Dutch finally found a permanent Chinese home in 1624, when the Fujian authorities allowed them to build a fortress on the south-west coast of Formosa – Fort Zeelandia. The VOC used Formosa as its base of operations in China until 1662, conducting trade with Japan first at Hirado and then at Nagasaki. This proved to be a lucrative trading opportunity as only Chinese and Dutch ships were allowed into Japan. The Spanish were also allowed to occupy the northern end of Formosa in 1626 and built their own fort (San Domingo); however, they were excluded from trade with Japan. This colony persisted until 1642, when it was attacked and overrun by the Dutch.

From their base on Formosa, the Dutch operated an illicit trade network between Fujian and Japan, joining a large group of Chinese pirates regularly plundering legitimate trade vessels moving between Fujian and Manila. These pirates had operated off the coast of China for many decades, after the Ming had given up their trade monopoly via the tribute system. These bands of pirates also smuggled contraband Japanese silver into the south-eastern provinces of China for raw silk, porcelain and other trade goods. Since the Ming had outlawed trade with Japan, the Japanese silver could only reach China through illegal activities.

Greatest of the Chinese pirates was Zheng Zhilong (1604–1661), who became such a problem for Chinese commercial trade that the authorities in Fujian wound up hiring him to police the waters along the south-eastern coast of China. He was named admiral of a fleet stationed in Amboy in 1628, from which he established a trade monopoly in Fujian Province. He reigned supreme until the Ming dynasty began to crumble in south-east China under pressure from the Qing from the north. Zheng Zhilong surrendered his forces in exchange for the position of governor of the Fujian and Guangdong provinces.

When Zheng Zhilong turned sides, the Ming loyalist Koxinga (1624–1662) took over as head of the Zheng family. In 1650, he pledged allegiance to the only claimant left to the throne of the Ming Dynasty, the Yongli Emperor. However, he was unable to save him and was forced to relocate to Taiwan in 1661, when the Qing emperor evacuated the whole coast of Fujian and Guangdong to block all trade. He seized the Dutch Fort Zeelandia and effectively ended the VOC's direct role in the China trade.

From this point on, the Dutch involvement in the China trade was through Batavia. They actively encouraged the establishment of a Chinese community of immigrants there, who were granted a certain level of autonomy and were given the responsibility to buy the cargo of the Chinese junks arriving with raw silk and porcelain. The VOC, in a sense, began outsourcing their Chinese trade to the Chinese themselves.

Dutch Take Over Ceylon

In 1602, Admiral Joris van Spilbergen was the first Dutch envoy to make contact with the rulers of Ceylon at Kandy (Gaur, 2015; Anonymous, 2020). Spilbergen and King Vimaladharmasuriya hit it off and developed quite cordial relations. The king saw the Dutch arrival as an excellent opportunity to gain naval support against his Portuguese adversaries and Spilbergen made lavish promises about military assistance. The king was so impressed by his visitors that he began to learn Dutch and Spilbergen left a couple of musicians behind to entertain the king.

A few months later another Dutch official Sebald de Weert arrived with a fleet of six ships and a concrete offer of help. The king agreed and they launched a joint attack on the Portuguese at Batticaloa on the eastern coast. During the attack, Weert took four Portuguese ships but allowed the officers and crews to go free. The king was furious about this action and after Weert insulted the queen at a drunken dinner party, he and all 47 Dutchman with him were slaughtered. This tragedy spooked the Dutch, and it was another three decades before they made another serious attempt to work with the locals to expel the Portuguese from Ceylon.

The Dutch returned in force in 1637, after Rajasinha II sent emissaries to meet the admiral of the Dutch fleet, Adam Westerwolt, who was then blockading Goa, India. After decimating the Portuguese fleet, the victorious Westerwolt took a fleet of four ships and 800 men and attacked the Portuguese fort at Batticaloa aided by Singhalese forces. The coalition conquered the fort on 18 May 1638.

Five days later, Westerwolt signed a new treaty with King Rajasinha in his palace in Batticaloa (Kandyan Treaty of 1638). Under the treaty the Dutch would make war with the Portuguese and in return would be given a monopoly over all trades except elephants, the king would pay all expenses incurred by the Dutch, and any forts captured from the Portuguese would be garrisoned by the Dutch or destroyed, depending on the king's wishes (de Silva and Beumer, 1988). This last stipulation would lead to bitter disagreement between the two parties in the future.

Slowly but surely the Dutch and Kandyan forces pushed the Portuguese out of Ceylon. In May 1639 the Dutch fleet captured Trincomalee, and in February 1640 the Dutch and the Kandyans combined to take Negombo. In March 1640 Galle was also taken, but the progress was temporarily halted by a truce declared in Europe between the Dutch Republic and Spain. In 1645 the boundaries between Portuguese and Dutch territory in Ceylon were demarcated and Jan Thijssen was appointed the first Dutch governor. Galle would remain the centre of Dutch power in Ceylon for the next 18 years.

In the 1650s, the Dutch resumed their open warfare against the Portuguese, who still had forts at Kalutara, Mannar and Jaffna. Gerard Pietersz Hulft was sent to Ceylon in 1656 with 11 ships and 1120 soldiers. The Dutch took the fort of Kalutara by surprise and then laid siege to the city of Colombo. The Portuguese surrendered on 12 May 1656, ending 150 years of Portuguese presence in the city. In 1658 the Dutch advanced north and captured Jaffna and Mannar, the last Portuguese strongholds on the island.

Once the Dutch had possession of Colombo, they barred Rajasinha's troops from taking over the city. The king was furious and demanded its delivery. The Dutch then presented the king with their bill for the costs of the war totalling 7,265,460 guilders, a sum they knew the king could not pay. They vowed they would keep the Portuguese possessions as collateral security until the Kandyan king repaid the costs of the war. Disappointed, Rajasinha had to conclude that the 'assistance' by the Dutch had changed nothing for Ceylon – he had simply exchanged one occupier for another.

To prevent the Portuguese or the English from ever recapturing Ceylon, and to obtain a valuable source of pepper, the VOC went on to conquer almost the entire Malabar Coast from the Portuguese, driving them forever from the west coast of India by 1663. They then established their own series of fortifications along the coast, with their headquarters set up in Cochin. Only Goa now remained in Portuguese hands. The subjugation of Malabar meant that the VOC was less dependent on Indonesian and Malaysian pepper producers. It gave the company the premier position in pepper commerce 'and Amsterdam became the new pepper and spice Mecca of Europe' (Brouwer, 2004, p. 215).

References

Anonymous (2020) Dutch period in Ceylon. *World Heritage Encyclopedia*. Available at: http://www.worldheritage.org/Find/Dutch%20period%20in%20Ceylon (accessed 28 January 2021).

Ashfaque, N. (2006) The Dutch East India Company and the Mughal court. *Proceedings of the Indian History Congress* 67, 271–274.

Barrow, I. (2017) *The East India Company: A Short History with Documents*. Hackett Publishing Company, Indianapolis, Indiana and Cambridge, Massachusetts.

Broadberry, S., Custodis, J. and Gupta, B. (2013) India and the great divergence: an Anglo-Indian comparison of GDP per capita, 1600–1871. *Explorations in Economic History* 55, 58–75.

Brouwer, C.G. (2004) Pepper merchants in the booming Port of al-Mukha: Dutch evidence for an oceanwide trading network. *Die Welt des Islams* 44(2), 214–280.

Bulut, M. (2020) The Ottomans and western Europeans during the mercantilist times: neutrality, competition and conflict. *Journal of Al-Tamaddun* 15(1), 13–30.

Chaudhuri, K.N. (1963) The East India Company and the export of treasure in the early seventeenth century. *The Economic History Review* 16, 23–38.

de Silva, R.K. and Beumer, W.G.M. (1988) *Illustrations and Views of Dutch Ceylon 1602–1796*. Serendib Publications, London.

Fischer, K. (1965) The beginning of Dutch trade with Gujarat. *Proceedings of the Indian History Congress* 27, 206–215.
Gaur, A. (2015) Kandy: historical kingdom, Sri Lanka. *Encyclopædia Britannica*. Available at: https://www.britannica.com/place/Kandy-historical-kingdom-Sri-Lanka (accessed 28 January 2021).
Goodman, B. (2010). The Dutch East India Company and the tea trade. *Emory Endeavors in History* 3, 60–68.
Heitzman, J. and Worden, R.L. (1995) *India: A Country Study*. Federal Research Division, Library of Congress, Washington, DC.
Krishna, B. (1924) *Commercial Relations Between India and England (1601–1757)*. George Routledge & Sons, Ltd, London.
Laarhoven, R. (1994) The power of cloth: the textile trade of the Dutch East India Company (VOC) 1600–1780. PhD thesis, Australian National University, Canberra.
Lunde, P. (2005) The coming of the Portuguese. *Aramco World: Arab and Islamic Cultures and Connections* 56(4). Available at: https://archive.aramcoworld.com/issue/200504/the.coming.of.the.portuguese.htm (accessed 28 January 2021).
Marshall, P.J. (1998) The English in Asia to 1700. In: Canny, N. (ed.) *The Origins of Empire*. Oxford University Press, Oxford, pp. 264–285.
Nadri, G.A. (2018) The English and Dutch East India Companies and Indian merchants in Surat in the seventeenth and eighteenth centuries. In: Clulow, A. and Mostert, T. (eds) *The Dutch and English East India Companies, Diplomacy, Trade and Violence in Early Modern Asia*. Amsterdam University Press, Amsterdam, pp. 125–150.
Nijman, J. (1994) The VOC and the expansion of the world system 1602–1799. *Political Geography* 13, 211–227.
Prakash, O. (2006) Empire, Mughal. In: McCusker, J.J. (ed.) *History of World Trade Since 1450*. Macmillan Reference USA, Farmington Hills, Michigan, pp. 237–240.
Richards, J.F. (1993) *The New Cambridge History of India*, Vol. V. *The Mughal Empire*. Cambridge University Press, Cambridge.
Sluiter, E. (1942) Dutch maritime power and the colonial status quo, 1585–1641. *The Pacific Historical Review*, 11(1), 29–41.
van Meersbergen, G. (2018) Diplomacy in a provincial setting: the East India companies in seventeenth-century Bengal and Orissa. In: Clulow, A. and Mostert, T. (eds) *The Dutch and English East India Companies: Diplomacy, Trade and Violence in Early Modern Asia*. Amsterdam University Press, Amsterdam, pp. 55–78.

The Ottoman and Safavid Silk Trade   21

Setting the Stage – Ottomans in the Centre

In the early modern world, the most powerful empire in Europe was that of the Turkish Ottomans. As Celetti describes:

> The sixteenth century brought a shift in the geopolitical landscape of the Eastern Mediterranean. With the conquest of Egypt and Syria in 1516–1517, the Ottomans established uncontested hegemony in the region. Extending from the eastern Adriatic coast to the Persian Gulf, the empire controlled the main routes of trade between the economies of the Mediterranean and the Indian Ocean, effectively becoming the hinge of Western Eurasia.
>
> (Celetti, 2020, p. 386)

Until the end of the sixteenth century, the Ottomans had almost total control of the profits from the spice and silk routes connecting the East to the West, trading primarily with Venice. This situation began to change when the Northern Europeans found the route to the Far East around the Cape of Good Hope into the Indian Ocean (Bulut, 2020). This pushed the Ottomans to develop new strategies to maintain the profitability of their central attachment to the East–West trade routes. They began to encourage the Western European nations to directly trade in their territories by granting them privileges (capitulations), starting with the French, then followed by the Dutch and the English.

As European commercial power grew in trade with the East, the economic landscape of the Levant changed dramatically. East and West became completely integrated into a Western-dominated world economy that was pivoting around the great Ottoman Empire. Commerce soared as the European merchants jockeyed for position in a process that the French historian Braudel:

> dubbed a 'Northern Invasion' of the Mediterranean, which was followed by flows of merchants and diplomats trying to secure 'each nation's' positions in what seemed one of the most relevant economic and political chessboards of the time. As a result, commercial ties between East and West soared, the Ottoman Empire ultimately being integrated into the emerging Western-dominated world economy.
>
> (Celetti, 2020, p. 386)

© J.F. Hancock 2021. *Spices, Scents and Silk: Catalysts of World Trade* (J.F. Hancock)
DOI: 10.1079/9781789249743.0021

Silk Production and Movement in Safavid Persia

During the fifteenth and sixteenth centuries, the European silk-weaving industry was primarily fed with Iranian raw silk. Iran was second only to China in world raw silk production, supplying its own domestic industry and exporting to India, the Ottoman Empire and Europe. In fact, 'for Iran, contemporary observers all agree that raw silk was the most important export, playing a vital role in the economy and in state finances by bringing silver into the country, to supply the mints, swell the treasury, and offset the perpetual trade deficit with India' (Herzig, 1992, p. 61).

Silk was produced in many parts of Iran, but the main sources for the export to the West were Gilan and Shirvan/Karabakh near the south-western corner of the Caspian Sea. Traditionally, the Iranian silk that found its way to Europe was transported by Iranian and Armenian merchants by caravan to the Levantine ports on the Mediterranean. The Levant route had a number of branches through Aleppo and Smyrna (Izmir). Much of that raw silk then passed through the hands of the English Levant Company founded in the late sixteenth century and then the EIC. The Venetians, French and Dutch were also active in this trade but by the middle of the seventeenth century the value of English imports greatly exceeded them.

In the late fifteenth to sixteenth century only a limited amount of silk entered Europe through the Levant (a few tens of thousands of kilograms), but by the early seventeenth century hundreds of thousands of kilograms could pass through this route annually. This production was highly volatile, shifting year to year from hundreds of thousands to a few hundred kilograms. In the second half of the seventeenth century, production became much more consistent at 200,000–300,000 kilograms a year. Total European import $c.1600$ was 180,000–190,000 kilograms; in the early 1620s it was still about 200,000 kilograms a year, but in 1675 the overall volume of the silk export via Smyrna rose to 422,000 kilograms and in 1700, it may have been as high as 500,000 kilograms.

Iranian silk also got to Europe via Russia in a route that went across the Caspian Sea, up the Volga River to Moscow and then overland to Central Europe or via the Baltic to Holland and England. The main collection point for this trade was Astrakhan. This particular route was used only sporadically in the sixteenth and seventeenth centuries, and the quantities were small, but its importance increased dramatically in the 1690s when the Levant trade was disrupted by the Nine Years' War and the Baltic was opened. The silk that previously made it to Europe through Russia had been transported all the way to Archangel on the White Sea for onward shipment to Holland. Levels of about 20,000 kilograms a year were recorded in the early 1690s and then in 1696 they rose to 84,000 kilograms. In the single month of April 1712, 44,000 kilograms of raw silk passed through Astrakhan (Herzig, 1992). By the eighteenth century the Russian route was carrying enough of Iran's silk export that both the East India companies and the Ottomans offered the Armenian silk merchants a very low single duty of 5% in an attempt to win them back to the Levantine route.

France and the Ottomans

In the early sixteenth century, a new player entered the silk and spice trade in a big way: France, who established itself as the leading Western power in the growing commercial and diplomatic relations of the Ottomans. In 1536, a 'treaty of capitulation' was negotiated between Francis I and Suleiman the Great that gave France jurisdiction over all Christian traders in the Levant. Anyone who wanted to do business there 'was obliged to do business under the French flag and under the exclusive surveillance and representation of the French ambassador and consuls' (Horniker, 1946, p. 302). This was the first time that Christian sovereigns had formed an alliance with the Muslims.

This capitulation was not a treaty, but a unilateral document, a grant or concession by the sultan to his friend, the king of France. It permitted to French subjects the rights of residence, trade and local jurisdiction. As Angell tells it:

> The substance of the concessions in the chief Capitulations was as follows: The Franks were to have the liberty to travel in all parts of the Ottoman Empire. They were to carry on trade according to their own laws and usages. They were to have liberty of worship. They were to be free from all duties save customs duties. They were to enjoy inviolability of domicile. Their ambassadors and consuls were to have exterritorial jurisdiction over them. Even if they committed a crime, they were to be arrested by an Ottoman official only in the presence of a consular or diplomatic official of their own country. The Ottoman officers, if asked by a consular or diplomatic officer to aid in the arrest of a French subject, must render such service. The Franks had the full right of making wills. If they died intestate in Turkey, their own consul must take possession of their property and remit it to their heirs.
>
> (Angell, 1901, p. 256)

Under this trade agreement, the French established merchant communities in the Levant that were regulated in hierarchical order by a local assembly, a consul, an ambassador, the Chamber of Commerce of Marseille, and the minister of the navy himself. 'Consuls interceded with the Ottoman authorities, resolved commercial disputes, acquired permits to export theoretically prohibited goods – such as wheat or olive oil – and helped French merchants or captains with the avoiding, easing, or lifting of sanctions and other avanies' (Celetti, 2020, p. 398). Only authorized merchants could reside in the Levantine French communities, their residence limited in time and scope, their contacts with the local population strictly confined to the necessities of trade, mediated by official dragomans, and controlled by the diplomatic institutions and the community itself.

Enter the Levant Company

Until late in the sixteenth century England had no direct commercial and diplomatic relations with the Ottoman Empire. It was not until the reign of Queen Elizabeth I that the English themselves became interested in that trade. English ships had the sultan's permission to trade in Ottoman ports, but only under the French flag. Once this trade became significant, it was inevitable that the English would find this arrangement unbearable; thus in 1583, under Elizabeth, the English were able to obtain their own treaty of peace and friendship which gave them the privilege of trading under their own flag.

Trade between England and the Ottoman Empire occurred mostly through 'The Governor and Company of Merchants of England trading into the Levant Seas', generally shortened to the 'Levant Company' or 'Turkey Company'. It was founded by an English royal charter in 1581 that was renewed in 1606 and 1661. The Levant Company was not chartered as a joint-stock company like the EIC or VOC – its members traded as independent merchants, subject to the restrictions imposed on all Levant Company members. They did, however, often pool capital to share risk and shared any resulting profits.

The members did not leave England themselves, but sent an ambassador to Istanbul (the Ottomans renamed Constantinople after its conquest in 1453), appointed consuls to the major trade centres and sent factors to all of the company's factories in the Levant (Epstein, 1908). The ambassador resided in Istanbul and had a staff including interpreters, a treasurer and a preacher. The ambassador received his salary from the company, but besides his mercantile duties he also served the political needs of the Crown. He had the ear of the sultan, but in the eyes of the other ambassadors in Istanbul his not being paid by the Crown diminished his status. The consuls were not traders themselves but were charged with administrating the affairs of the English communities and maintaining a positive trade atmosphere. Most served for three to five years, although some of the most successful ones served for decades in the eighteenth century when it became common for consuls to have their families with them.

The factors were generally young men who entered into the Levant trade after spending two or three years working at English ports or the Royal Exchange in London. Once they were schooled in the business of trade, they were sent to the Levant, where they tracked incoming and outgoing shipments to England and kept tabs on caravans moving across Mesopotamia. A major headache for the company was the factors doing independent trading. They were prohibited from this practice, but the distance from home was great and many profited handsomely, being 'able to live in grand style in consequence' (Epstein, 1908, p. 148).

The primary centres of the Levant Company's trade were Aleppo in north-eastern Syria and Smyrna in eastern Turkey on the Aegean Sea. Of these two Aleppo was the most important, being at the nexus of the caravan trade

(Silk Routes) from Persia, Mesopotamia, Africa and Asia. The primary trading commodities were wool broadcloth from England for Persian silk and currants. Currants dominated trade in the early period of the Levant Company and silk in the later stages. Most transactions were by barter, although when barter prices were not advantageous or shipments of English goods were lost or delayed, the factors sometimes had no choice but to purchase silk directly.

Dutch in the Levant

As early as the Middle Ages, the Dutch had 'been active in the Levant as pilgrims, crusaders, maritime operators, merchants and travelers' (Umunç, 2011, p. 373), but it wasn't until the seventeenth century that they became particularly active as traders. They began their activities under the French and English colours and received their own capitulation in 1612. During the Twelve Years' Truce between the Dutch Republic and Spain (1609–1621), the Dutch greatly outplayed their Venetian and English rivals and became the major player in the Levant trade. The States General of the Dutch Republic set up resident communities and consuls in the Levant, much like the French and English before them, to generate a vigorous trade atmosphere and protect the rights and privileges of the merchant colonies.

The Ottomans were particularly interested in establishing political relationships with the Dutch to support their ongoing battle with their common enemy, Spain. In fact, the Ottomans had been particularly grateful when in 1604 the Dutch captured the Spanish naval base on the coastal town of Sluis in the Netherlands and freed 1500 Turkish galley slaves. This led to the capitulations and protection from the North African corsairs. The Dutch also helped the Turks modernize their navy to better equip it to fight the Spanish in the Mediterranean.

During the early decades of the 1600s, Dutch trade in the Levant increased enormously and the Dutch Republic became the dominant economic power in the region (Umunç, 2011). Their primary point of contact became Smyrna, where the Dutch brought from Turkey mohair yarn, cotton and cotton thread, raw silk, carpets, figs, currants, raisins and hides. In return, the Dutch sold fine cloth and other textiles, furs, Spanish silver, dyestuffs, Swedish copper and spices.

Umunç quotes a petitioner to the States General in 1611 who states:

> that the Levant trade had latterly emerged as one of the most vital plied by the Dutch in any part of the globe ... that their commerce with Turkey, Cyprus, Egypt and The Aegean now, compared in value with the trade that the VOC had with the East Indies, was potentially even more important for the future wellbeing of the fatherland than the East India traffic ... that the Levant furnished an abundance of valuable raw materials, in particular, silks, cotton and mohair, which were indispensable to the manufacture of a wide range of luxury and middle quality textiles in the west.

(Umunç, 2011, p. 378)

French Impact Grows in the Levant

French trade in the Levant kicked back into high gear in the early eighteenth century, as that of the Dutch and English began to diminish (Table 20.1). France's position as a country adjacent to the Mediterranean put it in good position for Ottoman trade in the Levant. Marseille, being one of the largest French and Mediterranean ports, afforded easy flow of goods across Europe from the Mediterranean and became a centre for re-exporting colonial items in high demand. As Celetti describes:

> This commerce involved a wide array of commodities, ranging from cloth to coffee and sugar imported from the Caribbean. For many branches of growing French industry, the Eastern Mediterranean also constituted a potential source of raw materials, with imports of wool and raw cotton soaring in the course of the [eighteenth] century ... Both the variety of products and the scale of commercial exchange underpinned Marseille's role as one of the most relevant hubs of Levantine trade.
>
> (Celetti, 2020, p. 388)

Textiles played a central role in the exchange, either as raw materials or finished products, and were the most significant category among both importers and exporters. Textiles, most of them produced by the Manufacture du Languedoc, increased from 46% of total exports in 1700–1702 to 68% in 1785–1789, and raw materials increasingly dominated imports over the same period, rising to 75%. The colonial commodities – sugar, coffee and dyestuffs – also grew in importance as French exports. France was operating in a truly global market.

At the end of the eighteenth century, raw silk imports into France from the Levant dropped back again from 16% of the value of all imports in 1700–1702 to only about 3% in 1789. This drop can be attributed to several factors: (i) Ottoman domestic demand increased, reducing the amounts available for export; (ii) a higher amount of Italian silk, considered of higher quality, was now being imported; (iii) the Safavids' collapse disrupted the caravan trade of Armenian merchants; and (iv) the huge rise in the popularity of cotton. There was, however, an uptick in raw wool imports by the French during the eighteenth

Table 21.1. Shares of the major European nations in the Levant trade, 1604–1784. (From Bulut, 2020.)

Year	Share of Levant trade (%)			
	Venice	France	England	Dutch Republic
1604[a]	50	32	12	6
1613[a]	25	52	7	15
1686	7	16	43	3
1750	16	45	25	14
1784	12	37	9	18

[a]Just Aleppo.

century, who returned textiles back to the Ottoman market in the form of light woollen 'draps'. These were used for garments and accessories, such as veils produced in a variety of colours.

Raw Silk Around the Horn

Soon after it arrived in the Indian Ocean, the EIC made a bold play to divert some of the Iranian silk trade south via the Cape of Good Hope (Steinmann, 1987; Razzari, 2019). The opportunity arose in 1619 when Shah Abbas I (1587–1629) offered to give the English Ormuz if they would help him with naval support to regain the Persian Gulf from the Portuguese. The EIC accepted his request and began by attacking and ransacking the island of Qeshm in the Strait of Hormuz. The English then used their fleet to drop off 3000 Safavid troops about two miles from the town, while they remained at sea to provide artillery support. The Safavid force soon took the city, but not before the Portuguese begged the English for protection, fearing the Safavids would seek harsh reprisals. The English came to their aid and ferried a number of Portuguese away under the cover of dark, many of them charged with defending the fort. When the city fell the next day, the Safavids and English scoured the whole island of Hormuz for anything of value and split the booty.

The assault on Ormuz was a major blow to the Portuguese, but it was not without major consequences to the EIC, as it had overstepped its bounds. King James I was displeased with the siege both because the English were officially at peace with the Spanish (which then included Portugal) and the company had taken plunder without sharing it with the Crown. It was also messing with the business of the Levant Company. The EIC eventually wound up paying a £20,000 fine 'while enduring the Duke of Buckingham's bitter criticism of the Company as an organization of *piratts*' (Razzari, 2019, p. 500).

The victory at Ormuz represented a major maritime victory for the English and made the Indian Ocean safer for the EIC's ships travelling between London and the Far East, but it did not provide the financial and commercial rewards the company had hoped for. Razzari explains why:

> The English were unable to maintain the prestige they earned after their assault on the Persian Gulf, and as a consequence their ability to sustain favor within the Safavid state waned. This was particularly true as the English were unable to quench the shah's need for gold and silver specie. The English also found it extremely hard to guard the Iranian coastline while the Safavids were preoccupied with the Ottomans and peace with the Portuguese all but sealed their fate in Iran ... Unable and unwilling to maintain the naval offensive in the Persian Gulf, the second half of the decade marked a rapid decline of the Company in Iran. The Company struggled to employ valuable commodities and capital for trade.
>
> (Razzari, 2019, pp. 506–507)

In comparison, the Armenians plying the caravan routes continued to offer the Persians bullion, commodities and information in exchange for silk and a favoured status. The English could not compete and as the strength of their trade with India continued to grow, the EIC gave up on the Iranian silk trade all together in 1640.

The VOC Gets into the Silk Market

The VOC was quick to take advantage of the new opportunities opened by the English in Iran. It was able to sign its own treaty with the shah granting it the right to import into Persia selected products at fixed prices and toll-free, while the shah agreed to supply the VOC with a set amount of silk at higher-than-market prices (Floor, 2012). The VOC subsequently opened a rest house in Lār for Dutch caravans moving between Isfahan and the coast and over the years established a number of additional trading posts across Iran.

Between 1626 and 1640, silk was the main commodity exported from Persia by the VOC, although high prices, irregular quality and competition from Bengal silk kept the profits modest at best. In addition, despite their treaty, the shah supplied silk only sporadically and in smaller quantities than had been stipulated. The Dutch improved their profitability by going behind the shah's back and buying silk from private individuals, without paying customs duties on it. The VOC hung on in Iran with variable success for 136 years from 1623 to 1759, as the most important single foreign trading firm in Persia.

Bengali Silk Trade

A large silk industry had long existed in India when the English and Dutch arrived in the early 1600s, and it was clear to these early merchants that the establishment of a silk trade with Europe could be lucrative. However, the European trading houses were in general ignorant of Bengal silk and most of the Bengal silk trade was carried on by Gujaratis who operated in Agra, Delhi, Lahore and Surat. Bengal silks got a much wider market between 1660 and 1757 once the VOC and EIC set up permanent factories in Bengal and extended mulberry cultivation. The Dutch began trading Bengali silk first in Japan and then moved to Europe, while the English pointed towards the European market from the beginning.

Bengali silk soon largely replaced Persian silk in the European market due to its cheaper price, more consistent quality and more dependable availability. Until 'the eighteenth century, silk had long constituted a mainstay among imports from Levantine ports. However, in the subsequent period, its volume and importance declined sharply, from 15.54 percent of the value of imports in 1700–1702 to a mere 3.21 percent in 1789' (Celetti, 2020, p. 389). European weavers still desired raw silk from Italy and France because of its superior quality, but Bengali silk was much cheaper, and money spoke.

Over the latter half of the seventeenth century and into the eighteenth, exports of Bengali silk rose dramatically. The EIC exported 470–570 bales (73 pounds or 33 kilograms each) in 1673/74 but by 1679/80 was exporting 1200 bales and in 1680/81 it was 1800 bales (Guha, 1984). By the 1780s, exports had jumped to over 10,000 bales annually. VOC exports had greatly exceeded those of the EIC in the seventeenth century, but by the mid-eighteenth century were about the same (Chaudhury, 1995).

Luxury Silks in Europe

During the 1300s the city states of Italy took over production of most of the European luxury silks and by the fifteenth century, 'Venice and Florence were renowned for their sumptuous velvets incorporating gold and silver threads with large floral patterns after the pomegranate motif' (Walters, 2017). A significant silk-weaving industry also emerged in France in the fifteenth century, centred in the city of Tours which was close to Paris and the French court, the primary consumers of luxury textiles.

Throughout the sixteenth and seventeenth centuries, French royalty strongly supported the production of their own raw silk. Henri IV (1553–1610) ordered that mulberry trees be planted in each parish and that silk culture begin. He even brought in the services of the great French agronomist Olivier de Serres, whose book, *Théâtre d'Agriculture* (1600), was the textbook of French agriculture in the seventeenth century. Henri had more than 20,000 mulberry trees planted in the royal gardens and at the end of his reign over four million mulberry trees had been planted across Provence and Languedoc. Many of the farmers who had traditionally grown chestnuts and olives shifted to mulberries.

When Louis IV (1643–1715) became king, he invited Italian and Greek craftsmen to settle in Tours, and by 1646 this city had become the leading centre of sericulture. In the late sixteenth century the centre of silk manufacture moved from Tours to Lyon, which for a long time had been the chief French importer of Italian textiles. More than one-third of the population of Lyon, almost 15,000 workers, became involved in the silk industry (Walters, 2017). Lyon bustled with all the activities supporting silk weaving including spinning, dyeing and selling the raw materials and finished products.

A considerable English weaving industry also emerged in the late seventeenth century, when hundreds of thousands of French Huguenots, skilled as weavers, migrated to England after the revocation of the Edict of Nantes. 'In 1685 the influx of the Huguenots from all parts of France placed the industry for the first time among great English enterprises' (Hertz, 1909). The English also tried to establish a silk industry in the Americas, but this enterprise failed.

From the mid-seventeenth century until the late nineteenth century, the European silk textile industry was supplied with a significant amount of European-produced raw silk. Provence led the way followed by Italy, Sicily and Spain. From 1760 to 1780 cocoon production in Provence amounted to about 7000 tons per year, and by 1852 it had reached about 26,000 tons (Gaston, 2017). In England, a little over half of its raw silk came from the Levant. Of the 4650 bales (160 pounds or 72.5 kilograms each) imported in 1715, 2500 came from the Levant, 1300 from Italy and 850 from India and the East.

The silk industry of Lyon almost completely died during the French Revolution from 1789 to 1797. Thousands of workers were guillotined or shot, and the rest stayed hidden. Drawings, fabrics and designs were destroyed, and the silk industry was reduced by 90% in under a decade. It wasn't until the rule of Napoleon that the industry was restored.

Death of the Worms

The European sericulture industry collapsed in the 1850s when the silkworm disease called pebrine came to devastate the sericultural crops in France and Italy. In 1853 France produced 2100 tons of silk, the next year about 1790 tons, but in 1855 production dropped dramatically to about 600 tons (Ma, 1996). In Italy, raw silk production, which had been averaging about 3500 tons a year, fell to 1607 tons in 1863 and only 826 tons in 1865. The disease also significantly hit the sericultural industries in Turkey and Syria.

The French government desperately sought to find a solution to the disease problem and asked Louis Pasteur for help, even though he had no experience with silkworms or animals in general (Smith, 1999). He was able to identify the exact cause of the disease, determined the importance of keeping colonies clean, and came to the conclusion that the disease could be overcome by the selection of healthy, resistant breeds. By 1870 resistant European breeds were developed and the Italian industry recovered, with its output rebounding to 3200 tons in 1883, almost 90% of its former level. French raw silk production never did regain its former level and rose to only about a quarter of its pre-disease amount.

The death of the silkworms in Europe proved to be a boon for the EIC, which greatly increased its import of raw silk from China and then re-exported most of it to France. Between 1850 and 1860 Chinese exports to Britain more than quadrupled. 'It is no exaggeration to claim that British Far East commerce spared the French and European silk manufacturers a major raw material supply crisis' (Ma, 1996).

This dependence on British re-exports greatly concerned the French, who soon established direct access themselves to the silk of China and Japan. The French shipping giant Messageries Maritimes began moving silk directly from Calcutta in 1862 and China and Japan by 1864. The opening up of the markets and the first direct imports from China to France were made in 1863 and from

Japan in 1866. The level of these imports grew steadily and by the 1880s France was getting about half of its raw silk directly from China and Japan; by the 1900s, British re-exports to France were negligible.

References

Angell, J.B. (1901) The Turkish capitulations. *The American Historical Review* 6(2), 254–259.
Braudel, F. (1972) *The Mediterranean and the Mediterranean World in the Age of Pnilip II*, Vol. II. University of California Press, Berkeley, California.
Bulut, M. (2020) The Ottomans and western Europeans during the mercantilist times: neutrality, competition and conflict. *Journal of Al-Tamaddun* 15(1), 13–30.
Celetti, D. (2020) France in the Levant: trade and immaterial circulations in the 'Long Eighteenth Century'. *Journal of Early Modern History* 24, 383–406.
Chaudhury, S. (1995) International trade in Bengal silk and the comparative role of Asians and Europeans. *Modern Asian Studies* 29(2), 373–386.
Epstein, M. (1908) *The Early History of the Levant Company*. George Routledge & Sons Ltd, London.
Floor, W. (2012) Dutch–Persian relations. *Encyclopædia Iranica* 7, 603–613. Available at: http://www.iranicaonline.org/articles/dutch-persian-relations (accessed 28 January 2021).
Gaston, G.O. (2017) History of sericulture in France. *European Journal of Research in Social Sciences* 5(2), 43–58.
Guha, S.C. (1984) The silk market – a short note on external and internal trade. Available at: http://14.139.211.59/bitstream/123456789/311/8/08_CHAPTER_04.pdf (accessed 28 January 2021).
Hertz, G.B. (1909) The English silk industry in the eighteenth century. *The English Historical Review* 24, 710–727.
Herzig, E.M. (1992) The volume of Iranian raw silk exports in the Safavid period. *Iranian Studies* 25(1–2), 61–79.
Horniker, A.L. (1946) Anglo-French rivalry in the Levant from 1583 to 1612. *Journal of Modern History* 18(4), 289–305.
Ma, D (1996) The modern silk road: the global raw-silk market, 1850–1930. *The Journal of Economic History* 56, 330–355.
Razzari, D. (2019) Through the backdoor: an overview of the English East India Company's rise and fall in Safavid Iran, 1616–40. *Iranian Studies* 52(3–4), 485–511.
Smith, T. (1999) Pasteur and insect pathogens. *Nature Structural Biology* 6(8), 720.
Steinmann, L.K. (1987) Shāh 'Abbās and the royal silk trade 1599–1629. *British Scciety for Middle Eastern Studies Bulletin* 14(1), 68–74.
Umunç, H. (2011) The Dutch in the Levant: trade and travel in the seventeenth century. *Belleten* 75(273), 373–386.
Walters, S. (2017) The silk industry in Lyon, France. *Museum of the City*. Originally available at: http://www.museumofthecity.org/project/the-silk-industry-in-lyon-france/ (accessed 6 October 2017). Archived (23 October 2017) at *Wayback Machine*. Available at: https://web.archive.org/web/20171023063733/http:/www.museumofthecity.org/project/the-silk-industry-in-lyon-france/ (accessed 28 January 2021).

End of the Spice Era   22

Setting the Stage – European Tastes Change

As the Dutch and English battled at home and abroad, the major trade commodities also underwent a dramatic shift. Changes in tastes and political climates in Europe caused the profitability of the spices to fall precipitously. This led the VOC and the EIC to seek new markets including cotton, coffee, opium and tea.

It was in the middle of the seventeenth century that European interest in spices began to wane. In fact, there was an oversupply of pepper by mid-century, which dropped prices by about 40% compared with that which the Portuguese and then the VOC had long been able to maintain (Lunde, 2005). After a peak of seven million kilograms of pepper imported in 1670, levels fell to about three-and-a-half million kilograms in 1688 (Krondle, 2007). Pepper had lost its status as an exotic luxury in Europe and was now more or less a mundane commodity. The other spices held their high status longer, but they too began to lose their glow by the end of the seventeenth century.

The decline in spice use in Europe was due to many causes. As Freedman describes:

> A whole new group of beverages, stimulants and flavors had arrived including tea, coffee, chocolate and tobacco that offered new taste sensations but also produced psychological effects that proved to be mildly, or in the case of tobacco, quite seriously addictive ... Spices became cheaper with colonialism and the opening of new trade routes, so their consumption no longer conveyed an adequate sense of privilege and exclusivity.
>
> (Freedman, 2008, p. 221)

Their popularity as medicines had greatly diminished and they had 'lost their healthy glow'. The origins of the spices were now well known, and they did not seem so mysterious and exotic anymore. Perhaps most importantly there 'was a seismic shift in tastes. The wealthy people of Europe no longer liked fiery and perfumed food' (Freedman, 2008, p. 224).

A culinary revolution had sprouted in France in the mid-1600s that took the rest of Europe by storm. As Krondl tells it:

> Gone were the generous helpings of sugar and exotic spice of the Italian Renaissance masters, replaced by local herbs and mushrooms ... The French chef still includes nutmeg and cloves in plenty of his recipes, though he does so in stingy quantities. A typical recipe will call for two or three cloves and a grating of nutmeg. Pepper and ginger are mostly absent, and cinnamon has been quarantined in the dessert chapters.
>
> (Krondl, 2007, p. 254)

The old sweet and sour blends were now despised in favour of more delicate combinations of local herbs. Sauces were based on butter and egg yolk combinations that were flavoured with capers, anchovies, mushrooms and scallions rather than nutmeg and cinnamon.

The Shift Away from Spices

The shifts in European tastes led to a seismic shift in VOC imports between the seventeenth and eighteenth centuries. In 1650, the relative percentages of commodities shipped to Europe were about 32% pepper, 27% fine spices and 18% textiles, but by 1700 they had shifted to 42% textiles, 25% fine spices and 12% pepper (Bernstein, 2008). In 1750, the Dutch imports had further diversified to 28% textiles, 25% fine spices, 25% tea and coffee, and 12% pepper.

Commodity imports of the EIC were much less oriented towards spices than the VOC's, but their relative portions also changed dramatically between the mid-1600s and 1750. Between the years of 1668 and 1670, the averages were 25% pepper, 57% textiles and just a trace of tea; between 1738 and 1740, pepper was only 3%, while textiles had risen to 70% and tea to 10% (Barrow, 2017). Textiles were now 'the bread and butter' of the EIC's trade.

As European tastes shifted, the VOC would hold tenaciously on to its Indonesian sources of pepper, clove and nutmeg, and gain a monopoly on Ceylon cinnamon, while steadily expanding its reach towards the cotton of India and the tea of China. The EIC would largely give up on its Indonesian pepper and spice trade and focus on India, first to acquire cotton textiles and later to trade opium for tea.

The Shift to Cotton

So by the early eighteenth century, pepper and the other spices were no longer the primary commodity being shipped to Europe from the Indian Ocean. Thus ended a run that had lasted over 1500 years, starting with the Romans and passing to the Venetians, Portuguese, Dutch and English. The spices would be replaced first by cotton in the late 1600s, and then by tea and coffee in the

1700s. A large portion of Dutch and English trade had also shifted to the Atlantic Ocean with its sugar, tobacco and slaves.

As spice demand began to decline in Europe in the mid-1600s, the English viewed as potential alternatives Persian silk, indigo and calicoes from Surat, and calicoes from the Coromandel Coast (Marshall, 1998). These they shipped out from their fortified settlement in Madras. Persian silk did not prove to be generally profitable, and indigo could be acquired from cheaper sources in the Americas, which left cotton as the most viable alternative. This proved to be the case and cotton became the EIC's main source of income from East Asia until the eighteenth century, when tea took off.

What many historians have called a 'chintz craze' began in Europe in about 1600 and reached its peak at the end of the century (Hancock, 2017). Originally chintzes were used as wall coverings, pillowcases, curtains, bed covers and rugs, but by the late 1700s were widely used for clothing by people from all walks of life. The great English writer Daniel Defoe of *Robinson Crusoe* fame described 'the chintz was advanced from their floors to their backs, from their footcloth to the petticoat' (Yalfa, 2005, p. 32). The poorer classes were actually the first to wear the cheap but colourful chintzes, but the rich soon followed. Another satirical writer attested 'it became difficult for the better folk to know their wives from their chambermaids' (Hancock, 2017, p. 65).

By 1684, the EIC's textile import averaged between 60 and 70% of its total trade and amounted to more than one million pieces (Lemire and Riello, 2008). Imports of calicoes rose from 240,000 pieces in 1663/64 to 861,000 in 1699–1701. As the seventeenth century progressed the commodities imported by the VOC also shifted from predominantly pepper and fine spices to textiles. Like the English, they had been trading Indian cotton throughout South East Asia to secure spices, so it was not a great stretch to begin shipping significant quantities of those textiles to Europe as well (Laarhoven, 1994).

Cotton Politics

As the chintz craze gathered steam, the local English textile industries began to howl in protest (Hancock, 2017). Literally hundreds of pamphlets were published that decried the widespread sale of Indian textiles. They were defiled as being produced by pagans, encouraging increased spending by the already poor, putting struggling artisans out of work and damaging the wool industry. Daniel Defoe wrote: 'About half the woolen manufacture has been entirely lost, half of the people scattered and ruined, and this by the intercourse of the East India trade' (Yafa, 2005, p. 31).

France came to ban cotton importation of all kinds in 1686, and England started with a partial ban in 1702, followed by a full ban in 1721. However, these laws had little real impact, were flagrantly ignored by almost everyone and the bans were little enforced by the authorities. A few unfortunate women were stripped naked on the streets by mobs of angry textile workers, but the Indian chintz craze continued largely unabated.

British Weave Their Own

As the demand for cotton textiles exploded in Europe, it was only a matter of time until their own manufacturing industry would develop. This started as a cottage industry but by the 1770s an Industrial Revolution had emerged. British ingenuity led the way, stimulated by the ever-growing thirst for more cotton yarn and finer textiles.

The first major technological innovation came from John Kay in 1733. He invented a mechanical loom with a 'flying shuttle' which was hurled back and forth, weaving the horizontal 'weft' yarn through the vertical 'warp' threads. As the faster, more efficient looms became widespread, so did the demand for cotton yarn. The man who rose to the occasion was James Hargreaves in 1764, who invented the 'spinning jenny' (named after his daughter).

While the spinning jenny revolutionized the rate at which yarn could be produced, the product was only strong enough for weft and not warp. This meant that linen or wool had to be used as warp by the English weavers. Richard Arkwright and John Kay (not related to the inventor of the flying shuttle) were the first to develop a mechanical spinning machine that produced cotton thread strong enough for warp.

Through the eighteenth century, the energy used to run the textile machines passed through a number of transitions from human crank to horse drawn, then water driven and finally steam powered. The great textile plants of the mid-1700s were powered by water, but water wheels were replaced by steam power when Cartwright invented the 'power loom' in 1785. He was the first to put James Watt's 1769 invention of the steam engine into large-scale manufacturing.

Richard Arkwright was the man who built the first cotton mill and started the Industrial Revolution. He built his first factory in Cromford, Derbyshire in 1771 and within a decade he had expanded to 12 factories and almost 1000 employees; by 1790 he had 200 factories and 5000 workers. As the cotton revolution spread across England, the spinning of wool essentially disappeared. Between 1775 and 1783, calico production vaulted from 56,000 to 3,000,000 yards in England; and by 1841, there were more than 1000 mills spread across north central England, employing over 350,000 people.

The Shift to Tea

Cotton remained the primary export item from the East until the Industrial Revolution made England the world's leading textile manufacturer. The EIC then shifted to tea from China as its primary export crop. Tea popularity had been rising in England from about 1660 and had become widely popular by the early 1700s. Its popularity received a great boost during the Industrial Revolution, when it began being consumed during refreshment breaks that kept minds sharp. Workers were much more productive and safe drinking tea than the previous beverage of choice, beer.

In the first half of the 1600s, most of the tea in Europe came from Japan via the VOC. The Dutch gained a monopoly when the Portuguese were thrown out of Japan for pushing Christianity too hard. This left the door open to the VOC, who wound up with a stranglehold on Japanese trade that lasted until the VOC was dissolved in 1795 after the Fourth Anglo-Dutch War.

The EIC began to bring tea into England from China in 1689. Its trade was centred in Canton, where the Chinese had allowed English and French 'foreign devils' to establish factories. The amount of tea moved by the EIC was initially modest – in 1678 it shipped only about 5000 pounds (2270 kilograms) to England. Its highest profits were still coming from cotton textiles from India. However, when the Industrial Revolution shut down the profitability of Indian cotton, tea turned the EIC into an economic powerhouse. At the company's peak, 60% of its trade was in tea and 10% of the British government's revenue came from duties on tea. The company continued to export its tea from China until the 1800s, when it shifted to plantations it had established in India and Ceylon (Hancock, 2017).

Opium Wars

The EIC had a major problem in acquiring tea: the Chinese were really not interested in anything in trade except silver (Hancock, 2017). The EIC solved this problem by building a demand for opium in China, selling it for silver and then using that silver to buy tea. There was a little problem with this plan, however, as opium use in China was banned by the emperor and the British government certainly couldn't advocate the trade of an addictive substance. The EIC developed a circuitous route around this problem by growing poppies in India, processing them into opium, and then selling the opium for silver to private merchants called 'country agents'. These agents then smuggled the opium into China, bribed the necessary officials and sold it for silver.

What started as a minor trade item grew by leaps and bounds to gigantic proportions and opium become the most profitable trading commodity in the world. In 1760 only about 200 chests weighing 140 pounds (63.5 kilograms) each were exported to China but by 1770 the number had risen to over 1000. By 1800 this level was at 4000 chests and by 1830 had reached over 18,000 annually. By this time, there were likely over ten million opium addicts in China.

The emperors in China became increasingly concerned with the large number of Chinese who had become addicted, and opium's import was ultimately banned. This led the British, with help from the French, to fight two major wars with China to allow this trade. These wars greatly undermined the authority of the Qing Dynasty and confidence in the emperor plummeted. As described by Philip Allingham:

> What had begun as a conflict of interests between English desires for profits ... resulted in the partitioning of China by the Western powers (including the ceding of Hong Kong to Great Britain), humiliating defeats on land and sea by technologically and logistically superior Western forces, and the traditional values of an entire culture undermined by Christian missionaries and rampant trading in Turkish and Indian opium.
>
> (Allingham, 2006)

China would be the 'Celestial Empire' no more.

The EIC Moves Heavily into India

In 1686, the EIC felt the time was right to embark on a direct war with the now fading Mughal Empire to obtain broad trading privileges across the entire continent and more specifically to get permission to build a fortress in Bengal. A fortress in Bengal was seen as a critical step to protect the company's burgeoning trade there from the Dutch and interlopers.

The First Anglo-Mughal War (1686–1690) began at the Hugli River at Calcutta and ultimately was fought on both coasts (Barrow, 2017). Often called 'Child's War', it was driven by one of the company's major stockholders, Josiah Child, and proved to be a great embarrassment to the English nation. Twelve British battleships were involved, and a number of battles raged across the continent including a siege of Bombay and burning of the city of Balasore. The British navy blockaded the Mughal ports on the western coast and attacked its army on land. A number of major cities were significantly damaged including Bombay, Madras, Calcutta and Chittagong.

A pivotal point in the conflict came when the English confiscated a convoy of ships carrying a grain shipment to Yakut Khan, a regional leader who was allied with the Mughals but had stayed out of the fight. In response, Sidi Yakur landed a force at Bombay and began a siege of the city that lasted almost a year and a half. The war was finally ended in 1690, when the Mughal emperor Aurangzeb issued a *farman* and allowed the English to build a fortress at Calcutta, but not before requiring them to pay a huge fine, return seized property and comply with a number of conditions.

The reaction in England to the war was one of vehement disgust. Barrow (2017, p. 18) quotes a pamphleteer who described how the war had ruined the good name of England: 'Thus has the English Nation be made to stink in the Nostrils of that people; when before, from the time we set footing on that Golden Shoar, we were the most loved and esteemed of all Europeans.'

VOC Loses Steam

The profits of the VOC peaked in the 1670s and then began a slow, gradual decline. Its vigorous trade with China largely ended when the company was removed from Formosa in 1663. This meant VOC could no longer trade Chinese silk for the Japanese gold needed to acquire Asian goods. The Dutch compensated by focusing on the Bengal silk market, but the profits were not nearly as high. From 1675 to 1683 the VOC and the EIC got into a price war that brought both companies near bankruptcy. The VOC got a bit of a boost in 1685, when it was able to force the British competition out of Bantam, Java, but this was greatly offset by the Japanese decision to limit the export of gold and silver in that year, depriving the Dutch completely of their key source of precious metals. They faced a rebellion in Java from 1741 to 1743, after a massacre of 10,000 local Chinese in Batavia. New competition from the Danes and most notably the French arose in the late seventeenth century, putting their Indian holdings at risk. The Dutch strongholds along the Malabar Coast and the Persian Gulf were lost in the eighteenth century. The British greatly decimated their East Asian navy during the Fourth Anglo-Dutch War from 1780 to 1784. In the late 1780s, only the core of the company's 'empire' remained: the Indonesian Archipelago and Ceylon.

The VOC collapsed fully after the Napoleonic Wars in Europe, when it was nationalized in 1796 and its charter was allowed to expire in 1799. The company died quietly after almost 200 years of largely ruckus existence. Indonesia became a colony of the Kingdom of the Netherlands, and Ceylon was ceded to the British.

Last Century of the EIC – From Traders to Rulers

In the eighteenth century, the EIC carried its invasion of India to its final conclusion and made the startling transition from merchants to rulers. It came to control a colonial empire whose territory far exceeded the British Isles.

The company's shift to ruler began in Bengal in 1756 when after decades of benevolent rule, Siraj ud-Daulah became the *nawab* (Mughal governor) and decided to exert his authority. He demanded exorbitant amounts of cash from all the East India companies and launched an attack on the English fortifications in Calcutta. These he easily took, along with 146 prisoners who were forced to spend the night in the fort's prison called the 'Black Hole'. Many died in the heat and humidity, enraging the British colonial community.

A force was sent from Madras to retake the city that was led by Robert Clive, who quickly recaptured it and put Siraj ud-Daulah to death. A period of turmoil and court intrigue then followed until the Treaty of Allahabad was signed in 1765 with the Mughal emperor. This treaty allowed the company to collect revenue in the provinces of Bengal, Bihar and Orrisa for an annual stipend and in essence made the company a sovereign state.

Many other wars followed that ultimately gave the British control of most of India. Beginning in the 1740s, the company fought at least one major war a decade until about 1850. The Anglo-Mysore Wars (1766–1799) were fought over southern India, the Anglo-Maratha Wars (1772–1818) over central India, and the final Sikh Wars (1845–1849) over the Punjab of northern India.

European Competitors in India

The Danes decided to enter the lucrative Indian Ocean trade in the seventeenth century. The first Danish East India Company was chartered in 1616 under King Christian IV and was focused on India (Stow, 1979). Between 1624 and 1636, the Dutch had outposts in Surat, Bengal, Java and Borneo, and factories in Masulipatam, Surat, Balasore and Java. From 1643 to 1669 their trade with India was disrupted during the European wars and they lost almost all their Indian possessions. Two new Danish East India Companies were charted after the wars and between 1670 and 1750, a total of 27 ships were sent to India of which 22 returned.

France was the last of the major European maritime powers to enter the East India spice trade. The French did not get into the game until the EIC and VOC were well established, a full six decades after the foundation of these companies (Furber, 1976; Greig, 2018). The *Compagnie Française des Indes Orientales* was established by Jean-Baptiste Colbert, finance minister to King Louis XIV, with the goal to purchase gold, pepper, cinnamon and cotton directly from Indian Ocean merchants. As it stood, a considerable amount of French wealth was being transferred to England and the Dutch Republic, as about 30% of the EIC and VOC exports were moving into France.

Colbert's company was given a 50-year monopoly of trade east of the Cape of Good Hope, as well as the rather ambitious right to colonize Madagascar, Réunion and the islands of Mauritius, and establish trading centres in India, Ceylon and Indonesia. Unfortunately, the initial support of the other French merchants was lukewarm and only about half as much money was raised as expected, with the king being the primary investor. The first French expedition made its way to Madagascar in 1664 and in 1668 a second expedition reached Surat and established the first French factory in India. In general the early expeditions were commercially disastrous, and it took a number of years for the company to become modestly successful. The French were constantly plagued by financial woes and were in constant conflict with the Dutch and English.

Pondicherry (Puducherry), lying about 85 miles (137 kilometres) south of and not far from the EIC's trading centre at Madras on the Coromandel Coast, became the centre of French India in 1674. Here silk and cotton textiles were readily available that were produced by the surrounding villages. The French, however, found themselves in constant conflict with the Dutch and the English. The jealous VOC even drove them out in 1693, but the company returned in 1699 and for the next 100 years Pondicherry served as the Indian capital of the French, who eventually built factories in Surat, Chandernagor (French name; formerly Chandernagore, now Chandannagar), Calicut, Dhaka, Patna, Kasim Bazar, Balasore and Jodia. After the treaty of Utrecht in 1713, the VOC was forced to withdraw to Indonesia, leaving England and France the only rivals in India.

The most famous governor of French India was Joseph François Dupleix, who tried to build a French territorial empire in India even though the French government was not particularly interested in provoking the British. Dupleix's army came to control the area between Hyderabad and Cape Comorin but was subjected to constant political intrigues and military skirmishes with Great Britain. The ambition of Dupleix to create a French empire in India was dashed when British Major-General Robert Clive arrived in India in 1744, took possession of Bengal and then smashed the French forces. Dupleix was summarily recalled to France and dismissed in 1754.

End of the EIC

The EIC ruled India until 1858, when its handling of the Indian Rebellion led the British government to take over control of the country. At about the same time as the Opium Wars, the company began witnessing a rapidly increasing amount of insurgence in its Indian territories. The company's conquest of the subcontinent during the eighteenth and early nineteenth centuries had left many scars. Many of the rebels were Indians within the EIC's own army which by this time had grown to over 200,000, of whom 80% were Indian recruits.

The rebels caught the British off guard and managed to kill many British soldiers, civilians and Indians loyal to the company. In retaliation, the company brutally murdered thousands of locals, both rebels and anyone thought to be sympathetic to the rebellion. It ended in Delhi where 1400 were murdered. As quoted by William Dalrymple:

> 'The orders went out to shoot every soul,' recorded Edward Vibart, a nineteen-year-old British officer. 'It was literally murder ... I have seen many bloody and awful sights lately but such a one as I witnessed yesterday I pray I never see again. The women were all spared but their screams, on seeing their husbands and sons butchered, were most painful ... Heaven knows I feel no pity, but when some old grey bearded man is brought and shot before your very eyes, hard must be that man's heart I think who can look on with indifference ...'
>
> (Dalrymple, 2006, p. 7)

In the wake of this bloody uprising, the British government effectively abolished the EIC in 1858, taking away all of its administrative and taxing powers. The Crown assumed control of all its territories and armed forces. Thus began the British Raj and the direct British colonial rule over India which continued until independence in 1947.

Epilogue – Spice Trade After the East India Companies

With the end of the East India companies, so went the centralization of the spice and silk trade. No longer would the spices be grown solely in restricted geographical regions under the control of a specific trading company. They became scattered all over the world. Many now are grown and important in countries far from their South East Asian origins, in particular Africa, Central America and South America. So, after thousands of years as remote, localized commodities controlled by a few powerful trading empires, the spices became more or less routine commodities open to many international entrepreneurs.

After the collapse of the VOC, the Dutch kept close tabs on their Spice Island clove trees until 1770 when the French missionary, Pierre Poivre, smuggled seed off the islands to France and from there cloves were introduced to Mauritius and later to Zanzibar. Dutch profits in Indonesia came to rely on plantations of sugar, coffee, tea and rubber in Java and other Indonesian islands. Today Indonesia still produces the most cloves in the world, but its production is closely followed by Madagascar, Tanzania and Kenya. India, Pakistan and Sri Lanka are also important clove producers.

The British invaded and temporarily took control of the Banda Islands soon after the Napoleonic Wars and from there they spread nutmeg trees first to Sri Lanka, Malaysia and Singapore, then to Zanzibar and Grenada, and later to Brazil and Tanzania. Today, Guatemala, Indonesia and India are the leading world producers.

In the late eighteenth century, the Dutch introduced cinnamon trees to Indonesia, and from there the French took them to Mauritius, Réunion and Guyana. In the nineteenth century, cinnamon was also introduced from the Jardin des Plantes of Paris to Egypt. Today, world production has further spread to Indonesia, China and Vietnam, with Sri Lanka now fourth.

For centuries after the Europeans first found their way to the Indian Ocean, the pepper-producing countries remained static and this situation remained unchanged until the modern era, when pepper was successfully introduced to Brazil in the 1930s and to Africa, Vietnam and southern China after World War II. Vietnam leads world production today, followed by Brazil, Indonesia and India.

Ginger was introduced into Europe as early as the ninth century and from there was spread to all the tropical and subtropical countries of the world (Nayar and Ravindran, 1995). India is still the largest producer by far, but ginger is also very important in Nigeria, China, Indonesia and Nepal.

The centre of silk production shifted back and forth in the later part of the twentieth century. The completion of the Suez Canal in 1869 greatly reduced the price of importing Asian silk to Europe, and Japan became the world's foremost silk producer, producing about 60% of the world's raw silk. During World War II, silk supplies from Japan were cut off and silk prices skyrocketed, so Western countries began finding synthetic substitutes such as nylon. When the war ended, silk was not able to regain many of its previous markets and Japan never regained its export status. Today, China has returned as the world's leading silk producer with close to three-quarters of the market, followed by India with most of the rest.

References

Allingham, P. (2006) England and China: The Opium Wars, 1839–60. *Victorian Web*. Lakehead University, Thunder Bay, Ontario. Available at: https://victorianweb.org/victorian/history/empire/opiumwars/opiumwars1.html (accessed 13 April 2021).

Barrow, I. (2017) *The East India Company, 1600–1858: A Short History with Documents*. Hackett Publishing, Cambridge, Massachusetts.

Bernstein, W.J. (2008) *A Splendid Exchange: How Trade Shaped the World*. Grove Press, New York.

Dalrymple, W. (2006) *The Last Mughal: The Fall of a Dynasty: Delhi, 1857*. Bloomsbury, London.

Freedman, P. (2008) *Out of the East: Spices and The Medieval Imagination*. Yale University Press, New Haven, Connecticut and London.

Furber, H. (1976) *Rival Empires of Trade in the Orient, 1600–1800*. University of Minnesota Press, St Paul, Minnesota.

Greig, J.A. (2018) French East India Company. *Encyclopedia.com*. Available at: https://www.encyclopedia.com/history/modern-europe/french-history/french-east-india-company (accessed 13 April 2021).

Hancock, J.F. (2017) *Plantation Crops, Plunder and Power: Evolution and Exploitation*. Routledge, London.

Krondl, M. (2007) *The Taste of Conquest: The Rise and Fall of the Three Great Cities of Spice*. Ballantine Books, New York.

Laarhoven, R. (1994) The power of cloth: the textile trade of the Dutch East India Company (VOC) 1600–1780. PhD thesis, Australian National University, Canberra.

Lemire, B. and Riello, G. (2008) East & West: textiles and fashion in early modern Europe. *Journal of Social History* 41(4), 887–916.

Lunde, P. (2005) The coming of the Portuguese. *Aramco World: Arab and Islamic Cultures and Connections* 56(4). Available at: https://archive.aramcoworld.com/issue/200504/the.coming.of.the.portuguese.htm (accessed 28 January 2021).

Marshall, P.J. (1998) The English in Asia to 1700. In: Canny, N. (ed.) *The Oxford History of the British Empire, Vol. I. The Origins of Empire: British Overseas Enterprise to the Close of the Seventeenth Century*. Oxford University Press, Oxford, pp. 264–285.

Nayar, N.M. and Ravindran, P.N. (1995) Herb species. In: Smartt, J. and Simmonds, N.W. (eds) *Evolution of Crop Plants*. Longman Scientific and Technical, Harlow, UK, pp. 491–495.

Stow, R. (1979) Denmark in the Indian Ocean, 1616–1845: an introduction. *Kunapipi* 1(1). Available at: https://ro.uow.edu.au/kunapipi/vol1/iss1/3 (accessed 15 April 2021).

Yafa, S. (2005) *Cotton: The Biography of a Revolutionary Fiber*. Penguin Books, New York.

Index

Note: Page numbers in **bold** type refer to figures
Page numbers in *italics* type refer to *tables*

al-Abbas 139
Abbas I, Shah of Persia 284, 285
Abbasids 139
Abd al-Malik, caliph 158
Abd-al-Rahman 139
Abreu, António de 219
Abu Hafs 158
Abu Zayd 144
Abu-Lughod, J. 176
Academy of Gundishapur 128
Acapulco 241
Aceh, Kingdom of 198, 226, 253, 266
Achaemenid Empire (Persia) 68, 70
Achinese 227
Acre 183
Adelson, H.L. 148, 150
Aden (*Arabia Emporion*) 57, 194–195, 218
Aden–Cambay trade 195
Adler, E. 150–151
Adriatic Sea, route control 177
Adulis 130
Aegilops tauschii 13
Aegyptus 46
Aelius Gallus, prefect of Egypt 62
Africa 67
      Portuguese navigator missions 207–208
      route from Red Sea ports 75, **76**
      Swahili coast 196
Agatharchides, *On the Erythraean Sea* 72
Age of Discovery, European 206
Aila 62
Akananuru 77
Akbar the Great, Mughal Emperor 228–229, 265

Akhbar al-Sin wa'l-Hind (Accounts of China and India) 143
Aksum Kingdom 129–130
Alagakkonara, King of Ceylon 203
Albuquerque, Afonso de 215, 217, 222–223
      cartaz system 223–224
Aleppo 281–282
Alexander VI, Pope, bull to Spain 208–209
Alexander the Great 23, 28–29, 43–44, 52, 70–71, 94
      aftermath after India 108
Alexander of Tralles, *Twelve Books on Medicine* 31
Alexandria 3, 43–44, 57, 77, 160
Alexandros 119
Alexios I Komnenos, Emperor of Byzantine Empire 162, 165, 167
      Chrysobull Golden Bull (*chrysobull*) 162–163, 167
Alexios II Komnenos, Emperor of Byzantine Empire 168
Alexios Angelos, Prince of Byzantine Empire 169
Allahabad, Treaty of (1765) 295
Allingham, Philip 294
Almangor, Raja of Tidore 200–201
Almeida, Francisco de (Viceroy of Portuguese India) 214, 217
Almeida, Lourenço de 214, 216
alpine passes of medieval trade 149–150
Amalfi 163
amber 42
Ambon Island 254, 259
Amboyna massacre (1623) 259

301

amphorae, from shipwrecks 148
Amsterdam 248
Anaxicrates 71
ancient world
    frankincense and myrrh 5–6
        culture 7–8
        natural history 6–7
    luxury products 1–2
    silk 32–35
    spices 24–31
Andronikos I Komnenos, Emperor of Byzantine Empire 168
Angell, J.B. 280
Anglo-Mughal War, First (1686–1690) 294
Anglo–Dutch War
    First (1652–1654) 272
    Second (1665–1667) 272
    Third (1672–174) 272–273
Anthimus 147
antiquity see ancient world
Antoine plague 103–104
Antwerp 227, 247–248
Anuradhapura Kingdom 108–109
Apicius 147
    cookbook 21, 26, 27
        Excerpta 147
aquila 95
Arab agricultural revolution 138
Arab converts 139
Arab expansion, infighting 138–139
Arab merchants 158
Arabia, Dutch spices sales 271
Arabia Felix 54
Arabian Peninsula 53
    early seventh century 135
    Incense Route 3, 5, 53–54
Arabian Sea, transport across 67
Arabians
    Indian trade monopoly 72
    worldwide trade role 71
Arabs
    dirhams 152
    of Gaza 23
    'Ta-shis' 143
Archaemenid Persian Empire 94
Archimedes 44
Arctic exploration 232–233
Ardashir I (the Unifier), Emperor of Sasanian Persia 95, 102
Arkwright, Richard 292
    and Kay, John 292
Armada de Molucca 238
Armada del Nord 223
Armada del Sud 223

Armenians 285
Arsaces I, Chief of the Parni 94
Artabanus IV, King of Parthia 102
asbestos cloth 101
Asbridge, T. 146, 159, 165–166
Ashoka 108, 109
Asia
    broomcorn millet 67
    Dutch East Asian trading network 254–255, **255**
    Dutch imports from, to Bengal 269
    East, Dutch trading network 254–255
    spice production, sixteenth century 228
    see also Central Asia; South East Asia
Asia Minor 13
Assyria, Kingdom of 3
Assyrian cuneiform texts 23
Assyrians 39, 52, 66
Astrakhan 278
Atil 153
Atlantic trade, European 261–262
Augustus, Emperor of Rome 73, 95
Aurangzeb, Mughal Emperor 265
Aurel, Sir A. 9
Aurelian, Emperor of Rome 97–98
Austronesians 10
Ayurvedic medicine of India 21
    cinnamon use 24
    cloves use 30
    pepper use 27, 28
    saffron use 30
    texts 25
Aztec 236
Aztec Empire 235

Babylonians 39, 66
Bactria 85, 98
Baghdad 153–154, 158
    building and features 139–140
Baghdad–Canton express (Bernstein) 143
Bahadur Shah 217
Balboa, Vasco Núñez de 235
Bali, early trade evidence 108
Banda Islands 16, 25, 114, 199, 254, 298
    massacre (1621) 259
    orang kaya 257
    Portuguese in 218–219
bandits 55, 56, 74
banquets 22
Bantam 250, 255
Barbados 262

barbarians 146
    Germanic 147
Barbosa, D. 192–194
Barda, Rus' attack on (943) 154–155
Barents, William 233
Baros (Panchur/Pansur) 198
Barrow, I. 267, 294
barter 128
Basil I, Emperor of Byzantine Empire 162
Basil II, Emperor of Byzantine Empire 155, 160, 162
    Bulgarian campaigns 160
Batavia 107, 254, 258, 259, 274
Bavaria-Landshut, Duke of 172
Begram (Bagram) 99
    Old Royal City/New Royal City 99
Belisarius, General Flavius 157
Bendersky, G., and Ferrence, S.C. 14
Bengal 11, 295
    Dutch and English trade 269
    fortress 294
    silk 285–286
    VOC and EIC trade centre 269
Berenike 74, 77
Bergreen, L. 238
Berlin Ethnological Museum 92
Bern Acoarala, Sultan of Ternate 200
Bernstein, W.J. 34, 72, 143, 209
Best, Thomas 266
Bhuvanaikabahu 220
Bible 22, 39–40
    Book of Exodus 24
    Book of Proverbs 25
    First Book of Kings 52, 54
    Gospel of Mark 22
    Gospel of Matthew 23–24
biological warfare 186
black bile 32
Black Death 186–187
black pepper *see* pepper
Black Sea
    alternative route 184–185
    control of 184
    Middle Ages, commercial region 164–165, 184
    Venice move into 185–186
bloody flux 256
Bolivia, Potosí 242
Bombay 268
Bombyx mori 17
Boniface VIII, Pope 184
Bostock, J., and Riley, H.T. 8, 55, 57, 68
Boswellia sacra 6
Boxer, C.R. 190, 206

Braudel, F. 278
Brazil 212
    sugar plantations 262
bread wheat 13
breadbaskets
    Black Sea surroundings 137
    Egypt 97
    sea routes to 43
Brierley, J.H. 68
British East India Company 233
    see also English East India Company (EIC)
British Museum 92
British Raj 297
Brito, Antonio de 238
Broadhurst, R. 166
broomcorn millet 67
Brouwer, C.G. 276
Brouwer, Hendrik, Africa to Java route 255, 257
Brownstone, D.M., and Franck, I.M. 82
Bruges 180
Brunel, Oliver 232
Bryce, T. 60, 96
bubonic plague 186
Buddha, statues of 90
Buddhism 90–91, 91–92, 99
    Borobudur Temple 111
    cave complexes, artefact removal 91–92
    pilgrims from China 90
    use of saffron in textiles 30
Buddhist/Confucianist and Muslim boundaries 142
Buddhists 108
    communities, Islam spread and Silk Road loss 180
    emissaries 109
    monastic communities/institutions, Indonesia 108–109, 109–110
Bukhara 89
Bulgarian campaigns, Basil III of Byzantine Empire 160
bullion 270
Bulut, M. 271–272
Byzantine clergy 18
Byzantine Commonwealth 160
Byzantine Empire 4, 122–123
    after Justinian 157–158
    Golden Age 160
    pepper 27
    resurgence, tenth and eleventh centuries 159–160
    struggles with Venice 167–168

Byzantine Empire (*contiuned*)
    trade
        overlapping spheres 160
        redirection in seventh
            century 135–146
        with Sasanian Empire 128–129
        in silk and spices with
            Muslims 160–161
    under Justinian 125–127, **126**
Byzantine medical texts 31
Byzantine merchants 148
Byzantines 122, 146
    trade with Rus, through Khazaria 153–154
Byzantine–Holy Roman Empire tensions 168–169
Byzantine–Sasanian War (602–628) 138
Byzantine–Venetian animosity 167–168
Byzantium (Constantinople) 122, 123, 159–160
    capture by Ottomans 189
    contempt for trade 127
    'eparch' 127
    exotic luxuries of 124–125
    grain trade 124
    Hagia Sophia 169
    importance of Black Sea 164
    Justinian's plague 132
    kommerkiarioi 127
    medieval, food 125
        navicularii 127
    rebuilding by Constantine 123
    main features 123
    Rus' attack on (941) 154
    sacking by Crusaders (1204) 169–170
    Venetian commercial quarter in 167

Cabot, John 230
Cabot, Sebastian 239
Cabral, Pedro Álvarez 212–213
Cabrillo, Juan Rodríguez 240
Cacafuego 243
Cadamosto, Alvise 207
Caffa 185
Calcutta 269
calico 291
    cloth 192
Calicut 192, 211, 212
Caligula, Emperor of Rome 35, 48
Cam, Diego 208
Cambay 194, 195, 197
camels 56, 60, 103
    Bactrian (*Camelus bactrianus*) 88, 89
    caravans 89, 94
    domestication 54–55, 88–89
    dromedary (*Camelus dromedarius*) 54, 88
    Silk Routes 88–89
    see also caravan routes
canal, Nile–Red Sea 52–53
cannons 189
canoes
    kora-kora 114
    outrigger 68
Canton 293
Cape of Good Hope 224–225
    VOC way station 269
Capitulation Treaty (1535, Francis I and Suleiman the Great) 280
caravans
    camel 89
    routes 55–59, 68, 74, 77
        Royal Road 94
    state-sponsored by China 103
Caribbean, sugar production dominance 262
Carolingians 146, 159
Carrhae, Battle of (40 BCE) 21, 95
cartaz system 223–224
Carthage 41, 45–46
Cartier, Jacques 230–232
Cartwright, Edmund, power loom 292
Cartwright, M. 19, 41–42, 128, 179, 189, 196, 201–202, 203
Casa da Índia 226–227
cassia 2, 8–9, 24, 67–68
Cassia cassia, flower 39
Casson, L. 38, 42, 47, 48
Catalonian trade networks 178
Catholic Church *see* Roman Catholic Church
cavalry 82, 83
cave complexes, Buddhist 90
Cavendish, Thomas 243–244
Cebu 237, 240
Celetti, D. 278, 280, 283
Central Asia
    balance of power, and horses 82–83
    Islamization of 142
    Silk Routes 104
Cerón, Álvaro de Saavedra 239
Ceuta 207
Ceylon 108, 119, 198–199, 214
    Dutch takeover 275–276
    Manthi 108–109
    Portuguese in 220
Champagne, international trade fairs 179–180
Chancellor, Richard 232
Chandragupta 108
Charlemagne, King of the Franks 146, 159
Charles I, King of Spain 236

Chen Zuyi 203
Cheras 108
Chevallier, J. 147
Child, Josiah 294
Childeric II, King of the Franks 147
Child's War 294
China 103
    ancient dynasties 18
    cassia use 24
    cloves called *hi-sho-hiang* (bird's
        tongue 30
    division 104
    early 83–84
        and Xiongnu 84–85
    Great Wall 84–85
    Guangzhou 104, 143
    Han Dynasty 28, 30, 84, 84–85, 104
        borderlands control 86–87
        maritime trade 110
        power coalescence after
            collapse 142
        silk currency 33
        and Xiongnu 84–85, 86–87
    and horses 82–83
    isolationism 203–204
    Ming Dynasty 197, 201, 203–204
        tributary trade system 201–202, 204
    Muslim maritime trade with 142–143
    'Nanhai' trade 110
    pepper use 28
    silk 17–18, 19, 21, 34, 241–242, 298
    South East Asia trade network
        entry 110
    state-sponsored caravans 103
    Sui Dynasty 142
    Sung Dynasty 28
    Tang Dynasty 28, 104–105, 142
        foreign traders influx 143, 144
        silk scroll of Emperor Taizong 33
    trade with Portugal 229, 230
    Zhou Dynasty 110
China–South East Asia mutual trade 204
Chinese herbals 21
Chinese medicine
    cloves 30
    ginger 25
    saffron in texts 30
    traditional, nutmeg 31
Chinese pilgrims 111–113
Chinese smugglers 204
Chinese society, silk wearing 32–33
Chinese–Parthian merchants, trade 101
chintz craze 291
Chittick, H.N. 196

Chlothar, King of Franks 147
Choniates 167
Christian, D. 80, 82, 102–103
Christian God 18
Christianity 122, 130
    Eastern Orthodox 152
    foothold in Muslim Middle East
        165–166
    Jesus Christ 23, 128
    see also Bible; Roman Catholic Church
Christian–Islamic Orient 160
Chrysobull Golden Bull (*chrysobull*), given to
    Venice 162–163, 167
Church 187
    Byzantine Empire 122–123
    see also Roman Catholic Church
Cilicia 12
cinnamaldehyde 39, 66
Cinnamomum 39, 66
    cassia (cassia or bastard/Chinese
        cinnamon) 8
    verum/zealanicum (true/Ceylon
        cinnamon) 8
cinnamon 2, 3, 8, 21, 66
    in antiquity 24–25
    bales handling and stowage 214
    natural history and ancient culture
        of 8–9
    properties/effects from *Livre des simples*
        médicines 174–175
    trees 298
        bark strips/powder and flowers 9, **9**
Circa Instans (Platearius) 174–175
city centres, Mycenaean 38
city states
    Greek 42–43, 45
    Phoenician 40–41, **40**
civilizations, great 1
Cleopatra 29
Clive, Robert 295, 297
cloth
    asbestos 101
    calico 192
    purple-dyed, Phoenician 41
cloves 2, 14–15, 15–16, 30–31, 109, 113,
    114, 146, 259
    in *Apicius*, excerpta 147
    cultivation 15–16
    harvest 16
    production 298
    properties/effects from *Livre des simples*
        médicines 175
    trade, Ternate–Tidore play off 200
    trees 15, 298

Cochin 213
Coen, Jan Pieterszoon 258–259
coins
    Byzantine Empire 127–128, 149–150
    Islamic silver 152
    Ostrogoth 150
    Roman 119
Colbert, Jean-Baptiste 296
Colombo 199, 220, 276
Columbus, Christopher 1, 206, 208
Commiphora myrrha 6
commodities
    invisible 39
    traded on Red Sea–Africa/India routes 75
    see also luxury commodities
Compagnie Française des Indes Orientales 296
Company of Merchant Adventurers to New Lands 232
Confucianist/Buddhist and Muslim boundaries 142
Confucius 19, 26
Constantine I (the Great), Emperor of Rome 31, 122, 123
Constantine V, Emperor of Byzantine Empire 138
Constantinople *see* Byzantium (Constantinople)
copper 65
Coptos 74
coral 101
Corbie Monastery (Picardy) 147
Coromandel Coast
    Dutch trading posts 268, **268**
    EIC settlement, Masulipatam 269
Córtes, Hernán 235, 239
Cortesão, A. 192, 194, 194–195, 195–196, 197, 198, 199–201
cosmetics 22
cotton
    Indian, importance 225–226
    politics 291–292
    revolution, England 292
    role in Portuguese spice trade 270
    shift to 290–291
    textiles 225–226
country agents 293
country trade 270
Courthope, Nathaniel 258
Coutinho, Marshall Fernão 217
Coutre, Jacques de 223
crack-porcelain 252
Crassus 95
Creel, H.G. 82

Crete 38, 158
    Minos Palace (Knossos), crocus harvest fresco 13–14
Crocus
    cartwrightianus 12
    harvest, Minos Palace (Knossos) fresco 13–14
    sativus (saffron) 12, 13, **13**
crop rotation 138
crops, assemblage of, introduction by Muslim armies 138
Crowley, R. 163, 183–184, 185, 190, 209, 210, 213, 216
Crusader States 165–166, 168
Crusades 4, 165, 184, 206
    First (1096–1099) 164, 165
    Third (1189) 168–169, 207
    Fourth (1202) 169
Ctesiphon 13, 95
cuisine (food)
    medieval, spices in 173
    of medieval Constantinople 125
culinary revolution 290
cultural diffusion, along Silk Routes 90–91
cura annonae 124
Cyprus 158, 161

da Cunha, Tristão 215
da Gama, Vasco 209–212
    João's goals for 209
    routes 209–211, **210**, 213–214
Dalby, A. 8, 24, 26, 27, 30, 31
Dale, Sir Thomas 258
Dalmatian coast 177
Dalrymple, William 297
Damascus 139
Danish East India Company 296
Dar es Salaam 77
Darius the Great, King of Persia 70
    Royal Road building 68
    Suez Canal attempt 53
Darius, King of Persia 23
Dark Ages, of Europe 140, 141, 146–147
Davis, John 232
Dayuan 85, 86
de Menezes, Jorge 219
De observatione Ciborum (On the Observance of Foods) (Anthimus) 147
de Orta, G. 9, 10, 11, 16, 17
de Ruyter, Michiel 272
de Serres, Olivier 286
Debin, M.A. 241–242
Decker, M. 148

Dedan 58
Defoe, Daniel 291
Deir el-Bahri mortuary temples and tombs complex (Nile) 22–23
    temple bas-relief of Queen Hatshepsut's flotilla 50–51, **51**
Dejima 273
Demak Sultanate 198
Dewan, R. 29, 30
Dharmapala 220
dhows 107
Diamond Sutra, The 91
Dias, Bartolomeu 208, 212
Dido, Queen 45–46
Dilmun 66
Diocletian, Emperor of Rome 122
Diodorus 59
Dioscorides 21, 29, 31
    De Materia Medica 26
Diu 216
domestication
    camels 54–55, 88–89
    horses 83
donkeys 54
Donkin, R.A. 16
Drake, Sir Francis 243
dromedary *see* camels
Dunhuang 91–92
Dupleix, Joseph François 296–297
Dutch 4
    alternative northern routes for spices 230–233, **231**
    East Asian trading network 254–255
    first expeditions to East Indies 250–251
    in Levant 282
    merchants, sixteenth century 247
    monitoring and control in Japan 273–274
    and pirates 274–275
    world power path 248
Dutch East Asian trading network 254–255, **255**
Dutch East India Company (VOC) 233, 251, 253–254, 258, 269, 296
    compensation for Van Deynssen's belongings 266–267
    Coromandel Coast trading posts 268
    decline 294–295
    imports, seismic shift 290
    Indian Ocean control 264
    intra-Asian trade 271
    in North Moluccas 260
    silk exports from Persia 285
    trade with Japan 273–274
Dutch factors, sent to Gujarat 266
Dutch West Indies Company (WIC) 261–262
Dutch–Iberian conflict 261–262
Dutch–Portuguese confrontation 256
Dutch–Russians trade 232
dye, purple 41, 124–125
dye works, Phoenician 38

East Asian trading network, Dutch 254–255
East Indies, first Dutch expeditions to 250–251
Eastern Orthodox Christianity 152
Eastern Roman Empire, 'Fu-lin' 143
ebony 42
Echebarria, D. 162, 167
economic decline of Parthia 102
Edessa, County of 165
Edward Bonadventure 244
Egypt
    kingdoms split (c.1100 BCE) 52
    New Kingdom 22, 24
    Roman rule of 47–48, 73
    under Ptolomies 44–45, 47
    see also Deir el-Bahri mortuary temples and tombs complex (Nile)
Egyptian Dynasty 3
Egyptians
    ancient, international long-distance trade 3
    trade expeditions to Land of Punt 50–51
Egyptian–Indian coalition 216
Egyptian–Levantine, early trade 37–38, 39
Egypt–India journey of Eudoxus of Cyzicus 73
Elagabalus 35
Elcano, Juan Sebastián 237, 238
elephantegoi 72
elephants 46, 72, 108, 129, 199, 220, 257, 275
elite
    Chinese 201
    Romanized 148
Elizabeth I, Queen of England 243, 245, 281
Elmina 262
embalming of the dead 22
embroidery, Phoenician 41
Empedocles, fundamental elements of the universe 32
emperor, wearing of silk 32
empires 1
emporium/emporia
    Arabia 57
    Europe 149

emporium/emporia (contiuned)
　　Java and Sumatra 116
　　Marseille 149
　　Muziris 77
　　Red Sea 74
England, weaving industry 286
England–Ottoman Empire trade 281
English, alternative northern routes for spices 230–233, 231
English East India Company (EIC) 252–253
　　assault on Ormuz 284
　　commodity imports change 290
　　end of 297
　　first expedition 252–253
　　India
　　　　foothold in 264
　　　　move into 265–266, 294
　　　　Surat attention in 267–268
　　silk re-exports 287–288
　　tea imports 293
　　traders to rulers 295
English–Dutch confrontations, early seventeenth century 255–256
Epstein, M. 281
Erasistratus 29
Eratosthenes 44
Erythraean Sea 75
　　see also Periplus of the Erythraean Sea
Espinosa, Gonzalo Gómez de 237, 238
Estado da Índia 217, 227, 228, 229, 250
Euclid 44
Eudoxus of Cyzicus 73
Eugenia aromatica 14
Euphrates River 65, 66
Europe 4
　　agricultural productivity, high Middle Ages 172
　　competitors in India 296–297
　　Dark Ages 122, 140, 141, 146
　　　　spice use in 146–147
　　early medieval trade 149–150
　　luxury silks 286–287
　　medieval, silk in 175–176
　　population (1000–1300) 172
　　Portuguese distribution network 226–227
　　rule of, after fall of Rome 158–159
　　silk production 131
　　trade routes, thirteenth and fourteenth centuries 176, **177**
European Age of Discovery 206
European merchants 278
European nations, Levant trade 283, *283*
European settlements, in India **268**
European silk textile industry 287
European trade
　　East India Companies' effect on 261
　　influence of Crusader States on 165–166
　　spice, sixteenth century shifts 226
Europeans, Atlantic trade 261–262
European–Indian commercial exchange 264–265
exotic goods
　　from Europe 166
　　Italian cities trade, Middle Ages 179
　　movement through Islamic Empire 135–137
exotic luxuries
　　ancient world 5
　　of Byzantium (Constantinople) 124–125
　　desire for 65
　　for wealthy Romans 78

fairs, medieval, European 179–180
famines 186
Fan Ye, *Hou Hanshu* 101
Far East 3
farms, building in Greece 43
Fatimid–Byzantine trading relationship, tenth century 161
Favier, J. 172
Faxian (Chinese Buddhist monk) 25, 90, 113
　　sea travel in antiquity depiction 113
Feltham, H.B. 102
Feng, J. 103
Ferghana 85
Ferguson, D. 214
Fernandes, A.P. 192–194
Ferrence, S.C., and Bendersky, G. 14
Ferrer, M.T. 178
Fertile Crescent (Mesopotamia) 13, 66
Fischer, K. 266, 267
Fitch, Ralph 248–249
Fitzherbert, Captain Humphrey 257
Flanders, Middle Ages, trading network 179
Flecker, M. 143
Fleming, R. 175–176
Florence 286
follis 128
Foltz, R. 142
food *see* cuisine (food)
foreign devils, artefact removal from Buddhist cave complexes 91–92
foreign traders
　　into China 144
　　　　Tang Dynasty 143, 144
Formosa 274
　　Fort Freelandia 274

fortified way stations (*hydreumata*), Roman 74
France
    East India spice trade 296
    French impact, in Levant 283–284
    Lyon 286, 287
    Marseille 149, 283
    and Ottomans 280
    Rhône–Saône corridor 149
    silk industry 286
    Tours 286
Franck, I.M., and Brownstone, D.M. 82
frankincense 2, 3, 5–6, 21
    ancient culture 7–8
        resin collection 8
    in antiquity 21–23
    caravan route 55–57, **56**
    embalming use 22
    natural history 6–7
    profits 59
    scent use 21–22
        worship/funeral 22
    supply chain income 59
    trade expeditions to Land of Punt 50
    trees 6, **7**
        resin **7**, 22
Franks 122, 146
Frederick I, Emperor of Roman Empire 168–169
Freedman, P. 173, 183, 289
frescos
    Buddhist cave 90
    crocus harvest 13–14
Frobisher, Martin 232
Frumenius 130
Funan 115–116
    Oc Eo 116
funeral pyres 24
funerals 22, 24

Galen 9, 25, 104, 140
    humoral theory 32, 140, 172
Galerius, Emperor of Rome 122
Galle 275
galleon trade, trans-Pacific 241, **242**
galleys
    Minoan 38
    Portuguese 223–224
    Venetian 178
galliots 223, 224
Galvao, A. 16
garrisons (*phrouroi*), Roman 74
Gaza 23, 57
Gedrosian Desert 71

Genghis Khan, and Silk Routes 180–183
Genoa 163, 164, 166, 176, 184
    Caffa colony 184–185
Genoese quarter in Constantinople 168
geographical origins, silk, spices and scents **2**
Germanic governance 147–148
Germanic tribes migrations 122, 146, 147, 148
Gerrha 58
Gibbon, E. 97
Gil, M. 151
Gilboa, A., and Namdar, D. 39
ginger 2, 8, 66, 298
    in antiquity 25–26
    Chinese 125
    fresh rhizome **10**
    mahabbeshaj (great cure) 25
    medicinal properties 25
    natural history and ancient culture of 9–10
    properties/effects from *Livre des simples médicines* 174
giraffes 203
glass
    mould-blown 99–101
    Phoenician 41
    Roman ware 119
Goa 192–194, 217, 218, 224, 249
gold 101, 114, 235, 270
Golden Chersonese (Golden Peninsula) 13–115, 119
Golden Horde 180, **182**, 185, 186
Gomez, Diego 207–208
Goodman, B. 258, 270
Gorgan region, Rus' attack (913) 154
Gotland 152
Gotlanders 152
Gowa Kingdom 260
grain 46
    Byzantine own production 137
    feeding Byzantine masses 125
    from Egypt 45, 47
    from Greek city states 43
    merchants, Amalfi maritime republic 163
    trade, Constantinople focus 124
Great Banda Island 251
Great Saint Anna 244
Great Wall (China) 84–85
Greco-Roman *materia medica* 23
Greece
    city states 45
    saffron 12, 13
    silk clothing 34
    emergence 42

Greek civilization 3
Greek fire projectors 154
Greek gods 28
Greek medicine 26
Greek Orthodox Church of Byzantium 123, 124
Greeks
    ancestral 28
        saffron use 28–29
    city states (*poleis*) 42–43
    incense at funerals 22
Grijalva, Hernando de 240
Groot (Grotius), Hugo de 252
Guam 236–237
Guangzhou 143
    sea routes from 104
Gujarat 194
Gujaratis 143, 194, 197, 201, 266
Gujarati–Portuguese relationship 228
gynaecea 131

Hadhramaut 54, 63
Hairun, Sultan of Ternate 219
Hannibal 46
Hanseatic League 178–179, 261
Hansu 86
Harappa civilization 65, 66
    clay and porcelain tags 66
Hargreaves, James, spinning jenny 292
Harrapans 18
Hatshepsut 23
Hawkyns, William 265–266
healers, saffron use 30
heavenly horses 85–86
Hector 253
Heemskerck, Jacob van 251
    and Law of Prize and Booty 251–252
Hegra 58
Heitzman, J., and Worden, R.L. 265
Henri IV, King of France 286
Henrique, King of Portugal 208
    African coast navigator missions 207–208
Henry I, King of England 172
Heraclius, Emperor of Byzantine Empire 138
Hero 44
Herodotus 21, 22, 23, 29, 53, 70
    Histories 52
Herrin, J. 127
Herzig, E.M. 279
Heyn, Admiral Piet 262
Himyarite Kingdom 54, 61–62, 63
Himyarites 132

Hindu kingdoms 108, 109
    Sunda Kingdom 198
Hindu trading vessels 211
Hinduism 31
    Prambanan Temple Complex 111
Hippalus 75
Hippocrates 29, 32
Hirado 273
Hirst, K.K. 115
Historia Augusta 98
Hoffman, S. 28
Holy Roman Empire, and Byzantine tensions 168–169
Homer 43, 44
Horniker, A.L. 280
horse pastoralism 80
horseback riding 83
horses
    Central Asia, and balance of power 82–83
    China 82–83
    domestication 83
    'heavenly' 85–86
Hou Hanshu (*Book of the Later Han*) 120
Houtman, Cornelius de 250
Huangdi, Chinese 'Yellow Emperor' 18
Hudson, Henry 233–234
Hughli (Hooghly) 269
Huguenots, migration to England 286
humoral theory of Galen 32, 140, 172
Hunayn ibn Ishaq al-'Ibadi 140
Huns 122, 146
Huygen van Linschoten, Jan 249–250
    Itinerario 249–250, 270

Iberian Peninsula 148
Ibn al-Athīr 154–155
Ibn Amran, Isaac 31
Ibn Jubayr 166
Ibn Khordadbeh, *The Book of Roads and Kingdoms* 150–161
Ibn Majid, Ahmad, seagoing literature 191
Ibn Rustah 153
icon veneration 158
Iconoclasm 158
illegal trading
    Dutch, Fujian–Japan 274
    Portuguese 229
Immae, Battle of (272CE) 98
incense 22
    fall in demand 125
    maritime trade 61–62
    trees 6
    see also frankincense

Incense Route 3, 5, 53–54
    caravan 55–59, **56**
    land-based, length 77
    Roman invasion of 62–63
India 3
    ancient, ginger 15
    British control of 295
    Chola Empire 143
    Dutch spread into, early 266–267
    eastern trade cities 119
    English East India Company entry 265–266
    European competitors 296–297
    European settlements **268**
    global trade, in seventh century 143
    Gupta Empire 143
    Indian Ocean trade centre 192–194, **193**
    long pepper exports 11
    Madras 266, 269
    Mughals and Portuguese in 228–229
    Pondicherry (Puducherry) 296
    ports, route from Red Sea ports 75, **76**
    Portuguese 227
        conquest of 212–213
        seafarers, in South East Asia trading sphere 116
    textiles 291
    trade monopoly, by Arabians 72
    trading partners in 75
    use of saffron 30
    VOC foothold 264
    see also Ayurvedic medicine of India; Bengal; Coromandel Coast; Gujaratis
Indian Ocean 65, 264
    internal trade 270
    monsoon 73
    Monsoon Islam, before arrival of Portuguese **193**
    trade, description in sixteenth century (Pires) 191–192
    trade network
        discovery 248–250
        expansion in fifteenth century 190
Indian Rebellion (1858) 297
Indianization, of Indonesia 109–110
India–Arabia trade 67
India–Egypt trade route 71
India–Gulf trading 67
India–South East Asia trade, origins 107–108
Indonesia 298
    clove, nutmeg and mace 14–17
    demand for textiles 270
    early trade, in outer reaches 113–114
    Indianization of 109–110
    Java as nucleus of 110–111
    outer island bartering 110–111
    to Madagascar, return journey 68
Indonesian archipelago 107
    pepper producers 254
Indonesian seafarers
    early 67–68
    in South East Asia trading sphere 116
Indus River 65, 70–71
Indus Valley 18
    exports 66
Industrial Revolution 292
Inner Eurasian pastoralists 80
international trade fairs 179–180
Iran 132, 154
    silk
        EIC trade round Cape of Good Hope 284
        industry 279, 285
    VOC in 285
Iron Age Phoenician clay flasks 39
irrigation 138
Islam 105, 130
    converts to 139, 141, 142, 204
    European beliefs about 206
    expansion 135, **136**
        early seventh century 135
        towards China 142
    Golden Age of 140
    and loss of Silk Route Buddhist communities 180
    and medieval medicine 140–141
    in ninth century 158
    Qu'ran 54
    spread
        across South East Asia 141
        through Java and Sumatra 197
    see also Muslims
Islamic Empire 137
    Abbasid Dynasty 139
    Umayyad Dynasty 135
        rebellion against caliphate (747) 139
Islamization, of Central Asia 142
isolationism, China 203–204
Israel, Kingdom of 39–40
Italian cities
    exotic goods trade, Middle Ages 179
    trading colonies 149
Italian city states 161
Italian mercantile states 166
    rise of 163–164

Italy
  control by Eastern Roman
      Empire 157
  Genoa 163, 164, 166, 176, 184–185
  luxury silk production 286
  Pisa 163, 164, 166, 176, 184
  see also Venice
Italy–eastern Mediterranean link, thirteenth
      century 176
ivory 42, 129

Jacatra Kingdom 259
Jacoby, D. 164–165
Jade Gate 85
Jaffna Kingdom 220
Jahangir, Mughal Emperor 265, 266
  farman
      to Dutch for trade in Surat 267
      to English for trade and factories
          across Empire 266
Janibeg 186
Japan
  monitoring and control of Dutch
      273–274
  Portuguese discovery of 229
  Sakoku (closed door policy) 273
  silk production 298
  trade
      with Dutch East India
          Company 273–274
      with Portugal 229, 230
Java 110, 197, 198, 250
  as nucleus of Indonesia 110–111
  trade emporia 116
Jerusalem 168, 170
Jesus Christ
  crucifixion 23
  image on coins 128
Jewish merchants, Radhanites
      150–152
Ji Han, Nanfang caomu zhuang 31
João I, King of Portugal 207
João II, King of Portugal 208, 209
  goals for da Gama 209
John Tzimiskes, Emperor of Byzantine
      Empire 161
Johor Kingdom 251
junks, Chinese 201
Justin, Emperor of Eastern Roman
      Empire 132
Justinian I, Emperor of Byzantium 124,
      125–127, 157
  Empire after 157–158

Justinian II, Emperor of Byzantium 128,
      138, 158
Justinian, Emperor of Eastern Roman
      Empire 132
Juvenal 34

Kan Ying 120
Kang Dai (K'ang T'ai) 115
Karakorum Mountains 88
Kashgar 87–88
Kay, John
  and Arkwright, Richard 292
  flying shuttle 292
Kelder, J.M., et al. 45
Kerak fortress 166
Key, J. 44
Khaganate, term 152
Khalil, al-Ashraf 183
Khan, Toqta 186
Khan, Yakut 294
Khareef 6
Khasneh 61
Khazar, trans-Caucus trade routes 137
Khazar Khaganate 137–138
Khazaria 138
  Rus trade with Muslims and Byzantines
      through 153–154
Khoikhoi 209, 210
Khvalkov, E. 185
Kiev 152
Kilwa 196
Kingdom of Jerusalem 165
kings, medieval, silk clothes 175
Kingsley, S. 149
Kinoshita, S. 181–183
Knox, J.S. 40
Ko-ying 111
kora-kora canoes 114
Kotte 220
Koxinga 274
Kra, Isthmus of 115
Krishna, B. 223–224, 261, 270
Krondl, M. 176, 177, 225, 290
Kuhn, D. 18, 19
Kuk, N.J. 174–175
K'ung Yin-ta 19
Kusha, tariff payments 101
Kushan Empire 99
  trade
      resources 99
      routes crossroads 99–101
Kushans 104
  and Silk Routes 98–99, **100**

Laarhoven, R. 270–271
Lamb, A., and Skrine, C.P. 87
Lancaster, James 244, 245, 252
Land of Punt 3, 50–51
Langkasuka Kingdom 115
Lapita cultural complex 10
Latin Christendom 160
Le Coq, A. von 92
Legazpi, Miguel López de 240
Leizu, Lady 18
Lelis, A.A. 147, 150
Lendering, J. 52
Leo III, Emperor of Byzantine Empire 158
Leo III, Pope 159
Leo IV, Pope, defence of 163
Leonidas 23
Leuke Kome (white port) 57, 62
Levant
    Dutch in 282
    French impact in 283–284
    French merchant communities 280
    Italian mercantile states trade with 166
    route, for Iranian silk 279
    trade, European nations 283, *283*
Levant Company 248, 261, 281–282, 284
Levantine
    early trade with Egypt 37–38, 39
    Muslims 165–166
Libya 52
Liggio, L.P. 179
Lisbon 247, 261
Liu, X. 97
    The Silk Road in World History 60, 61
Liu-ye 115
Livre des simples médicines (Plateatius - French version) 174–175
Loaísa, García Jofre de 238, 239
Lombards 157, 161
Longobards (long beards/Lombards) 157, 161
Lonthor 259
Lost Murals of Bezeklik Thousand Buddha Caves 92
Louis IV, King of France 286
lovemaking 22, 25, 29
Lunde, P. 212, 216, 223, 224, 229–230, 269
luxury commodities 21, 102, 172
    ancient world 1–2, 5
    see also exotic luxuries
Lyon, silk industry 286, 287

Macao 273
mace 2, 14–15, 16–17, **16**, 31, 113, 114

properties/effects from *Livre des simples médicines* 175
    tree 16
McLaughlin, R. 59, 88
Madagascar 68
Madden, T.F. 169
Madras 266
    Fort St George 269
Madre de Deus 244–245
Magan (now Oman) 66
Magellan, Ferdinand 236–237
    fate of ships and crew 237–238, 239
    search for western sea route to Spice Islands 236
Magellan Strait 236
Mago of Carthage 46
Ma'in, Kingdom of 54, 55
Mairs, R. 99
Majapahit Kingdom 197
Makassar 260
Makassar War (1666–1669) 260
al-Makha (Mocha), port of 271–272
Malabar 11
Malabar Coast 192, 276
Malacca (Melaka) 196–197, 217–218, 229, 254
    cotton cloth trade 225–226
    Portuguese, Dutch control 259–260
    region history 197
    sixteenth century sieges 222
    Verhoeff attack on 257
Malacca Strait 111, 196, 197
Malay Archipelago 114
Malay Peninsula 114, 119
Malay sultans 222
Maldives 214
Malindi 196, 211, 225
Maluku Islands *see* Moluccas (Maluku Islands/Spice Islands)
Mamluks 181, 183, 194, 216
Mandeville, Sir J. 207
Manichaeism 90–91
Manila 240, 241
al-Mansur, Caliph 139
Manuel I, Emperor of Byzantine Empire 167–168
Manuel I, King of Portugal 212, 213, 214, 215
    Casa da Índia 226–227
Manufacture du Languedoc 283
Marculfe (monk) 147
Marcus Aurelius, Emperor of Rome 25
Mare Librum (Grotius) 252
Ma'rib, dam 57

maritime trade 65
    of Anuradhapura Kingdom 108–109
    Arab trade routes, seventh century 142–143
    China's 110
    high Middle Ages 164–165
    incense, Nabataeans 61–62
    Muslim
        with China 142–143
        with South East Asia 142–143
Mark Antony, General 96
Mark, J.J. 46, 98
markets
    medieval, European 179–180
    Phoenician 42
Marseille 149, 283
Masts, Battle of the (655) 135
Mataram Kingdom 111
Mauryan Empire 107–108
Maximian, Emperor of Rome 122
Mayacunne 220
medical books, Arabic 174
medicinal use
    cassia and cinnamon 24
    frankincense and myrrh 23
    ginger 25
    nutmeg 31
    pepper 27
    saffron 29–30
medicine
    Ayurvedic 21, 24, 25, 27, 30
    Islamic medicine 140–141
    medieval
        and Islam 140–141
        spices in 174–175
    Roman, *theriac* 25
medieval cuisine 173
medieval trade
    early
        in Europe 149–150
        in Mediterranean 148–149, 150
        Western 147–148
Mediterranean
    Middle Ages, commercial region 164
    as Roman world 47
    supremacy, Rome and Carthage rise and fight 45–46
Mediterranean trade
    in early medieval period 148–149, 150
    routes 166
        Middle Ages 179
Mehendale, S. 99
Mehmed I, Ottoman sultan, Period of Great Expansion 189

Mehmed II, Ottoman sultan 189
Melaka *see* Malacca (Melaka)
Meliqueaz (Malik Ayyas) 216–217
Meluhha 66
merchant colonies, Islamic 158
Merchant Royal 244
merchant vessels
    Phoenician 41
    Roman 47–48
merchants
    Arab 158
    Byzantine 148
    Dutch, sixteenth century 247
    European 278
    foreign, in China 143, 144
    Jewish, Radhanites 150–152
    middlemen, Venice 161–162
    Muslim 141
    Parthian–Chinese 101
    Silk Routes 89
Mesopotamia
    early vessels 65–66
    exports 66
    Fertile Crescent 13
    maritime links, Persian Gulf–Indus Valley 66
Messageries Maritimes 287
metals 271
    imported by Phoenicians 41–42
Metamorpheses of Apuleius, The 25
Mexico
    silk trade 243
    Spanish arrival 235–236
Mexico City 236
Michael VIII Palaeologus, Emperor of Nicaea 184
Michalopoulos, S., *et al.* 141
Middle Ages, high, maritime trade 164–165
Middle East
    greatest trading centre, Palmyra 95–97
    Mongol control 181
    Muslim, Christian foothold in 165–166
    Roman intrusions into 95
    Roman-controlled 97
middlemen 62, 120
    'of Nanhai' 143
    Phoenician 41
    primary merchant, Venice 161–162
Middleton, Henry 255–256
Miksic, J. 116
miliaresion 128
Minaean traders 55
Minoans 12, 38
    civilization 38

Minos Palace (Knossos, Crete) 13–14
Mintz, S.W. 138
Mithridates I, King of Parthian Empire 94
Mogadishu 203
Mogao Grottoes (Cave of the Thousand
    Buddhas) 91
Moluccas (Maluku Islands/Spice Islands) 14–
    15, **15**, 113–114, 199–201, 236
    found by Spanish 237
    Portuguese in 218–219
    Portuguese–Spanish claims conflict 237
    routes to 201
    Spanish fleet sent to 239
    Ternate 200, 219, 239, 256, 260
    Tidore 200–201, 239, 254, 256, 260
Mombasa 196
monasteries
    Buddhist 90
    Nalanda 113
Mongol Empire 181, **182**
Mongols
    and Genghis Khan 180–181
    routes, trans-Asiatic 186
monks
    Byzantine 160
    Chinese Buddhist pilgrims 111–113
    Faxian (Chinese Buddhist) 25, 90, 113
    Nestorin 131
    Yijing (Chinese Buddhist) 111, 113
monsoons 75, 107, **193**
    Indian Ocean 73
    winds 118, 143, 224
    winter 225
Montezuma, Aztec emperor 236
moralists, restriction of silk wearing 35
Mostert, T. 260
Mozambique channel 225
Mozambique City, Muslim traders 210
Mozambique Island 224
Muawiyah I, founder/caliph of Umayyad
    Caliphate 135
Mughal Empire 228, 265
    Indian Ocean territory 190
Mughals 264
    and Portuguese in India 228–229
    in seventeenth century 265
mulberry trees 2, 17, 18, 19, 131, 242, 285, 286
murex shellfish 41
Musa 95
Muscovy Company 232–233
Muslim and Buddhist/Confucianist world
    boundaries 142
Muslim merchants 141

Muslim Middle East, Christian foothold
    in 165–166
Muslim navy 135
Muslim physicians 32
Muslim world, shift 138–139
Muslims 4
    control of Indian Ocean trading network,
        fifteenth century 190
    'Hui' 143
    Levantine 165–166
    maritime trade
        with China 143–144
        with South East Asia 142–143
    trade with Rus, through Khazaria 153–154
    traders, Mozambique City 210
    trading vessels, unarmed 211
Muslim–Christian trade, tenth century 161
mutineers 240
mutual trade, South East Asia–China 204
Muziris 75–77
Mycenaeans 38
    trading goods 38
Myos Hormos 74, 77
Myristica fragrans 16
myrrh 2, 3, 5–6, 21
    ancient culture 7–8
        resin collection 8
    caravan route 55–57, **56**
    embalming use 22
    natural history 6–7
    profits 59
    supply chain income 59
    trade expeditions to Land of Punt 50
    trees, resin/tears 6, 8
    use in antiquity 22–23

Nabataea, Kingdom of 59–60
    ships 77
Nabataeans 45, 59–60
    maritime incense trade 61–62
    Petra 60–61
Nadri, 264–265
Nagasaki 229
Najran oasis 58
Nalanda Monastery 113
Namdar, D., and Gilboa, A. 39
Navigation Act (1651) 272
Nearchus 71
Nebuchadnezzar, King of Babylon 23
Necho II, Pharaoh 52
    Nile–Red Sea canal attempt 53
Neira 257

Neolithic farmers 6
Nestor 154
Nestorian monks 131
Nestorianism 91
New Spain
  route to 240–241
  silk industry 242–243
Nicaea 184
Nicholas V, Pope, *Romanus Pontifex* (1455) 208, 209
Nicol, D.M. 162, 168
Niebuhr, B.G., *Lectures on the History of Rome* 104
Nijman, J. 271
Nile, River 65
  papyrus 45
  see also Deir el-Bahri mortuary temples and tombs complex (Nile)
Niya 91
nomadic pastoralists 84
nomadic people 59–60
nomisma (solidus) 128
North African amphorae 148
North America, exploration for spice route 230
North Moluccas, Dutch East India Company in 260
North Sea, maritime trade routes, Middle Ages 179
Northeast Passage, search for 233
Northern Black Polished Ware 108
Northwest Passage, search for 230–232, 233, 243
Nubians, southern 52
nutmeg 2, 14–15, 16–17, **16**, 31, 113, 114
  Byzantine use 125
  harvest 17
  properties/effects from *Livre des simples médicines* 175
  trees 16, 298

oasis towns, Silk Route 105
Oc Eo 116
Odoacer, King of Italy 157
Ogedei 181
Olinda 262
Ophir 52, 238
opium 25, 293–294
  addicts 293
Orhan 189
Ormuz 195–196, 215, 218, 223, 284
Osman I, leader of Ottoman Turks 189
Ostrogoths 122, 146
  coins 150

Ottoman Empire 189
Ottoman Turks 222
Ottomans
  Aden as portal to Indian Ocean trade 194
  East to West spice and silk route profits control 278
  Egypt takeover 226
  and France 280
  Indian Ocean territory 190
Ottoman–Dutch political relations 282
Ottoman–Venetian trading links, sixteenth century 226
outrigger canoes 68
Ovid, Crocus story 28
Özbeg 186

Pacific Ocean 236
  Spanish expeditions to 239–240
Palakka, Arung 260
Palembang 111
Palermo 144
'Palermo stone' 50
Palmyra ('Bride of the Desert') 87, 95, 98
  greatest trading centre in Middle East 95–97
  Liu's description 97
  as Roman *colonia* 96
Palmyraeans 96
Palmyrene Empire 97
Panama, Isthmus of 235
Panchur (Pansur) 198
Pandyas 108
Papal States 159
paper 45
papyrus 45
Parameswara (Iskandar Shah) 197
Pararajasinghe 220
Parthian Empire 94, 101
Parthians 86
Parthian–Chinese merchants, trade 101
Parthian–Roman hostilities 95
Pasteur, Louis 287
pastoralism, horse 80
pastoralists, nomadic 84
Paul of Aegina 31
Pearson, M.N. 228, 229
pebrine 287
Pegolotti, Francesco 183
Penghu 273
Pepin the Short, King of the Franks 159
pepper 2, 8, 21, 27–28, 228, 253, 298
  black (*Piper nigrum*) 11, 28, 39, 77, 109
  cultivation 11, 12

Byzantine use 125
exports/imports, Portugal 227
Far Eastern trade 222
long 11, **11**, 28
natural history and ancient culture of 11–12
oversupply 289
properties/effects from *Livre des simples médicines* 174
trade routes 226
tree 11
white 11, 12
perfumes
frankincense and myrrh 22
intoxicating, cinnamon 25
saffron in 29
Periplus of the Erythraean Sea 75–77, **76**, 119, 129
Ormuz mention 195
Persia 4
Achaemenid Empire 68
Royal Road 68–70, **69**
Persian Gulf
ports 94
route to Rome 117, **118**
Persian Gulf–Red Sea route, via southern Arabia 70
Persian Gulf–Syrian Desert route to Rome 117, 118, **118**, 119
Persian Sasanian Empire 95
Persians 52
Medic dress 34
'Po-ssi' 143
saffron use in textiles 29
Perur, S. 77
Petra 58, 60–61, 95
pharmacology texts, medieval 174
Pharos, Island of 44
Philippines, profitable colonization by Spanish 241, **242**
phlegm 32
Phoenicians 40–42, 52
city states 40–41, **40**
clay flasks, Iron Age 39, 66
markets 42
saffron trade 38
as superpower 42
physicians, Islamic 140
Pidgin Portuguese 224
Pigafetta, A. 15–16
pilgrims
Buddhist 90
Chinese Buddhist monks 111–113
Piper
longum (long pepper) 11
nigrum (black pepper) 11, 12

pirates
Arab 149
Chen Zuyi 203
Chinese 274
and Dutch 274–275
English 243–244
English/Dutch, Spanish galleon convoy escorts 242
Timoji 217
Zheng Zhilong 274
Pires, Tomé 191–192, 198, 229
on Aden 194–195
on Banda Islands 199–200
on Ceylon 199
on Malacca 197
on Ormuz 195–196
Suma Oriental 191–192, 225
on Ternate and Tidore 200–201
Pisa 163, 164, 166, 176, 184
plague
Antoine 103–104
Justinian's 131–132
Pleyn delit: Medieval Cookery for Modern Cooks (Butler *et al.*) 172
Pliny the Elder 8, 21, 22, 24, 27, 29, 55, 57–58, 68, 129
Natural History 26, 30
nutmeg named *comacum* 31
Plutarch of Chaeronea 23
Life of Alexander 43–44
Poivre, Pierre 298
political relations, Ottoman–Dutch 282
Polo, Marco 10, 28, 91, 206
Description of the World 183
Pondicherry (Puduchery) 296
Poppaea, funeral of 22
Portugal
conquest of India 212–213
emergence in fifteenth century 207–208
Lisbon 247, 261
lust for wealth 207
pepper exports/imports 227
Tordesillas Treaty (1494) 209
trade
with China 229, 230
with Japan 229, 230
loss in East 261
Portuguese 4
in Ceylon 220
and Mughals in India 228–229
removal attempts from Ceylon 275–276
trading missions, India run (*Carreira da Índia*) 224–225

Portuguese distribution network, in
    Europe 226–227
Portuguese India 227
Portuguese spice trade 227
    role of cotton 270
Portuguese–Mamluk Naval War (1498) 215–217
Potosí 242
pottery sherds, stamped 108
Potts, D. 89
precious metals 271
Prester John, myth of 206–207, 209
Principality of Antioch 165
privateering 230
privateers, English 243–245
Ptak, R. 201
Ptolomies, Egypt under 44–45, 47
Ptolomy I, King of Egypt 44–45, 71
Ptolomy II, King of Egypt, attempt to dig Suez canal 53
Ptolomy (Cartographer) 119
    Golden Chersonese 114
    second century map 119
Pulicat, Fort Geldria 268
Punic Wars I/II/III (264–146 BCE) 46
purple dye 124–125
purple-dyed cloth
    Phoenician 41
    silk 124
Puteoli 58

al-Qadisiyah, Battle of (637 CE) 135
Qataban, Kingdom of 54, 57, 63
Qeshm 284
Qu'ran 54

Radhanites 150–152
    age, ending 151–152
    origin 151
    trade routes 150, 151
rafts 65–66
Rajasinha, King of Ceylon 275, 276
Raleigh, Sir Walter 244–245
Rameses II, Pharaoh of Egypt, mummy 39
Rameses III, Pharoah of Egypt 50
Ravenna 161
Razzari, D. 284
Recife 262
Red Dragon 252–253, 256, 266
Red Sea 51, 67
    incense trade, Nabataean 61–62
    maritime routes

    to Africa 75, **76**, 77
    to India 75, **76**, 77
    ports 75
    under Roman rule 73–74
    trade
        after Rameses III 52
        Askum hold over 130
    trade routes, Ottoman control 190
Red Sea coast 3
Red Sea–Nile route to Rome 117, 118, **118**, 119
reeling silk 18
religions
    and South East Asian trade 107
    see also Buddhism; Christianity; Hinduism; Islam
religious conflict, Byzantine Empire 158
religious ritual, saffron use 30
religious services
    cinnamon burning 24
    incense burning 22, 24
religious-political struggle, Spanish–Dutch (1568) 247
Renaissance, European 128–129
Rhône–Saône corridor 149
rice basket, Java precolonial 110
rice belt, Funan 115
Richter, G.M.A. 34
Riello, G. 225
Riley, H.T., and Bostock, J. 8, 55, 57, 68
Roaring Forties 257
Rochwulaningsih, Y., and Sulistiyono, S.T. 110–111
Roe, Sir Thomas 261, 266
Roman Catholic Church 123, 146, 158
    structure 158–159
Roman coins 119
Roman Empire
    decline and fall 104
    Eastern 165
        control of Italy 157
        Golden Age of 125–126
    fall, rule of Europe after 158–159
    maritime link with South East Asia, end of 132–133
    reunification attempt 122
    Völkerwanderung/Migration Period 122, 146
    Western, fall 124, 146, 148
    Western and Eastern 122
Roman-controlled Middle East 97
Romans
    goods, trade through Red Sea ports 73–74

incense at funerals 22
invasion of Incense Route 62–63
lifestyles, after the fall 148
major trade routes 74–75
medicine 26
    theriac 25
pepper use in food 27
saffron use 29
ships, in first century Indian ports 116
and silk 21, 34
silk clothes 18
trade 3–4
    through Aksum Kingdom 129
wealthy, exotic luxuries 78
Roman–Parthian hostilities 95
Rome
    ancient, cinnamon use 24–25
    and Carthage, Mediterranean
        supremacy 45–46
    and China, first direct contact 119–120
    founding of 45
    international trade network 77–78
    military power 45
    routes for goods from South East Asia
        to 117–119
    sacking (410 CE) 146
    third century flux 122
Romulus Augustulus, Emperor of Western
    Roman Empire 157
rouletted ware 108
Royal Road (Persia) 68–70, **69**
    caravan routes 94
royalty, silk wearing 32
rugs 101
Run 258–259
Rus 152
    attack on Islamic and Byzantine
        worlds 154–155
    trade with Muslims and Khazars,
        through Khazaria 153–154
    trade routes 153, **153**
Rus' Khaganate 152
Russia 232–233
    route for Iranian silk to West 279–280
Russian Primary Chronicle, The (Nestor) 154
Russian–Dutch trade 232

Saba (Sheba), Kingdom of 54, 57, 63
Safavid Dynasty, Indian Ocean territory 190
Safavids 284
    Ormuz as portal to Indian Ocean
        trade 195
Saffarids 143

saffron 12–14, **13**, 28–30
    Mycenaean trade 38
saffron-pigmented textiles 29
sago palm (*Metroxylon sagu*) 114
Saladin, Ayyubid sultan 168
saltpetre 269
Samarkand 89
Samos, sanctuary of Hera 39
Samudera Pasai 198
Sanskrit
    black pepper names 27
    text *Charaka saṃhita*, cloves 30
Santa Catarina 251
    porcelain auction 252
Santa Maria de la Victoria 239
Santo Domingo 235
Santorini 12
Sardinia 158
Sarmatians 88
Sasanian Empire 102
    Byzantine trade with 128–129
    world luxury trade domination 102
Sasanians 101–102, 135
    defeat by Muslims 142
    power in Persian Gulf and Indian
        Ocean 132
    and Romans, trade struggle 129
    weavers 102
Sayhad desert (*Arabia deserta*) 57–58
scent
    of saffron 29
    tributes of 23–24
Schoff, W.H. 71–72, 195
scurvy 209, 211
Scylax of Caryanda 70
Scythians 82
Sea of Azov 185–186
'Sea Peoples, The' 38
sea power, Venetian 162
sea routes
    from Guangzhou (Canton) 104
    to breadbaskets 43
sea snail (*Bolinus brandaris*) 124–125
seasonings
    spices used 147
        medieval cuisine 172
seduction with cinnamon 25
Seland, E.H. 119
Seleucid Empire 96
Seleucids 94, 108
Seleucus I, General 94
Seljuk Turk Empire 189
Seljuk Turks 165
Seller, W., and Watt, M. 5–6

Sen, T. 108–109, 113
Seneca the Younger, *Da Beneficiis* 34–35
Serapis Temple (Alexandria) 44
serfdom system 187
sericulture 34, 131
    Chinese 19, 131
    European industry 287
    industry, Spain 242
Serrão, Francisco 219, 236
Shabwa 55
Shaffer, M. 206
Shah Jahan, Mughal Emperor 265
Shah Shuja, Bengal Governor 269
Sheba, Queen of 40, 54, 130
Sheba (Saba) 54, 57, 63
Shennong (Shen-nung), Emperor of China 24, 25
Shiji 89
ships/boats
    canoes
        kora-kora 114
        outrigger 68
    convoys (*flota*) to limit English/Dutch pirates and privateers 242
    dhows 107
    galleys 38, 178, 223–224
        Minoan 38
    galliots 223, 224
    Jewish-owned 150
    junks 201
    merchant vessels 41, 47–48
    Phoenician war 41
    Roman 116
    trading vessels
        Hindu 211
        Muslim 211
    VOC 254–255
shipwrecks, Mediterranean 148
shrines, Buddhist 90
Sicily 158
    Carthage–Rome warring over 46
Sidebotham, S. 74, 130
Sig (gorge) 60
silk 2, 17–19, 119, 124, 269
    annual ceremony in China 34
    in antiquity 32–35
    archaeological evidence in China 18
    Bengali 285–286
    Byzantine 124
    Chinese 17–18, 21, 34, 241–242
    cloth, from Syria 101
    imports into France 283
    Iranian, trade round Cape of Good Hope 284
    luxury, Europe 286–287
    making 104
    in medieval Europe 175–176
    northern steppes route 137
    production 34, 298
        Europe 131
        as object of worship 34
    raw, European 287
    secret, escapes 131
    as tax payment 34
    trade
        after 400 CE 104–105
        importance of 18
    trading network 34
    VOC exports from Persia 285
silk industry
    France 286
    Iran 279
    New Spain 242–243
Silk Road
    central Asia (2000 BCE) **81**, 82
    Han Dynasty, routes 87–88
    northern hub 137
Silk Routes 4, 5, 18, 21, 87, 99, 131
    camels 88–89
    Central Asian 104
    cultural diffusion 90–91
    cut off, due to Tibetan tribes revolt 143
    and Genghis Khan 180–183
    length 102
    merchants 89
    Mongol garrisons 181
    Muslim intermediaries, seventh century 142
    Northern and Southern 151
    oasis towns 105
    Sasanian control 102
    Tarim 103
    trade ebbs and flows 102–103
    trade revival and expansion under Mongols 181–183
silkworm
    cocoon 17–18, **17**
    disease 287
silver 101, 227, 241, 270
    Japanese 274
    mines, Japan 229–230
Sinae 119
Singer, C. 73, 78
Sirrullah, Bayan (Sultan of Ternate) 219
Skrine, C.P., and Lamb, A. 87
slaves 42
Slavs 153
Sluiter, E. 259–260

Smith, R. 39, 83–84
smugglers, Chinese 204
Smyrna 282
Snell, M. 131
social class
    differences, European in Middle
        Ages 172
    Swahili city states 196
Socotra 215
Sogdiana (Sogdia) 85, 98
Sogdians 89
Solomon, King of Israel 39–40, 52
    visit of Queen of Sheba 40, 54
South Asia, Europe trade, by
        100 CE 116
South East Asia
    Islam spread across, mid-600s 141
    trading spheres, early first century
        (CE) 116
South East Asian trade 107
    with India 107–108
    maritime, with Muslims 142–143
    network, China's entry 110
South East Asia–China mutual
        trade 204
Spain 4
    acquisition of territories by papal
        bull 208–209
    colonists to Hispaniola 235
    expeditions to Pacific (1530s and
        40s) 239–240
    second expedition to Spice Islands,
        goals 238
    sericulture industry 242
    Tordesillas Treaty (1494) 209
Spanish Armada (1588) 248
'Spanish Fury' 247
Spanish–Malay struggle 240
Speelman, Cornelis 260
spice centre, first major, Muziris 75–77
Spice Islands *see* Moluccas (Maluku Islands/
        Spice Islands)
Spice Routes 3, 4, 65
    from India
        through Persian Gulf 151
        through Red Sea 151
spices
    definition 8
    northern steppes route 137
    profitability 289
    profits, Portuguese 227–228
    shift away from, European 290
    trade
        earliest in India–Mesopotamia 66

East–West 109
Spilbergen, Admiral Joris van 275
Sri Lanka 119
    see also Ceylon
Srivijaya Kingdom 111, 197
steam power in mills 292
Stein, Sir A. 91–92
Stephen II, Pope 159
Steppe Roads 80, **81**
Stewart, R.T. 57
stone beads, origin-specific 108
storax 101
Strabo, *Geography* 62, 73
Strassler, R.B. 70
Strogonov family 233
sugar
    industry, Portuguese 262
    plantations, Brazilian 262
Sulistiyono, S.T., and Rochwulaningsih,
        Y. 110–111
Suma Oriental (Pires) 191–192, 225
Sumatra 110, 197
    Samudera Pasai 198
    trade emporiums 116
Sumerians 66
superpower, Phoenicians 42
Surat, Dutch given *farman* to trade in 267
Swahili, term 196
Swahili coast of Africa 196
    city states, social structure 196
Syria 96
Syriac Book of Medicines, The 27
Syzygium caryophyllata 145
    see also cloves
Szczepaski, K. 140

Tahirids 143
Táino 235
Taklamakan Desert 87
Talas 142
Talip, Ömer 271–272
Tana 185–186
Tanaka, Y., *et. al.* 86
Tang Dynasty 104–105
Tao Hongjing 25
Tarim Basin 86, 87
Tarim Silk Routes 103
taxes, caravan route 55, 57, 59
tea 292–293
tenate 254
Tenochtitlán 236
Ternatans–Portuguese relationship 219
Ternate 200, 219, 239, 256, 260

textiles 283, 290
    demand from Indonesia 270
    Indian 291
    saffron-pigmented 29
    silk, Sasanian 102
    see also cotton
Theophilos, Emperor of Byzantine
        Empire 127
Theophrastus 21
    Cn Odors 23
Theotokis, G. 162, 167
theriac 25
Tidore 200–201, 239, 254, 256, 260
Tigris River 65, 66
timber 41
Timna 57
Timoj. (pirate) 217
Tlaxcalans 236
Tordesillas Treaty (1494) 209, 237
Tours 286
trade
    Byzantine attitudes to 127
    early Egyptian–Levantine 37–38
    India–South East Asia, origins 107–108
    maritime 65
        of Anuradhapura Kingdom
            108–109
        China's 110
    incense 61–62
    pastoralist 80–82
    South East Asian 107
    spices, East–West 108
    state control of, Byzantine Empire 127
trade centres, Phoenician 40–41, **40**
trade networks
    Catalonian 178
    early China 84
    Persian, frankincense, myrrh, cassia and
        cinnamon 94
trade routes
    Dutch, early seventeenth century 254,
        255
    European, thirteenth and fourteenth
        centuries 176, **177**
    Portuguese and Spanish, sixteenth
        century **241**
    Roman major 74–75
    western Asia–China 82
traders, Palmyraean 96
trading, silk 33
trading centres, Middle Eastern greatest,
        Palmyra 95–97
trading colonies, Italian cities 149
trading empires, Muslim 190

trading networks, frankincense and myrrh
        traders 6
trading powerhouse, Constantinople
        as 125–127
trading spheres, early first century (CE), South
        East Asia 116
trading vessels see merchant vessels
treasure ships 202
Trebizond 160–161, 184
Trento, R. 125
tributary trade system 201–202, 204
tributes, of scent 23–24
Tripoli, County of 165
Triticum
    aestivum (bread wheat) 13
    dicoccum (emmer wheat) 13
Tschanz, D.W. 140
Tudella, Rabbi B. 10
Turfan 92
Turkish Ottomans 278
Turner, J., *Spice* 27
Twelve Years Truce, Dutch Republic–Spain
        (1609–1621) 282
Tyre 124
Tyrian purple 124–125, 127
Tyson, P. 50, 51

Umunç, H. 282
unguent cones 22
Union of Utrecht (1579) 247
Urban II, Pope 165
Urdaneta, Andrés de, route to New
        Spain 240–241
Utrecht, Treaty of (1713) 296

Van den Broecke, Pieter 267
Van Deynssen, David 266
van Neck, Jacob 251
van Wyhe, J. 114
Varangian Guard 155
Varangians 152
Vedas 31
Venetians 216
    arrogance 167
Venetian–Byzantine animosity 167–168
Venice 4, 166, 176, 286
    Byzantine Empire struggles with 167–
        168
    Chrysobull Golden Bull
        (*chrysobull*) 162–163, 167
    move into Black Sea 185–186
    power in thirteenth century 176

rise of 161–163
sea power 162
    navy 169
    navy as Byzantine Adriatic
        buffer 167
    spice imports, fifteenth to sixteenth
        centuries 226
    trade in spices 178, 227
    trading empire 177–178
Venner, John 245
Veracruz 236, 241
Verhoeff, Admiral Pieter Willemszoon
    256–257
Via Egnatia 160
Via Fracegena 180
Vietnam 202
Vijayanagara Empire 266
Vikings 152
Villalobos, Ruy López de 240
Vimaladharmasuriya, King of Ceylon 275
Visigoths 122, 124, 146, 148, 149

Wadi Hammamat 67
Wadi Moussa 60, 61
Wagner, S. 175
Wallace, A.R., *The Malay Archipelago* 114
Walters, S. 286
Wang Yuan 91
warships, Phoenician 41
Warwijck, Wybrand van 251
watchtowers, Roman 74
water wheels 292
Watson, W.E. 154–155
Watt, M., and Seller, W. 5–6
wealthy, saffron in lives of 29–30
Weatherford, J. 181
Weert, Sebald de 275
Western Europe 149
    economic state, after Roman Empire
        fall 147
Western Roman Empire, collapse 146
Western trade, in early Medieval Ages
    147–148
Western world, luxury trade commodities 2
Westerwolt, Adam 275
wheat 109, 124
Wheeler, Sir M. 116
Willard, P. 29
    Secrets of Saffron... 38
Willoughby, Sir Hugh 232
'wise men' 23–24

wood, Phoenician trade in 41
wool 283–284
    Phoenician trade in 41
Worden, R.L., and Heitzman, J. 265
world system central hinge, thirteenth
    century 176
Wudi, Emperor of Han China 85, 86

Xiongnu 83
    and China 84–85, 86–87
Xuande, Emperor of Chian 203
Xuanzang (Chinese Buddhist monk)
    90, 91, 113
Xun, Master 33

Yafa, S. 291
Yakur, Sidi 294
Yangzhou 143, 144
Yarmuk 135
Yathrib 58
yellow bile 32
Yemen 6, 71, 132
    ancient, four states 53–54
    deserts of 53
Yijing (Chinese Buddhist monk) 111, 113
Yongle, Emperor of China 202
Young, G.K. 130
Yue 110
Yuezhi 85, 98–99

zamorin 192, 211
Zamorin of Calicut 192, 211,
    212–214, 216
    overthrow attempt 217
Zara 169
Zaragoza, Treaty of (1529) 239–240
Zebu cattle 67
Zenobia, Palmyraean queen 96
    power play 97–98
Zeus 28
Zhang Qian 85–86
Zheng He, Indian Ocean missions 202–203
Zheng He, 'Treasure Fleet of the Dragon
    Throne' 28
Zheng Zhilong 274
Zhu Ying (Chu Ying) 115
Zingiber officinale (ginger) 9–10
    see also ginger
Zoroastrianism 129

www.ingramcontent.com/pod-product-compliance
Lightning Source LLC
Chambersburg PA
CBHW040744020526
44114CB00048B/2908